Engineering Materials

More information about this series at http://www.springer.com/series/4288

Mohammad Jawaid · Abou el Kacem Qaiss
Rachid Bouhfid
Editors

Nanoclay Reinforced Polymer Composites

Natural Fibre/Nanoclay Hybrid Composites

Springer

Editors
Mohammad Jawaid
Universiti Putra Malaysia
Serdang, Selangor
Malaysia

Abou el Kacem Qaiss
Laboratory of Polymer Processing
MAScIR Foundation
Rabat
Morocco

Rachid Bouhfid
Laboratory of Polymer Processing
MAScIR Foundation
Rabat
Morocco

ISSN 1612-1317 ISSN 1868-1212 (electronic)
Engineering Materials
ISBN 978-981-10-9291-6 ISBN 978-981-10-0950-1 (eBook)
DOI 10.1007/978-981-10-0950-1

Printed on acid-free paper

This Springer imprint is published by Springer Nature
The registered company is Springer Science+Business Media Singapore Pte Ltd.

Dedicated to Our Parents

Preface

Natural fibres such as flax, hemp, kenaf, jute, coir, oil palm, cotton, etc. are abundantly available worldwide and already established in automotive, and construction industries due to its sustainable and renewable nature. Nowadays natural fibres are getting attention of researchers and academician to utilize in different application but it has limiting utilization due to moisture absorption, less thermal stability, and difficulty to get quality fibre supply. Nowadays nano filler based polymeric materials display better utilization in different applications. Recent studies reported about hybridization of nanoclay with natural fibre reinforced polymer composites which display better mechanical and thermal properties as compared to natural fibre composites. The proposed book is focused on nanoclay based natural fibre reinforced polymer composites fabrication, characterization and applications. It will also including the effect of Nanoclay on the mechanical, thermal, morphological, tribiological, and flammability properties of natural fibre composites. The readers found complete information about recent development, mechanical, morphological, thermal and rheological properties, processing techniques of natural fibre/nanoclay composites, mechanical performance of cellulose particulate, effects of micro- and nano- cellulose on tensile and morphological properties of Montmorillonite/Nanoclay, tribological properties of hybrid composites, effect of plasticizer on fracture toughness of Polylactic Acid (PLA)/Kenaf Fiber (KF)/Montmorillonite organoclay and dielectric properties, impedance spectroscopy and transport properties and its application in functional materials.

We are highly thankful to all authors who contributed book chapters in this edited book and supported by contributing valuable ideas and knowledge. We appreciate their proficiency on scattered information from diverse fields in nanoclay/natural fibers based hybrid composites and accepting editorial suggestions to finally produce this venture that will hope be a success. We greatly appreciate their commitment.

We thank Springer-verlag, Singapore team for their generous cooperation at every stage of the book production.

Mohammad Jawaid
Abou el Kacem Qaiss
Rachid Bouhfid

Contents

About the Editors

Dr. Mohammad Jawaid is currently working as Fellow Researcher (Associate Professor), at Biocomposite Technology Laboratory, Institute of Tropical Forestry and Forest Products (INTROP), Universiti Putra Malaysia, Serdang, Selangor, Malaysia and also Visiting Professor at Department of Chemical Engineering, College of Engineering, King Saud University, Riyadh, Saudi Arabia since June 2013. He is also a Visiting Scientist to TEMAG Lab, Faculty of Textile Technologies and Design at Istanbul Technical University, Turkey. Previously, he worked as Visiting Lecturer, Faculty of Chemical Engineering, Universiti Teknologi Malaysia (UTM) and also worked as Expatriate Lecturer under UNDP project with Ministry of Education of Ethiopia at Adama University, Ethiopia. He received his Ph.D. from Universiti Sains Malaysia, Malaysia. He has more than 10 years of experience in teaching, research, and industries. His area of research interests includes Hybrid Reinforced/Filled Polymer Composites, Advance Materials: Graphene/Nanoclay/Fire Retardant, Lignocellulosic Reinforced/Filled Polymer Composites, Modification and Treatment of Lignocellulosic Fibres and Solid Wood, Nano Composites and Nanocellulose fibres, Polymer blends. So far he has published 5 books, 12 book chapters, and more than 130 International journal papers (CIJ) and 5 Published review papers under Top 25 hot articles in science direct during 2013–2015. He is also the Deputy Editor-in-Chief of Malaysian Polymer Journal and Guest Editor of Special issue-Current Organic Synthesis & Current analytical chemistry, Bentham Publishers, UK. Besides that he is also reviewer of several high impact ISI journals of Elsevier, Springer, Wiley, Saga Publishers, etc. Presently, he is supervising 15 Ph.D. students and 5 Master students in the field of hybrid composites, Green composites, Nanocomposites, Natural fibre-reinforced composites, etc., 3 Ph.D. and 2 Master students graduated under his supervision in 2014–15. He has several Research grant at University and National level on polymer composites of around RM 575,000 (USD 159, 722). He also delivered Plenary and Invited Talk in International Conference related to composites in India, Malaysia, Turkey, Saudi Arabia, Thailand and China. Besides that he is also member of technical committee of Several National and international conference on Composites and Material Sciences.

Dr. Abou el kacem Qaiss is currently working as Director of polymer processing laboratory, at Moroccan Foundation for Advanced Science, Innovation and Research (MAScIR), Institute of Nanomaterials and Nanotechnology (Nanotech), Rabat, Morocco since October 2010. He received his Ph.D. from Université Laval, Québec, Canada. He has more than 20 years of experience in teaching, research, and industries. His area of research interests includes Hybrid Reinforced/Filled Polymer Composites, Advance Materials: Graphene/Nanoclay, Lignocellulosic Reinforced/Filled Polymer Composites, Modification and Treatment of Lignocellulosic Fibres, Nano Composites and Nanocellulose fibres, Polymer blends. So far he has published 4 book chapters and more than 30 International journal papers during 2011–2015. He is also a reviewer of several high-impact ISI journals of Elsevier, Springer, Wiley, etc., publishers. Presently, he is supervising 4 Ph.D. students in the field of hybrid composites, Green composites, Nanocomposites, Natural fibre reinforced composites, etc., 4 Ph.D. and 6 Master students graduated under his supervision in 2012–15. He has Research Grant at MAScIR and Minister of Higher Education, Scientific Research and Training, Morocco (MESRSFC) of around USD 690000. He was a co-chair of The International Polymer Processing Society Meeting (PPS-27) that hold for the first time in Africa and in the MENA region (Middle East and North Africa), in the City of Marrakech, Morocco, May 10–14, 2011.

Dr. Rachid Bouhfid is a researcher at Moroccan Foundation for Advanced Science, Innovation and Research (MAScIR), Institute of Nanomaterials and Nanotechnology (NANOTECH), Rabat, Morocco. He obtained his Ph.D. in organic chemistry from Mohammed V. Following his Ph.D. he joined Artois University, France as ATER, then Assistant Professor Dr. Rachid Bouhfid's research are mainly in the field of organic synthesis of functional molecules in order to use as modifier of inorganic fillers and the development of new polymeric nanocomposites based on clay and nanoclay, graphene, and natural fibers. He is the author/coauthor of more than 65 international scientific publications, 4 book chapters, and more than 12 patents. He is also a reviewer of several high-impact ISI journals of Elsevier, Springer, RSC, Bentham Science Publishers, etc. He co-organized several international conferences in the field of heterocyclic chemistry (Trans Mediterranean Colloquium on Heterocyclic Chemistry TRAMECH-7, Rabat, 2013) and polymer processing (The International Polymer Processing Society Meeting PPS-27, Marrakech, 2011).

Recent Advances in Nanoclay/Natural Fibers Hybrid Composites

N. Saba, M. Jawaid and M. Asim

Abstract The growing demand for continual improvement in the engineering applications of thermoplastic and thermoset polymer materials compared with metals in various applications led to the emergence of hybrid nanocomposites by the addition of different nano fillers, with advanced properties. Nano fillers such as carbon black, pyrogenic silica, nano oxides and metal particles are being used as additives in polymers from decades. However, the increasing stringent environmental legislation and consumer awareness highlights the importance of natural, low cost and abundant clay materials such as nanoclays. The hybridization of natural fiber with nanoclay is interestingly positive due to the tendency of nanoclay to upsurge both flexibility and rigidity of the natural fiber in one step. The most promising nanoclay involved in the modification of polymers and natural fibers reinforced polymer composite are montmorillonite, organoclay, saponite and halloysite nanotubes. Nanoclay/natural fibers hybrid nanocomposites have engrossed great attention since their discovery due to their wide variety of properties in food packaging, biomedical devices, automotive industries and other consumer applications with better thermal, physical, mechanical, optical and barrier properties. Present article designed to be a comprehensive source of recent literature and study on nanoclay fillers, its different classes, modification of polymers by nanoclay and their varied applications. This article also intended to covers the recent advances in natural fiber/nanoclay hybrid polymer nanocomposites research study, including their different commercial applications.

Keywords Natural fiber · Nano filler · Mineral clay · Nanoclay · Montmorillonite · Nanocomposite · Hybrid composites

N. Saba (✉) · M. Jawaid · M. Asim
Laboratory of Biocomposite Technology, Institute of Tropical Forestry and Forest Products (INTROP), Universiti Putra Malaysia, 43400 Serdang, Selangor, Malaysia
e-mail: naheedchem@gmail.com

M. Jawaid et al. (eds.), *Nanoclay Reinforced Polymer Composites*, Engineering Materials, DOI 10.1007/978-981-10-0950-1_1

1 Introduction

1.1 Hybrid Composites

The word "hybrid" is of Greek-Latin origin. Hybridization is a process of incorporating of two reinforcements (either synthetic fibers/nano fillers/natural/ or metallic fibers) in one polymeric matrix phase or the incorporation of single reinforcements in polymer blends in order to yield better properties such as (high mechanical strength, compressive strength, stiffness, thermal stability and lowered water absorption properties) which cannot be realized in conventional composite materials (Borba et al. 2014; Saba et al. 2014). Hybrid materials are very advanced composites materials consisting of two or more different constituents at the molecular or nanometer level. Properties of hybrid composites is a weighed sum of the individual components, thus it provide combination of properties such as tensile modulus, compressive strength and impact strength (Gururaja and Hari Rao 2012). Hybrid composite materials have extensive engineering application where strength to weight ratio, ease of fabrication and low cost are required. The hybrid composite properties with two different fibers are exclusively governed by the individual fibers length, fibers orientation and fibers content, extent of intermingling of fibers, fibers arrangement and fiber-matrix bonding.

Researchers concluded from their study that the addition of a very small amount of nanoparticle into a matrix can improve both thermal and mechanical properties significantly without compromising the weight or processability of the composite (Hossen et al. 2015). Until now many research works has been made on natural fiber hybrid nanocomposite materials by using nano particles such as nano tube, metal oxides, nanoclays, carbon nanofiber and other nano particles for varied advanced applications.

1.2 Natural Fibers

Currently researchers are concerning more towards the natural fibers as cost effective and light weight reinforced materials. A huge variety of different kinds of natural fibers including cellulosic, wool and animal fibers are abundantly available all around the world (Thakur et al. 2012, 2014). Natural fibers can be classified into different types depending upon their origin such as plant fibers and animal fibers (Eichhorn et al. 2010; Cherian et al. 2011). Natural fibers are of different types based on their source of origin from the plant such as leaf (PALF, sisal, banana etc.), bast (kenaf, jute, hemp etc.), fruit (OPEFB, coir) and seed (cotton, kapok). The mechanical properties of natural fibers (kenaf, jute) are found to be comparable with the commercial synthetic fibers, making the natural fibers highly preferable to synthetic fibers in automotive and other high end applications where weight and stiffness are of primary concerns (Saba et al. 2015b; Thakur et al. 2013). Natural

Fig. 1 Different types of natural fiber used as reinforcement

fibers exhibit a number of advantages in terms of lower densities, specific modulus, cost per weight, cost per unit length besides their renewable and biodegradable nature required for making composite and green composite with bio-resin (Dittenber and GangaRao 2012; Thakur and Thakur 2014). Some of the most common natural fiber that are used as reinforcement in polymer for fabricating hybrid polymer composite especially with nano sized fillers are displayed in Fig. 1.

1.3 Composites

Composite materials have been used by the people of earlier civilization since many centuries. One of finest example include the preparation of bricks for building construction using straw and mud (Akil et al. 2011; Eichhorn et al. 2010). Composite materials are the most adaptable engineering materials having the perfect combination of two different constituents, reinforcing natural/synthetic fibers and plastic polymeric matrix holding the perfect physical or chemical properties of each component (Yan et al. 2014). Moreover, both the constituent of composite materials remain separate and distinct on a macroscopic level within the finished structure (Bajpai et al. 2012). The term "natural fiber reinforced composite" refers to natural fibers reinforced in some polymeric matrix (thermoplastic or thermoset; natural or synthetic). A number of different polymers are used as matrix materials for composite fabrication depending upon the type of applications (Teacă et al. 2013). The matrix hold and transfer the loads along the reinforced materials (organic to inorganic) in polymer composites (Singh et al. 2013; Saba et al. 2015a) and the reinforcing material (usually fibers) determines the overall properties (Valença et al. 2015). Natural fiber composites are cost effective, possess tool wearing rates, thermal stability, acoustic properties, better formability and confers health safety for manufacturer (Saba et al. 2015b; Batouli et al. 2014).

1.4 Nanoscience and Nanotechnology

Nanoscience is the study of phenomena and manipulation of materials at atomic, macromolecular and at molecular scales, where properties differ significantly from those at a larger scale (Saba et al. 2014). The term "nanotechnology" was invented by Taniguchi in 1974 and is defined as "the science of nano materials (1–100 nm)" and systems with structures and components displaying improved chemical, physical, thermal and biological properties". Nanotechnology is the advanced technology that deals and describes the design, production, characterization, and application of nano based materials and systems. The field of nanoscience and nanotechnology has blossomed over the last 20 years (Galpaya 2012; Kuilla et al. 2010) and gaining significance in areas of biosensors, biomedical, nanomedicine, computing, aeronautics, polymer modification and many other applications owing to their high surface area with better tendency to amend materials properties. The current applications of nanotechnology in different industrial sectors are shown in Fig. 2.

1.5 Nanocomposites

The discovery of polymer nanocomposites by the Toyota research group has opened a new dimension in the field of materials science (Kuilla et al. 2010; Kojima et al. 2011; Ibeh and Bubacz 2008). A nanocomposite belong to the nanomaterials groups, where nanoparticles have been added to improve a particular property of the material (Saba et al. 2015a, b, c; Marquis et al. 2011). The nanocomposites material encompasses a large variety of systems and frequently regarded as amorphous multiphase solid material (such as the structure of the bone), made of distinctly different components at the nanometer scale having at least one or two or three of the

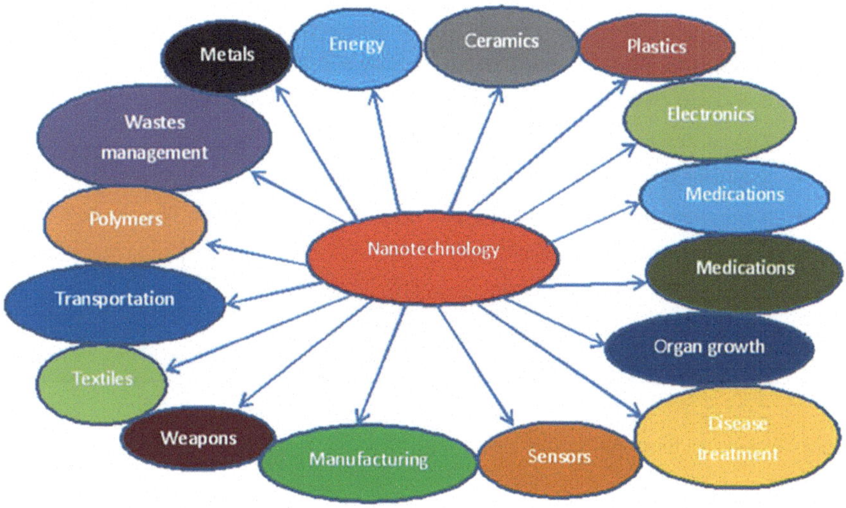

Fig. 2 Diverse applications of nanotechnology

phases dimensions in nano meter range (1–100 nm) (Saba et al. 2014). Moreover, according to the matrix materials, nanocomposites can be classified into three different categories (a) ceramic matrix nanocomposites (CMNC) such as SiO_2/Ni, Al_2O_3/TiO_2 (b) metal matrix nanocomposites (MMNC) such as Ni/Al_2O_3, Co/Cr and (c) polymer matrix nanocomposites (PMNC) such as thermoplastic/thermoset polymer/layered silicates (Raman et al. 2012). Nanocomposites emerged as appropriate substitutes to overwhelmed the limitations offered by monolithics and microcomposites, owing to the exceptionally high surface to volume ratio or high aspect ratio of the nano sized reinforcing phase (Boufi et al. 2014). Polymer nanocomposites often exhibit better chemical and physical properties dramatically different from the corresponding pure polymers. The polymer nanocomposites properties are governed by the incorporated nanoparticles shape, size, concentration and interactions with the polymer matrix (Karger-Kocsis et al. 2015; Armentano et al. 2010; Abdelrasoul et al. 2015). Nanocomposite however pretend some preparation challenges related with the dispersion and stoichiometry control in the nanocluster phase (Raman et al. 2012).

Currently, nanocomposites have attracted great interest, both in industry and in academia, due to their unique properties and multi-functionality that are well applicable in various fields (Saba et al. 2014) as compared with traditional composites (Babaei et al. 2014). Furthermore, despite of their nano dimensions, most of the processing techniques of nanocomposites remain almost the same as in microcomposites (Saba et al. 2015a, b, c). Several technique are been utilized to characterize or probe the structure and properties of polymer nanocomposites are shown in Fig. 3.

- **Transmission electron microscopy (TEM)** (Saba et al. 2015a)
- **Scanning electron microscopy (SEM)** (Hakamy et al. 2014)
- **Wide angle (WAXS) and Small angle (SAXS)** (Heeley et al. 2014)
- **X-ray diffraction (XRD)** (Najafi et al. 2012), (Ahmed et al. 2015)
- **Atomic force microscopy** (Fian et al., 2008; Zhu et al., 2002; Liu et al., 2003)
- **X-ray photoelectron spectroscopy** (Fian et al., 2008; Liu et al., 2003; Wang et al., 2002)
- **Fluorescence** (Jing et al., 2005)
- **Rheology** (Kashiwagi et al., 2008; Lee and Han 2003
- **Infrared spectroscopy (IR)**

Fig. 3 Techniques for characterizing nanocomposites

1.6 Nano Particles

Nanometer scale items (10^{-9} m) are defined by the term nano. A nanometer is 80,000 times thinner than a human hair and are equivalent to the billionth of a meter (Saba et al. 2015a, b, c; Raghavan et al. 2012). Nano particle or nano filler are the doping agent that are distributed in the polymer matrix of a composite having at least one of their dimensions in the nanoscale range (Dastjerdi and Montazer 2010). Currently nano fillers either from natural or synthetic source is of great interest and regarded as the most promising materials of the future owing to their unique properties in comparison with bulk counterparts. The most common nano fillers include carbon nanotubes, laminated alumosilicates (clays), nano fibers, ultra-disperse diamonds (nano diamonds), inorganic nanotubes, fullerenes, nano metal oxides, calcium carbonate, metallic nanoparticles, POSS and graphene (Henrique et al. 2009; Dey et al. 2016). Addition of nano fillers significantly improve or adjust the variable properties of the materials including physical, mechanical, optical, electrical and thermal properties, sometimes in synergy with conventional or traditional fillers (Saba et al. 2014; Shalwan and Yousif 2014). The nano materials can be synthesized by a variety of methods such as chemical, biological and mechanical techniques. The techniques involved in the synthesis are listed in Fig. 4.

Inorganic and organic nano fillers has gained considerable interest owing to their unique properties and numerous potential applications in the aerospace, automotive, electronics and construction industries (Kuilla et al. 2010; Potts et al. 2011). However, currently the majority of research are focused on polymer nanocomposites based on layered materials of a natural origin, such as a montmorillonite type of layered silicate compounds or synthetic clay (layered double hydroxide) (Li et al. 2009b; Kuilla et al. 2010; Paul and Robeson 2008). The nanoclay reinforced

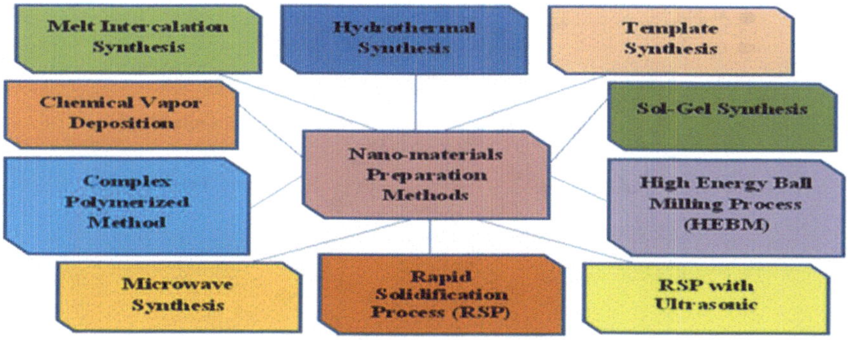

Fig. 4 The various methods for the synthesis of nano materials

polymer composites displays vast potential applications in the variety of areas such as electronics, sensors, information storage, catalysis and structural components (Kushwaha and Kumar 2011; Subramaniyan and Sun 2008). Research study illustrates that the usual content of lamellar nanoclay and organo-modified montmorillonite is in the range of 5–10 wt% due to their high aspect ratio (more than 1000), high surface area (more than 750 m^2/g) and higher modulus values (176 GPa) (Uddin 2013). Depending upon the processing conditions and characteristics of both the polymer matrix and organoclay, the in situ dispersion of organoclay inside the host polymer by melt blending can be more or less achieved, leading to intercalated or exfoliated nanocomposites. The general scheme of melt compounding and film casting methods involved in the preparation of nanoclay polymer composites are shown in Fig. 5.

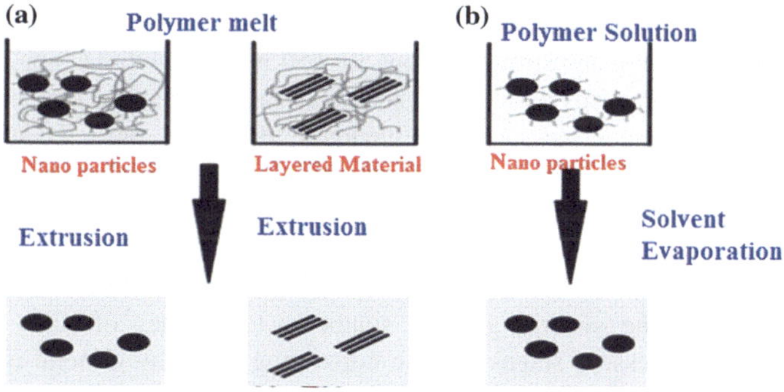

Fig. 5 Preparation methods for nanocomposites: **a** melt compounding, **b** film casting (Modified from: Li et al. 2010a, b)

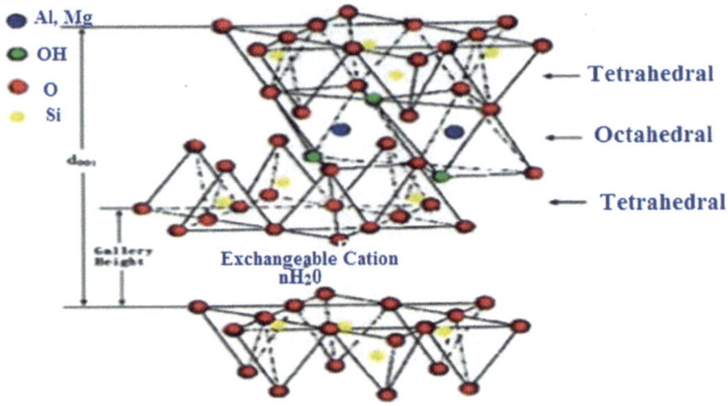

Fig. 6 General structure of clay mineral (Modified from: Nourbakhsh and Ashori 2009)

2 Clay Minerals

Clay deposits composed of clay minerals (phyllosilicate minerals) (Prasanth et al. 2013; Morsy et al. 2011). Clay is a naturally occurring mineral composed primarily of fine grained minerals. These minerals shows plasticity owing to variable water content and on drying or fired can be hardened (Zhang et al. 2015). The most typical structure of clay mineral are displayed in (Fig. 6).

2.1 Nanoclays

Nanoclays is the good example of naturally occurring nanomaterials and are generally used for a clay mineral having a phyllosilicate or sheet structure with a thickness of about 1 nm and surfaces about 50–150 nm in one dimension (Sedaghat 2013; Pavlidou and Papaspyrides 2008). Nanoclays are thus the general term for the layered mineral silicates nanoparticles having high aspect ratio. Depending on morphology of nano particles and chemical composition nanoclays are organized into various classes such as illite, halloysite, bentonite, kaolinite, montmorillonite, hectorite and chlorite (Pavlidou and Papaspyrides 2008). Researchers explored that the nanoclays can be obtained from raw clay minerals through fewer steps (Uddin 2013), instead of the several general synthesis techniques of nano particles. Descriptions of processing stages used in converting raw clay mineral into a nanoclay mineral are displayed in Fig. 7 (Uddin 2013).

The layered structure enables the material to either swell or shrink depending on its water uptake tendency. Moreover, the purity and cation exchange capacity of the nanoclay are one of the critical properties, as it provides the surface activity required for inclusion of modifiers and for surface treatments. The typical structure of the sodium based MMT are shown in Fig. 8 (Paul and Robeson 2008).

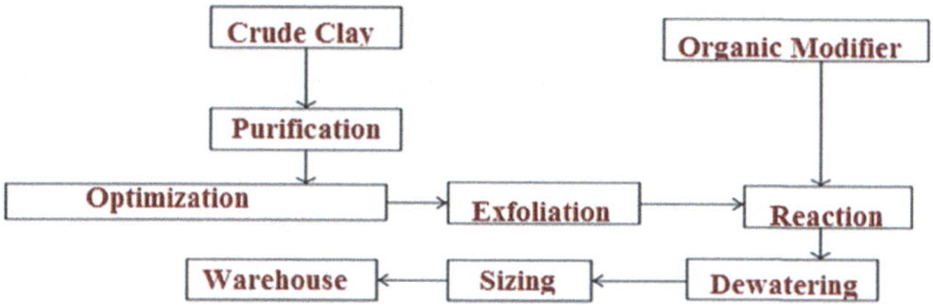

Fig. 7 Stages involved in the conversion of crude clay into nanoclays (Modified from: Uddin 2013)

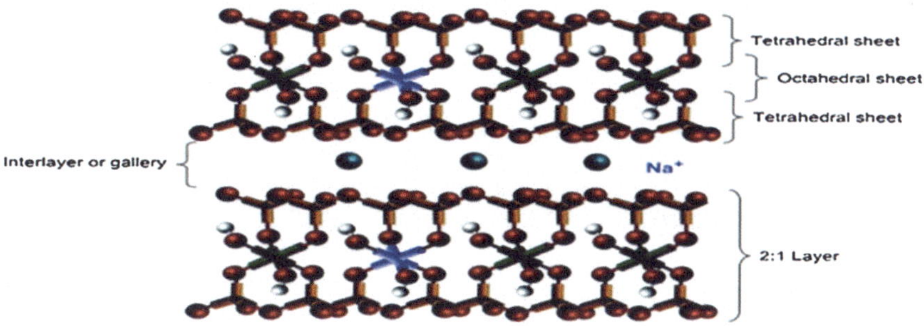

Fig. 8 Structure of sodium montmorillonite (Modified from: Paul and Robeson 2008)

Nanoclays products are typically modified with ammonium salts and are often called "Organically modified nanoclays or organoclays" (Zhang et al. 2015). Organoclays are one of the attractive and promising hybrid organic inorganic nanomaterials generally used for polymers and polymer based composites modification. The synthesis of organoclay is shown in Fig. 9. The clay surfaces modification render them organophilic to make them compatible with hydrophobic organic polymers (Jahanmardi et al. 2013).

Researchers stated that the structure of nanoclays or its dispersion in resins are of various types and can be characterized as phase separated, intercalated, or exfoliated displayed in Fig. 10 (Nabil et al. 2015).

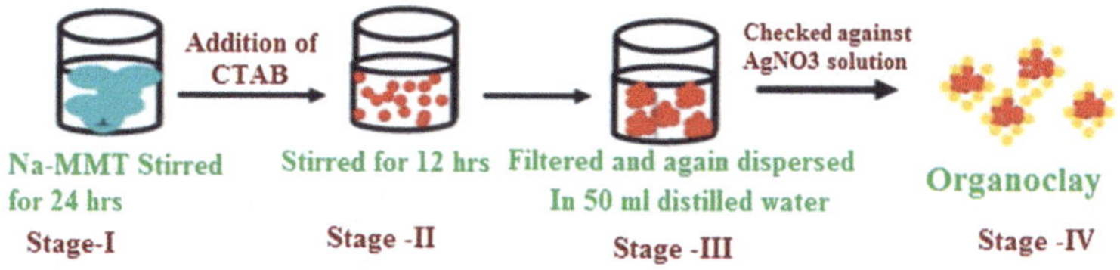

Fig. 9 Scheme for the preparation of nanoclay

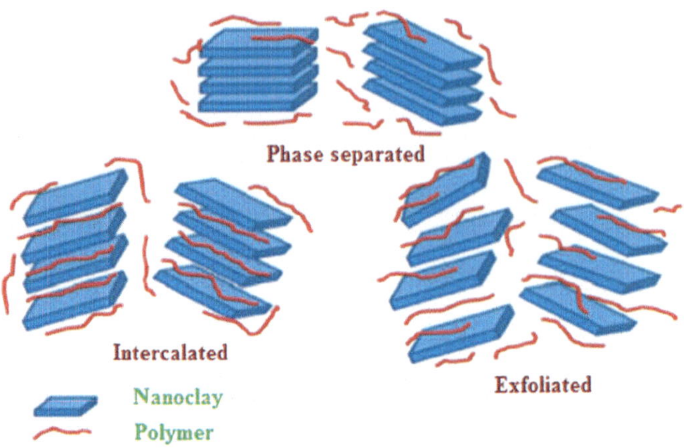

Fig. 10 Dispersion mechanism of nanoclay in resin (Modified from: Nabil et al. 2015)

2.2 Platy Nanoclays

The 2:1 phyllosilicates family of nanoclay is the most commonly used layered silicate in polymer nanocomposites (Jahanmardi et al. 2013). Their two-dimensional layers are made up of two tetrahedral coordinated silicon atoms fused to an edge-shared octahedral sheet of either aluminium or magnesium hydroxide. MMT, talc, hectorite, pyrophyllite, nontronite and saponite are the members of the layered smectite clay group (Sedaghat 2013; Henrique et al. 2009).

2.3 MMT Nanoclays

MMT nanoclays since 1847 have gained much research interests over the past decade. MMT was first labelled for its occurrence in montmorillon mineral in department of Vienne (France) more than 50 years before the discovery of bentonite in the US (Özcan et al. 2004; Ursache and Rodrigues 2014). However, chemically, it is hydrated sodium calcium aluminium magnesium silicate hydroxide $(Na, Ca)_{0.33}(Al, Mg)_2(Si_4O_{10})(OH)_2 \cdot nH_2O$ (Shirini et al. 2012; Sedaghat 2013). Most generally it is merged with muscovite, chlorite, illite, kaolinite and cookeite (Ehlmann et al. 2007). Iron, potassium and other cations are the common substitutes, but their ratios however varies greatly with source (Uddin 2008). MMT nanoclays are similar like saponite clay, belongs to extremely soft phyllosilicate group of minerals having extremely small nano sized particles, with fine layered grained structure (Shirini et al. 2012). MMT typically form in microscopic crystals and are derived from bentonite ore, forming anisotropic platy clay structure.

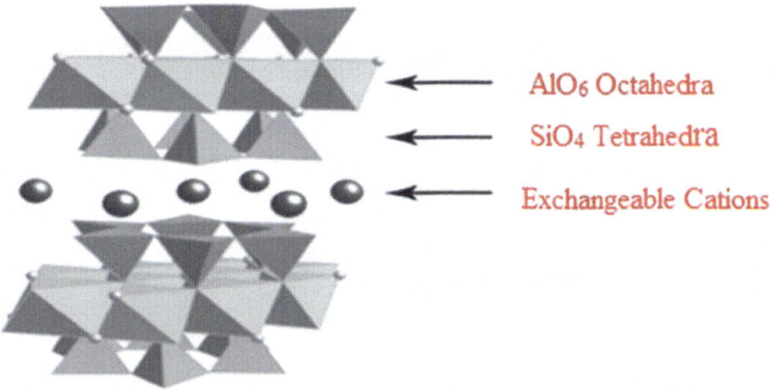

Fig. 11 Crystal structure of dioctahedral smectite type clay, MMT (Modified from: Nakato and Miyamoto 2009)

MMT is a subclass or member of the natural smectite group (2:1 clay), having 2 tetrahedral sheets sandwiching a central octahedral sheet. Crystal structure of 2:1 layered smectite nanosheet (MMT) are illustrated in Fig. 11 (Nakato and Miyamoto 2009). Thus MMT consist of aluminosilicate layers surface-substituted with metal cations stacked in ~ 10 μm-sized multilayer stacks. The particles are plate-shaped with an average diameter around 1 μm with the surface dimensions generally 300 to more than 600 nm, resulting in an unusually high aspect ratio (Nakato and Miyamoto 2009). MMT featured by having better cation exchange capacity and 50 % octahedral charge (due to isomorphous substitution of Mg for Al) (Srasra et al. 1994; Chan et al. 2011a, b).

Naturally occurring MMT is highly hydrophilic. The MMT water content tendency varies greatly and it increases greatly in volume when it absorbs water. Researchers also synthesized modified MMT through clay surface modification, since polymers are generally organophilic, the unmodified nanoclay disperses with great effort in polymers. After modification ("surface compatibilization or intercalation") the MMT can become compatible (organophilic) to the conventional petroleum polymers and dispersed readily in polymers for better properties in nanocomposites (Nakato and Miyamoto 2009).

2.4 Halloysite Nanotubes (HNTs)

Halloysite nanotubes were firstly reported by Berthier as a di-octahedral 1:1 clay mineral of the kaolin group in 1826, and its deposits are extensively found widely in the soils of countries such as France, China, Belgium, New Zealand and Australia (Du et al. 2010). HNTs are unique and versatile white and slight red nanomaterials (Fig. 12a), formed naturally within the earth over millions of years,

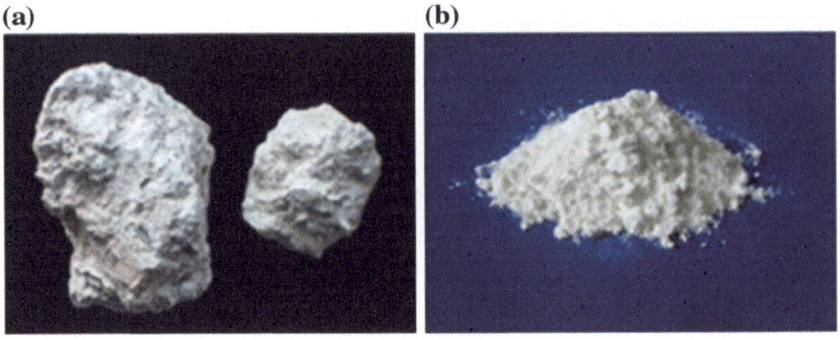

Fig. 12 Showing raw halloysite (**a**) and ground halloysite (**b**) (Liu et al. 2014)

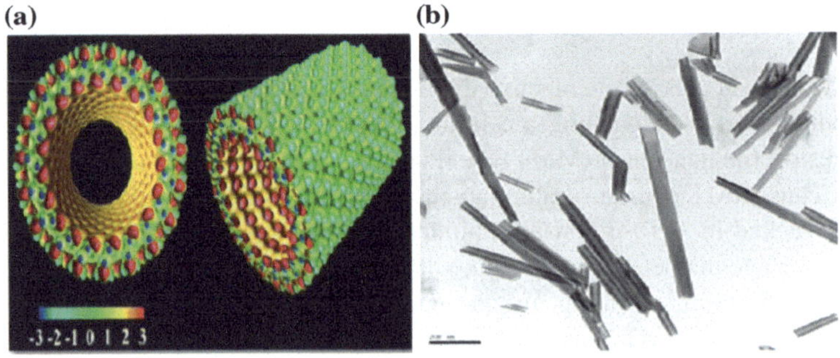

Fig. 13 Showing HNT model (**a**) and the TEM image of HNTs (**b**). **a** Kamble et al. (2012) **b** Du et al. (2010)

due to strain caused by lattice mismatch between adjacent silicon dioxide and aluminium oxide layers (Kamble et al. 2012). The stone-like raw halloysite is easily ground into powder (Fig. 12b) (Liu et al. 2014).

HNT is the naturally occurring double layered aluminosilicate (Al_2 Si_2 $O_5(OH)_4 \cdot 2H_2O$), ultra-tiny hollow tubular structure with diameters typically lesser than 100 nm and with lengths typically ranging from about 500 nm to over 1.2 μm (Kamble et al. 2012). HNT model and its SEM micrograph are shown in Fig. 13a, b. The HNTs rapidly emerged as a raw material for the future because of being relatively cheaper, abundant and sustainable nature compared to commercial and expensive CNTs. They are chemically similar to kaolin clay with several interesting applications in the construction industry (Yunsheng et al. 2008; Montes et al. 2015). Recently researchers developed two types of halloysite nanotubes (single walled and multi walled) models to amplify effectively its nanotechnology applications (Zhang et al. 2015).

3 Nanoclay Polymer Composites

Nowadays the growing concern towards the utilization of renewable materials as filler in the matrix highlights greatly the importance of layered silicates clay minerals as nano filler owing to its availability and versatility (Ray and Okamoto 2003). MMT and organoclays are widely used for dispersion in polymers due to the high aspect ratio and the large interface of the polymer nanoclay interaction in the polymer composites (Fareed and Stamboulis 2014). Toyota research group in the early 90s, declared the first use of nanoclays reinforcement in the manufacturing of nylon-6-based nanoclay composites. They concluded that addition of nanoclays perfectly influence the crystallization process and the structural changes (Sandler et al. 2004). Untill now a variety of polymers except a few hydrophilic polymers such as poly (vinyl alcohol) and poly(ethylene oxide) which are modified by organoclay are well reported in literature.

Depending on the process conditions and polymer/nanofillers bonding the layered silicates dispersed into the polymer matrix in different states of intercalation and/or exfoliation (Baniassadi et al. 2011). Polymer nanoclay composites involve the interaction of polymer matrix with the nano-plates of clay, formed by the dispersion of low weight percentages of nanoclay into polymers (Fareed and Stamboulis 2014). The three different types of nanocomposite usually results by the dispersion of nanoclay to the polymer matrix are shown in Fig. 14 (Baniassadi et al. 2011).

However, the exfoliated structures displayed far better properties compared to other two nanocomposites (Pollet et al. 2004). Nanoclay reinforced polymer composites and their laminates have excellent properties including enhanced physical (optical, dielectric, permeability, transparency and lower shrinkage), thermal (decomposition temperature, flammability, coefficient of thermal expansion and higher thermal stability) and mechanical properties (tensile toughness, tensile strength and tensile modulus) even at a very low filler loading with respect to pure polymers (Babaei et al. 2014; Hossen et al. 2015; Najafi et al. 2012). Untill now many research study were made, to investigate the effect of nanoclay loadings on the properties of different polymers. The nanoclays addition in very low weight

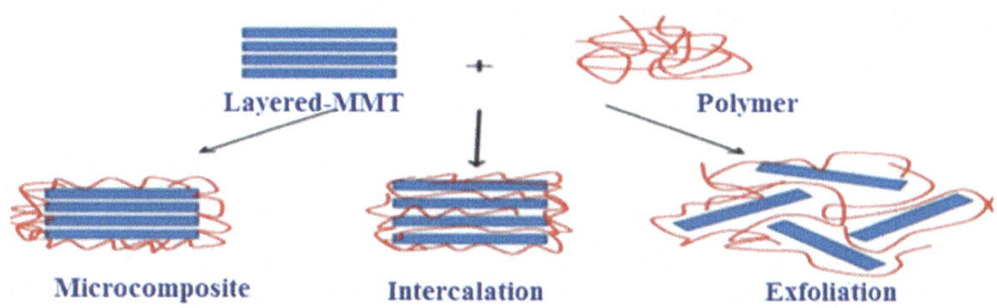

Fig. 14 Structural morphologies in polymer/MMT (nanoclay) nanocomposite (Modified from: Baniassadi et al. 2011)

Table 1 Reported study on nanoclay based polymer nanocomposites

Reinforcements/polymer	Properties improvement	References
Nanoclay/polyamide	Nanocomposites shows effective thermal conductivity, elastic modulus and mechanical properties	Baniassadi et al. (2011)
MMT aerogels/NR	Improved mechanical, rheological, swelling properties and viscoelastic response	Pojanavaraphan et al. (2010)
PLA/nanoclay	Improved mechanical properties (Peak stress, elongation at break, young modulus and degradation rate	Nieddu et al. (2009)
Nanoclay/PEU	Nanoclays acts as effective flame retardants coatings for polyester urethanes. It also improves the physical, mechanical and thermal stability of the composite	Patel and Patel (2014)

Notes Polyester urethanes (PEU), Natural rubber (NR), Montmorillonite (MMT), Polylactic acid (PLA)

percent (1–5 %) can significantly improves the properties of the polymer base. Some of the important research work on nanoclay based polymer composites were tabulated in Table 1.

4 Nanoclay Hybrid Polymer Composites

Many researchers have used MMT nanoclay as filler in hybrid polymeric composites and their laminates due to its well-known exfoliation/intercalation chemistry, surface reactivity, high surface area, cost effectiveness and easy availability (Hossen et al. 2015; Dewan et al. 2013). Some of the important research work on nanoclay based polymer composites were tabulated in Table 2. A variety of natural fibers (from jute to oil palm fibers, either in short fiber, nonwoven mat or in woven fabrics) are modified by the nanoclays.

5 Applications of Nanoclays and Polymer Nanoclay Composites

Clay materials are relatively commercially available, inexpensive, easily can be modified, non-corrosive and recyclable in nature (Shirini et al. 2012; Jahanmardi et al. 2013; Najafi et al. 2012). They have been explored from decades as they shows huge importance in various fields of agriculture, ceramics, construction and other industrial or domestic applications. Nanoclays (layered silicates) and organoclay are of considerable interest for bulk applications from many decades as they have high aspect ratio, large specific surface areas associated with platy layered

Table 2 Reported study on natural fiber/nanoclay hybrid nanocomposites

Reinforcements/polymer	Properties improvement	References
WSF/nanoclay/foamed HDPE	The addition of nanoclay up to 2 % improved the mechanical resistance. The water absorption and thickness swelling properties of HPDE/WSF composites also witnessed reduction by the addition of nanoclay	Babaei et al. (2014)
Glass fiber/nanoclay/polypropylene	Incorporation of nanoclay improves the thermal stability, flexural strength, flexural modulus and tensile strength of the hybrid material	Rahman et al. (2013)
Almond shell flour/OMMT/polypropylene	3 wt% OMMT filler addition decreases the water absorption and thickness swelling. It also improves the physicomechanical (tensile and flexural properties) properties of the developed hybrid composites	Zahedi et al. (2015)
Bamboo fiber/nanoclay/HDPE	Nanoclays addition significantly enhanced the modulus properties of the HDPE–clay hybrids	Han et al. (2008)
Wild cane grass fiber/MMT nanoclay/polyester	The addition of nanoclays improves the tensile strength, modulus and impact strength of the wild cane grass fiber composites. The weight loss of nanoclay filled wild cane grass fiber/polyester composites found to be 30 and 22 % less compared to composites without nanoclay at maximum percentage volume of fiber	Prasad et al. (2015)
Flax fiber/nanoclay/soy protein resin	Nanoclay addition considerably enhanced the mechanical (tensile and flexural) properties of the hybrid composite	Huang and Netravali (2007)
Jute/nanoclay/polyester	Enhancement of mechanical (compression, flexural, ILSS) and dynamic mechanical properties by the addition of nanoclay. It also reduces water absorption properties of JPC	Dewan et al. (2013)
Wood fiber/nanoclay/polypropylene	Nanoclay addition improves both the mechanical and physical properties. It also lowers the water absorption properties with relative to pure composite	Nafchi et al. (2015)
OPEFB/nanoclay/PU foam	The addition of nanoclays improves both, the thermal and barrier properties of BHPU composite	Ali and Ahmad (2012)

(continued)

Table 2 (continued)

Reinforcements/polymer	Properties improvement	References
WF/MMT/PLA	Intercalation of MMT in the PLA matrix enhances the tensile modulus and water resistance properties of the bio-composite.	Kumar et al. (2010)
WPI/Modified nanoclay (Cloisite Na^+, Cloisite 20A, and Cloisite 30B)	Nanoclay significantly improves the tensile and the water resistant properties of the developed composite films. The WPI/Cloisite 30B composite films showed a strong resistance towards bacteriostatic effect of Listeria monocytogenes and Gram-positive bacteria	Sothornvit et al. (2009)
Flexible PU/organoclay	Organoclay incorporation in PU foams leads to reduction in cell size. It also upsurge cell density and tensile modulus of the developed foams	Jahanmardi et al. (2013)
Sepiolite clay/epoxy-DGEBA	Sepiolite reduces the whole degradation process on the temperature scale. It improves the thermal stability and flame retardant effects of the hosting epoxy by fasten the char layer formation during the burning process on both condensate and vapor phase	Zotti et al. (2015)
HF/nanoclay/Portland cement	Improved the thermal stability, lowered the water absorption, porosity. The density, fracture toughness, flexural strength and impact strength get improved by the addition of nanoclays	Hakamy et al. (2014)
Treated and untreated Jute fibers/nanoclay/polyethylene	Nanoclays incorporation improves both physical and mechanical properties of the hybrid composite. Tensile strength and modulus improves by approximately 8 and 15 % respectively with respect to the composites without nanoclays	Hossen et al. (2015)
Poplar (Populus nigra L) sawdust/nanoclay/polypropylene	Nanoclay considerably inhibited the growth of wood-deteriorating fungi in the wood polymer composite	Bari et al. (2015)
Wood flour/nanoclay/ polypropylene	Both mechanical (flexural and tensile strengths) and physical properties improved by the incorporation of nanoclay. The water absorption properties significantly reduced in nanoclays filled composite	Yusoh (2015)
Wood flour/glass fiber nanoclay/polypropylene	Tensile modulus and tensile strength increases, while the water absorption tendency decreases considerably by the addition of nanoclay	Kord and Kiakojouri (2011)

(continued)

Table 2 (continued)

Reinforcements/polymer	Properties improvement	References
Nanoclay/low molecular weight PF	Addition of nanoclay into a phenolic matrix significantly increased the surface hardness, MOE, dimensional stability, water repellency and abrasion resistance	Nabil et al. (2015)
Nanoclay/glass fiber/polyester	Nanoclay improves mechanical properties and also accelerates the tool wear.	Prabhu et al. (2013)
OPMF/nanoclay/PLA/PCL	Nanoclay successfully improves the flexural properties, impact resistance, water resistance and compatibilizes the OPMF/matrix adhesion.	Eng et al. (2014)
Nanoclay/sisal fiber/epoxy	Tensile, dynamic properties and wear properties improves with nanoclay incorporation	Mohan and Kanny (2011)
Pine cone fibers/nanoclay/ polypropylene	Improvement in mechanical and thermal properties by the addition of nanoclay	Arrakhiz et al. (2013)
Poly(ester–urethane)/OMMT	Nanoclay addition increases the Tg, increased non-flammability (V-0 rating). It also improves morphology, mechanical and thermal properties	Kausar (2015)
Wood flour/nanoclay/ polypropylene	Improvements in thermal stability and resistivity towards water absorption by nanoclay addition	Tabari et al. (2011)
Reed fiber-MAPP/nanoclay/ polypropylene	Tensile modulus and strength of hybrid composites significantly increased. The water absorption property decreases with nanoclay loading.	Najafi et al. (2012)
Bamboo fiber/nanoclay/ epoxy/polyester	Addition of nanoclay enhances the elastic modulus, thermal stability, water absorption resistivity and wear resistance properties of hybrid composites	Kushwaha and Kumar (2011)
Bagasse/nanoclay/polypropylene	Nanoclay improves the physical and mechanical properties, lowered the water absorption tendency compared with conventional composites	Nourbakhsh and Ashori (2009)
Untreated and silane treated NaOH/woven coconut sheath/MMT/polyester	Dynamic factors and Tg value slightly increases with the incorporation of nanoclay in the composite system	Rajini et al. (2013)
Wood fiber/nanoclay/HDPE	Nanoclay enhance the mechanical properties and the flame retardancy of WPCs	Lee et al. (2010)
MMT/wood flour/HDPE	Mechanical properties of HDPE/wood-flour composites get considerably improved by MMT incorporation	Faruk and Matuana (2008)

(continued)

Table 2 (continued)

Reinforcements/polymer	Properties improvement	References
Industrial hemp/nanoclay/UPE-EMS	Nanoclays addition retains stiffness without sacrificing toughness. It also improves the thermal barrier properties	Haq et al. (2008)
Coir fiber/wood fiber/MMT/polypropylene	Physicomechanical and biodegradable properties subsequently increases by the incorporation of nanoclays	Islam et al. (2015)
HF/nanoclay/cement	Nanoclay addition increases the microstructure density, thermal stability, flexural strength, fracture toughness, and its impact strength. Beside this it reduced the porosity and water absorption tendency as compared to nanoclay unfilled HF cement composite	Hakamy et al. (2014)
Sisal fiber/nanoclay/metakaolin/Portland cement	The interfacial bond between cement matrix and fiber and flexural properties get improved significantly by the addition of nanoclay	Wei and Meyer (2014)
Carbon fibers/E-glass fiber/OMMT/epoxy	Mechanical properties get considerably improved by the incorporation of OMMT	Liu et al. (2004)
Carbon fibers/glass fibers/MMT/wood flour/polypropylene	The MMT addition enhanced the tensile modulus by 34 % and lowers the water absorption tendency of the hybrid composite	Turku and Kärki (2014)

Notes Wheat straw flour (WSF), Organically modified montmorillonite (OMMT), High density polyethylene (HDPE), Inter-laminar shear strength (ILSS), Unsaturated polyester (UPE), Epoxidized soybean oil (EMS), Hemp fabric (HF), Modified montmorillonite (OMMT), Phenol formaldehyde (PF), Polycaprolactone (PCL), Oil palm mesocarp fiber (OPMF), Natural rubber (NR), Whey protein isolate (WPI), Woven flax (WF), Polyurethane (PU), Oil palm empty fruit bunch (OPEFB), Melting temperature increases (Tm), Storage modulus (E′), Jute fabric reinforced polyester composites (JPC), Bio-nanocomposite hybrid polyurethane (BHPU), Modulus of elasticity (MOE), Glass transition temperature (Tg), wood fiber/plastic composites (WPCs), Polyester urethanes (PEU)

morphology, to results an improvement in the mechanical properties of polymers. Nanoclays also being used as flame retardants in the plastics industry to improve fire safety of their products. Integrating nanoclays into the plastic reduces drastically the amount of combustible material through a protective layer preventing the spreading of the fire. The transparency of nanoclay are also being used in films for achieving maximum mechanical properties and optimum clarity, as the presence of impurities act as stress concentrators, resulting in poor impact and tensile properties (Zhang et al. 2015). Nanoclays also conveys some practical potential biomedical applications such as in anti-bacterial efficacy, sterilizing phenomenon, toxins adsorption and in membrane coating (Dastjerdi and Montazer 2010; Wilson 2003).

Currently, one of the major applications of nanoclays is in the fabrication of polymer clay nanocomposites. These natural nanomaterials also acts as a rheological modifiers and additives in paints, inks, greases and cosmetics, pollution control bio-systems carriers, delivery systems for the controlled release of drugs (Floody et al. 2009). The incorporation of small amount (0.5–5.0 wt%) of MMT in the polymer matrix modified the polymer leading to a remarkable improvement of the physical, mechanical, fracture, wear resistance, thermal stability, peak heat release rate, flame retardancy, biocompatibility and chemical properties of the resulting composite as compared to conventional materials (Li et al. 2010a, b; Kushwaha and Kumar 2011; Uddin 2013). Currently the most particularly polymers that are modified includes epoxy, polyester, polyethylene, polypropylene, polystyrene and nylon using platy nanoclay mineral (Uddin 2013). MMT clays and acid treated MMT also been extensively used in catalytic cracking, acid-based catalysis processes and in materials applications from past 60 years. Organoclays are potentially being used as gas absorbents, rheological modifiers, polymer nanocomposites and as drug delivery carriers (Matsuda et al. 2013; Shirini et al. 2012).

In recent time, HNTs are the economically available and relatively cheaper nano sized raw material as compared to morphologically similar MWCNTs to apply in conjunction with various polymers such as nitrile rubber, natural rubber and polypropylene to form polymer-halloysite composite nanotubes (Rawtani and Agrawal 2012). Halloysite is a nontoxic and biocompatible material, which makes it a potential candidate for medical applications and household products (Zhang et al. 2015).

HNTs displays vast non-biological and biological applications such as drug and DNA carrier, tissue engineering scaffold, for cellular response, for osteoblasts and fibroblasts response (Rawtani and Agrawal 2012; Liu et al. 2014). HNTs are the main active ingredient in the "living nature deep cleansing mask" (New Zealand) which deeply purifies and refines the facial pores. The hollow HNTs tube can be filled with various active ingredients to meet the requirements for household, cosmetics, personal care and extended release products (Liu et al. 2014). HNT also being used for atom transfer radical polymerization, layer-by-layer coatings on wood microfibers, polyurethane coatings, scratch resistance coatings, also acts as a sorbents for various contaminants and pollutants (Li et al. 2013). Nowadays they are also being used for catalytic conversions, differentiation studies, hydrocarbon processing, molecular hydrogen storing, dispersion improvement and as an enzymatic nano reactor (Rawtani and Agrawal 2012).

Polymer nanoclay composites are one of the most promising alternatives to steel and wood based polymer composite, initially developed for the aerospace industry. Table 3 summarizes the different applications of nanoclays and their polymeric nanocomposites.

Table 3 Tabulates the different applications of nanoclays and their polymeric nanocomposites

Applications of nanoclay and its types				Applications of nanoclay polymer composite
Nanoclays	MMT	Organoclays (Cloisite® 30B)	HNT	Nanoclay based nanocomposites
Enhance properties of many polymers (thermoset and thermoplastics)	Fabrication of polymer nanocomposites with superior thermal, mechanical, gas-barrier and water resistant properties	Fabrication of polymer clay nanocomposites	Fabricating high performance polymer nanocomposites and green composites with natural rubber	MMT and HNTs nanocomposite application in optical, electronic, magnetic, thermosensitive and as biopolymer composites devices
Properties improvement of polypropylene spun bond fabrics	Catalytic cracking catalysts from past 60 years	Dimensional stability of flexible polyurethane foams	Bio-nanomaterial and formation of bio-nanocomposite film	MMT and HNTs perfectly modified the properties of epoxy resin, polypropylene, polyamide, natural rubber
Wide-spectrum mycotoxin binder	Modification of the polymeric materials	Rheological modifiers in paints, inks, grease and cosmetics	For corrosion prevention and fire retardant for nylon 6 and polymers as insulating coatings	Anti-microbial and green nanocomposites from HNTs. Formation of bio-nanomaterial and formation of bio-nanocomposite film
Wastewater treatment as a potential and cheaper adsorbent for cationic dyes	Materials applications	Color, shining retention and coverage for cosmetics (lipsticks, nail lacquers and eye shadows)	As adsorbent for cationic dyes such as neutral red (NR), methylene blue, malachite green from dye contaminated wastewater or aqueous solutions	Civil infrastructure
Drug delivery applications	Synthesis of commercial glass ionomer cement	Thicken the lubricating oils, greases	Cosmetics formulation, tissue engineering, macro molecular delivery systems and cancer cell isolation applications	Commercial packaging

(continued)

Table 3 (continued)

Applications of nanoclay and its types				Applications of nanoclay polymer composite
Nanoclays	MMT	Organoclays (Cloisite® 30B)	HNT	Nanoclay based nanocomposites
Rheological modifiers in inks, grease, paints, varnishes and cosmetics	Adsorb bacterial toxins associated with gastrointestinal disturbance and hydrogen ions produced in acidosis and steroidal metabolites related with pregnancy	Formulation of ink to amend the evenness of printing inks	Host for the polymerization host for several conductive and non-conductive polymers	Medicine
Bio-medical applications (sterilizing effect, adsorption of toxins membrane coating)	Carrier and in drug delivery applications	Removes humic acid from potable water	For immobilization of metalloporphyrins and silver nanoparticle	Domestic household appliances
Modification of textiles, UV-absorber	Pollution control and water treatment	Pollution control and waste water treatment	Manufacture of high quality white and ceramic ware	Mostly in Aerospace and industries automobiles. nanoclay composites in sensor technology for making sensor coating materials and in QCM sensors

Source Nakato and Miyamoto (2009), Sedaghat (2013), Floody et al. (2009), Jahanmardi et al. (2013), Jahanmardi et al. (2013), Li et al. (2009a), Floody et al. (2009), Fareed and Stamboulis (2014), Subramaniyan and Sun (2008), Rawtani and Agrawal (2012), Du et al. (2010), Marney et al. (2008), Dastjerdi and Montazer (2010), Mittal (2009)

They are now feature in almost every area of modern industrial production and enterprise, like automobiles, household appliances, medicine, weaponry, civil infrastructure and packaging (Floody et al. 2009). Polymer nanocomposites reinforced with HNTs possess highly increased tensile and flexural strength, toughness, elastic moduli, higher thermal stability, flame retardancy and unique crystallization behavior (Liu et al. 2014; Yuan et al. 2015).

6 Conclusion

Research activities in the area of nano-materials have been increased dramatically from past decades. Currently nanoclays are the most valuable nano fillers displaying great reinforcing effects due to their high thermal stability, improved mechanical strength, biocompatibility and great abundance. Natural tubules HNTs are unique, versatile and important class of nanoclays extensively used in modifying polymer in polymer composites formed by surface weathering of aluminosilicate minerals. We concluded that nanoclays minerals are the most promising alternative to expensive nano fillers such CNT's, POSS and graphene in improving or modifying the polymer and polymer composite properties with exciting potential applications. The incorporation of naturally existing nano materials as filler in nanocomposites exhibit distinctly improved properties such as improved thermal stability, higher flame retardancy, reduced CTE, better mechanical performance and improved clarity for optical applications. Organoclay, MMT and HNTs are of considerable interest in the field of polymer nanocomposites both for structural and functional applications. In this review article, we provided an overview of the recent research regarding nanoclay including MMT, HNT, organically modified nanoclays, nanoclays based polymer nanocomposites and an insight into their different potential applications. This review paper expected to delivers valuable evidence or literature information for further research and in the elaborative study of novel nanoclay based macro-micro hybrid nano polymeric composites as compared to other nanocomposites derived from commercially synthetic nano fillers.

Acknowledgments The first author acknowledges the International Graduate Research Fellowship (IGRF) UPM Malaysia grant to support this work. The authors also thankful to the Universiti Putra Malaysia for supporting this research study through Putra Grant Vot No. 9420700.

References

Abdelrasoul, G.N., Farkas, B., Romano, I., Diaspro, A., Beke, S.: Nanocomposite scaffold fabrication by incorporating gold nanoparticles into biodegradable polymer matrix: synthesis, characterization, and photothermal effect. Mater. Sci. Eng., C **56**, 305–310 (2015). doi:10.1016/j.msec.2015.06.037

Akil, H.M., Omar, M.F., Mazuki, A.A.M., Safiee, S., Ishak, Z.A.M., Abu Bakar, A.: Kenaf fiber reinforced composites: a review. Mater. Des. **32**, 4107–4121 (2011). doi:10.1016/j.matdes.2011.04.008

Ali, E.S., Ahmad, S.: Bionanocomposite hybrid polyurethane foam reinforced with empty fruit bunch and nanoclay. Compos. Part B Eng. **43**, 2813–2816 (2012). doi:10.1016/j.compositesb.2012.04.043

Armentano, I., Dottori, M., Fortunati, E., Mattioli, S., Kenny, J.M.: Biodegradable polymer matrix nanocomposites for tissue engineering: a review. Polym. Degrad. Stab. **95**, 2126–2146 (2010). doi:10.1016/j.polymdegradstab.2010.06.007

Arrakhiz, F.Z., Benmoussa, K., Bouhfid, R., Qaiss, A.: Pine cone fiber/clay hybrid composite: mechanical and thermal properties. Mater. Des. **50**, 376–381 (2013). doi:10.1016/j.matdes.2013.03.033

Babaei, I., Madanipour, M., Farsi, M., Farajpoor, A.: Physical and mechanical properties of foamed HDPE/wheat straw flour/nanoclay hybrid composite. Compos. Part B Eng. **56**, 163–170 (2014). doi:10.1016/j.compositesb.2013.08.039

Bajpai, P.K., Singh, I., Madaan, J.: Comparative studies of mechanical and morphological properties of polylactic acid and polypropylene based natural fiber composites. J. Reinf. Plast. Compos. **31**, 1712–1724 (2012). doi:10.1177/0731684412447992

Baniassadi, M., Laachachi, A., Hassouna, F., Addiego, F., Muller, R., Garmestani, H., Ahzi, S., Toniazzo, V., Ruch, D.: Mechanical and thermal behavior of nanoclay based polymer nanocomposites using statistical homogenization approach. Compos. Sci. Technol. **71**, 1930–1935 (2011). doi:10.1016/j.compscitech.2011.09.008

Bari, E., Taghiyari, H.R., Schmidt, O., Ghorbani, A., Aghababaei, H.: Effects of nano-clay on biological resistance of wood- plastic composite against five wood-deteriorating fungi. Maderas. Cienc. y Tecnol. **17**, 205–212 (2015). doi:10.4067/S0718-221X2015005000020

Batouli, S.M., Zhu, Y., Nar, M., D'Souza, N.A.: Environmental performance of kenaf-fiber reinforced polyurethane: a life cycle assessment approach. J. Clean. Prod. **66**, 164–173 (2014). doi:10.1016/j.jclepro.2013.11.064

Borba, P.M., Tedesco, A., Lenz, D.M.: Effect of reinforcement nanoparticles addition on mechanical properties of SBS/Curauá fiber composites. Mater. Res. **17**, 412–419 (2014)

Boufi, S., Kaddami, H., Dufresne, A.: Mechanical performance and transparency of nanocellulose reinforced polymer nanocomposites. Macromol. Mater. Eng. **299**, 560–568 (2014). doi:10.1002/mame.201300232

Chan, M.-L., Lau, K.-T., Wong, T.-T., Hoe, M.-P., Hui, D.: Mechanism of reinforcement in a nanoclay/polymer composite. Compos. Part B Eng. **42**, 1708–1712 (2011a). doi:10.1016/j.compositesb.2011.03.011

Chan, M.L., Lau, K.T., Wong, T.T., Cardona, F.: Interfacial bonding characteristic of nanoclay/polymer composites. Appl. Surf. Sci. **258**, 860–864 (2011b). doi:10.1016/j.apsusc.2011.09.016

Cherian, B.M., Leão, A.L., De Souza, S.F., Costa, L.M.M., De Olyveira, G.M., Kottaisamy, M., Nagarajand, E.R., Thomas, S.: Cellulose nanocomposites with nanofibres isolated from pineapple leaf fibers for medical applications. Carbohydr. Polym. **86**, 1790–1798 (2011). doi:10.1016/j.carbpol.2011.07.009

Dastjerdi, R., Montazer, M.: A review on the application of inorganic nano-structured materials in the modification of textiles: focus on anti-microbial properties. Colloids Surf., B **79**, 5–18 (2010). doi:10.1016/j.colsurfb.2010.03.029

Dewan, M.W., Hossain, M.K., Hosur, M., Jeelani, S.: Thermomechanical properties of alkali treated jute-polyester/nanoclay biocomposites fabricated by VARTM process. J. Appl. Polym. Sci. **128**, 4110–4123 (2013). doi:10.1002/App.38641

Dey, A., Bajpai, O.P., Sikder, A.K., Chattopadhyay, S., Khan, M.A.S.: Recent advances in CNT/graphene based thermoelectric polymer nanocomposite: a proficient move towards waste energy harvesting. Renew. Sustain. Energy Rev. **53**, 653–671 (2016). doi:10.1016/j.rser.2015.09.004

Dittenber, D.B., GangaRao, H.V.S.: Critical review of recent publications on use of natural composites in infrastructure. Compos. Part A Appl. Sci. Manuf. **43**, 1419–1429 (2012). doi:10.1016/j.compositesa.2011.11.019

Du, M., Guo, B., Jia, D.: Newly emerging applications of halloysite nanotubes: a review. Polym. Int. **59**, 574–582 (2010). doi:10.1002/pi.2754

Ehlmann, B.L., Mustard, J.F., Bishop, J.L., Griffes, J.L.: New secondary minerals detected by MRO CRISM and their geologic settings: kaolinite, chlorite, illite/muscovite, and the possibility of serpentine or carbonate in Nili Fossae. LPI Contrib. **1353**, 3270 (2007)

Eichhorn, S.J., Dufresne, A., Aranguren, M., Marcovich, N.E., Capadona, J.R., Rowan, S.J., Weder, C., Thielemans, W., Roman, M., Renneckar, S., Gindl, W., Veigel, S., Keckes, J., Yano, H., Abe, K., Nogi, M., Nakagaito, A.N., Mangalam, A., Simonsen, J., Benight, A.S., Bismarck, A., Berglund, L.A., Peijs, T.: Review: current international research into cellulose nanofibres and nanocomposites. J. Mater. Sci. **45**, 1–33 (2010)

Eng, C.C., Ibrahim, N.A., Zainuddin, N., Ariffin, H., Yunus, W.M.Z.W., Then, Y.Y.: Enhancement of Mechanical and Dynamic Mechanical Properties of Hydrophilic Nanoclay Reinforced Polylactic Acid/Polycaprolactone/Oil Palm Mesocarp Fiber Hybrid Composites (2014)

Fareed, M., Stamboulis, A.: Effect of nanoclay dispersion on the properties of a commercial glass ionomer cement. Int. J. Biomater. **2014**, 685389 (2014). doi:10.1155/2014/685389

Faruk, O., Matuana, L.: Nanoclay reinforced HDPE as a matrix for wood-plastic composites. Compos. Sci. Technol. **68**, 2073–2077 (2008). doi:10.1016/j.compscitech.2008.03.004

Floody, M.C., Theng, B.K.G., Mora, M.L.: Natural nanoclays: applications and future trends—a Chilean perspective. Clay Miner. **44**, 161–176 (2009). doi:10.1180/claymin.2009.044.2.161

Galpaya, D.: Recent advances in fabrication and characterization of graphene-polymer nanocomposites. Graphene **01**, 30–49 (2012). doi:10.4236/graphene.2012.12005

Gururaja, M.N., Hari Rao, A.N.: A review on recent applications and future prospectus of hybrid composites. Int. J. Soft. Comput. Eng. **1**, 352–355 (2012)

Hakamy, A., Shaikh, F.U.A., Low, I.M.: Thermal and mechanical properties of hemp fabric-reinforced nanoclay–cement nanocomposites. J. Mater. Sci. **49**, 1684–1694 (2014). doi:10.1007/s10853-013-7853-0

Han, G., Lei, Y., Wu, Q., Kojima, Y., Suzuki, S.: Bamboo-fiber filled high density polyethylene composites: effect of coupling treatment and nanoclay. J. Polym. Environ. **16**, 123–130 (2008). doi:10.1007/s10924-008-0094-7

Haq, M., Burgueño, R., Mohanty, A.K., Misra, M.: Hybrid bio-based composites from blends of unsaturated polyester and soybean oil reinforced with nanoclay and natural fibers. Compos. Sci. Technol. **68**, 3344–3351 (2008). doi:10.1016/j.compscitech.2008.09.007

Henrique, P., Camargo, C., Satyanarayana, K.G., Wypych, F.: Nanocomposites: synthesis, structure, properties and new application opportunities. Mater. Res. **12**, 1–39 (2009). doi:10.1590/S1516-14392009000100002

Hossen, M.F., Hamdan, S., Rahman, M.R., Rahman, M.M., Liew, F.K., Lai, J.C.: Effect of fiber treatment and nanoclay on the tensile properties of jute fiber reinforced polyethylene/clay nanocomposites. Fibers Polym. **16**, 479–485 (2015). doi:10.1007/s12221-015-0479-x

Huang, X., Netravali, A.: Characterization of flax fiber reinforced soy protein resin based green composites modified with nano-clay particles. Compos. Sci. Technol. **67**, 2005–2014 (2007). doi:10.1016/j.compscitech.2007.01.007

Ibeh, C.C., Bubacz, M.: Current trends in nanocomposite foams. J. Cell. Plast. **44**, 493–515 (2008). doi:10.1177/0021955X08097707

Islam, M.S., Ahmad, M.B., Hasan, M., Aziz, S.A., Jawaid, M., Haafiz, M.K.M., Zakaria, S.A.H.: Natural fiber-reinforced hybrid polymer nanocomposites: effect of fiber mixing and nanoclay on physical, mechanical, and biodegradable properties. BioResources **10**, 1394–1407 (2015)

Jahanmardi, R., Kangarlou, B., Dibazar, A.: Effects of organically modified nanoclay on cellular morphology, tensile properties, and dimensional stability of flexible polyurethane foams. J. Nanostructure Chem. **3**, 82 (2013). doi:10.1186/2193-8865-3-82

Kamble, R., Ghag, M., Gaikawad, S., Panda, B.K.: Halloysite nanotubes and applications : a review. J. Adv. Sci. Res. **3**, 25–29 (2012)

Karger-Kocsis, J., Mahmood, H., Pegoretti, A.: Recent advances in fiber/matrix interphase engineering for polymer composites. Prog. Mater Sci. **73**, 1–43 (2015). doi:10.1016/j.pmatsci.2015.02.003

Kausar, A.: Processing and properties of poly(ester-urethane)/modified montmorillonite nanocomposite foams derived from novel diol and tolylene-2,4-diisocyanate. J. Thermoplast. Compos. Mater. (2015). doi:10.1177/0892705715610401

Kojima, Y., Usuki, A., Kawasumi, M., Akane Okada, A., Fukushima, Y., Kurauchi, T., Kamigaito, O.: Mechanical properties of nylon 6-clay hybrid. J. Mater. Res. **8**, 1185–1189 (2011). doi:10.1557/JMR.1993.1185

Kord, B., Kiakojouri, S.M.H.: Effect of nanoclay dispersion on physical and mechanical properties of wood flour/polypropylene/glass fibre hybrid composites. BioResources **6**, 1741–1751 (2011)

Kuilla, T., Bhadra, S., Yao, D., Kim, N.H., Bose, S., Lee, J.H.: Recent advances in graphene based polymer composites. Prog. Polym. Sci. **35**, 1350–1375 (2010)

Kumar, R., Yakabu, M.K., Anandjiwala, R.D.: Effect of montmorillonite clay on flax fabric reinforced poly lactic acid composites with amphiphilic additives. Compos. Part A Appl. Sci. Manuf. **41**, 1620–1627 (2010). doi:10.1016/j.compositesa.2010.07.012

Kushwaha, P.K., Kumar, R.: Reinforcing effect of nanoclay in bamboo-reinforced thermosetting resin composites. Polym. Plast. Technol. Eng. **50**, 127–135 (2011). doi:10.1080/03602559.2010.512350

Lee, Y.H., Kuboki, T., Park, C.B., Sain, M., Kontopoulou, M.: The effects of clay dispersion on the mechanical, physical, and flame-retarding properties of wood fiber/polyethylene/clay nanocomposites. J. Appl. Polym. Sci. **118**, 452–461 (2010). doi:10.1002/app.32045

Li, B., Zhang, X., Gao, J., Song, Z., Qi, G., Liu, Y., Qiao, J.: Epoxy based nanocomposites with fully exfoliated unmodified clay: mechanical and thermal properties. J. Nanosci. Nanotechnol. **10**, 5864–5868 (2010a). doi:10.1166/jnn.2010.2500

Li, P., Kim, N.H., Bhadra, S., Lee, J.H.: Electroresponsive property of novel poly (acrylate-acryloyloxyethyl trimethyl ammonium chloride)/clay nanocomposite hydrogels. Adv. Mater. Res. **79–82**, 2263–2266 (2009a). doi:10.4028/www.scientific.net/AMR.79-82.2263

Li, P., Kim, N.H., Hui, D., Rhee, K.Y., Lee, J.H.: Improved mechanical and swelling behavior of the composite hydrogels prepared by ionic monomer and acid-activated Laponite. Appl. Clay Sci. **46**, 414–417 (2009b). doi:10.1016/j.clay.2009.10.007

Li, S., Lin, M.M., Toprak, M.S., Kim, D.K., Muhammed, M.: Nanocomposites of polymer and inorganic nanoparticles for optical and magnetic applications. Nano Rev **1**, 1–19 (2010b). doi:10.3402/nano.v1i0.5214

Li, X., Nikiforow, I., Pohl, K., Adams, J., Johannsmann, D.: Polyurethane coatings reinforced by halloysite nanotubes. Coatings **3**, 16–25 (2013). doi:10.3390/coatings3010016

Liu, M., Jia, Z., Jia, D., Zhou, C.: Recent advance in research on halloysite nanotubes-polymer nanocomposite. Prog. Polym. Sci. **39**, 1498–1525 (2014). doi:10.1016/j.progpolymsci.2014.04.004

Liu, Z., Erhan, S.Z., Calvert, P.D.: Solid freeform fabrication of soybean oil-based composites reinforced with clay and fibers. J. Am. Oil Chem. Soc. **81**, 605–610 (2004). doi:10.1007/s11746-006-0949-9

Marney, D.C.O., Russell, L.J., Wu, D.Y., Nguyen, T., Cramm, D., Rigopoulos, N., Wright, N., Greaves, M.: The suitability of halloysite nanotubes as a fire retardant for nylon 6. Polym. Degrad. Stab. **93**, 1971–1978 (2008). doi:10.1016/j.polymdegradstab.2008.06.018

Marquis, D,M., Guillaume, É., Chivas-Joly, C.: Properties of nanofillers in polymer. In: Cuppoletti, J. (ed.) Nanocomposites and Polymers with Analytical Methods, pp. 261–284. Intech Publishing (2011)

Matsuda, D.K.M., Verceheze, A.E.S., Carvalho, G.M., Yamashita, F., Mali, S.: Baked foams of cassava starch and organically modified nanoclays. Ind. Crops Prod. **44**, 705–711 (2013). doi:10.1016/j.indcrop.2012.08.032

Mittal, V.: Polymer layered silicate nanocomposites: a review. Materials (Basel) **2**, 992–1057 (2009). doi:10.3390/ma2030992

Mohan, T.P., Kanny, K.: Water barrier properties of nanoclay filled sisal fibre reinforced epoxy composites. Compos. Part A Appl. Sci. Manuf. **42**, 385–393 (2011). doi:10.1016/j.compositesa.2010.12.010

Montes, C., Joshi, A., Salehi, S., Lvov, Y., Allouche, E.: In: Sobolev, K., Shah, S.P. (eds.) Nanotechnology in Construction. Springer International Publishing, Switzerland (2015)

Morsy, M.S., Alsayed, S.H., Aqel, M.: Hybrid effect of carbon nanotube and nano-clay on physico-mechanical properties of cement mortar. Constr. Build. Mater. **25**, 145–149 (2011). doi:10.1016/j.conbuildmat.2010.06.046

Nabil, F.L., Zaidon, A., Jawaid, M., Anwar, U.K.M., Bakar, E.S., Paridah, M.T., Ridzuan, S.M.A., Aizat, G.M.: Physical and morphological properties of nanoclay in low molecular weight phenol formaldehyde resin by ultrasonication. Int. J. Adhes. Adhes. **62**, 124–129 (2015). doi:10.1016/j.ijadhadh.2015.07.012

Najafi, A., Kord, B., Abdi, A., Ranaee, S.: The impact of the nature of nanoclay on physical and mechanical properties of polypropylene/reed flour nanocomposites. J. Thermoplast. Compos. **25**, 717–727 (2012). doi:10.1177/0892705711412813

Nakato, T., Miyamoto, N.: Liquid crystalline behavior and related properties of colloidal systems of inorganic oxide nanosheets. Materials (Basel) **2**, 1734–1761 (2009). doi:10.3390/ma2041734

Nieddu, E., Mazzucco, L., Gentile, P., et al.: Preparation and biodegradation of clay composites of PLA. React. Funct. Polym. **69**, 371–379 (2009). doi:10.1016/j.reactfunctpolym.2009.03.002

Nourbakhsh, A., Ashori, A.: Influence of nanoclay and coupling agent on the physical and mechanical properties of polypropylene/bagasse nanocomposite. Polymer (Guildf) **112**, 1386–1390 (2009). doi:10.1002/app

Özcan, A.S., Erdem, B., Özcan, A.: Adsorption of acid blue 193 from aqueous solutions onto Na-bentonite and DTMA-bentonite. J. Colloid Interface Sci. **280**, 44–54 (2004). doi:10.1016/j.jcis.2004.07.035

Patel, R.H., Patel, K.S.: Synthesis of flame retardant polyester-urethanes and their applications in nanoclay composites and coatings. Polym. Int. **63**, 529–536 (2014). doi:10.1002/pi.4547

Paul, D.R., Robeson, L.M.: Polymer nanotechnology: nanocomposites. Polymer (Guildf) **49**, 3187–3204 (2008). doi:10.1016/j.polymer.2008.04.017

Pavlidou, S., Papaspyrides, C.D.: A review on polymer-layered silicate nanocomposites. Prog. Polym. Sci. **33**, 1119–1198 (2008). doi:10.1016/j.progpolymsci.2008.07.008

Pojanavaraphan, T., Schiraldi, D.A., Magaraphan, R.: Mechanical, rheological, and swelling behavior of natural rubber/montmorillonite aerogels prepared by freeze-drying. Appl, Clay Sci. **50**, 271–279 (2010). doi:10.1016/j.clay.2010.08.020

Pollet, E., Delcourt, C., Alexandre, M., Dubois, P.: Organic-inorganic nanohybrids obtained by sequential copolymerization of ε-caprolactone and L, L-lactide from activated clay surface. Macromol. Chem. Phys. **205**, 2235–2244 (2004). doi:10.1002/macp.200400180

Potts, J.R., Dreyer, D.R., Bielawski, C.W., Ruoff, R.S.: Graphene-based polymer nanocomposites. Polymer (Guildf) **52**, 5–25 (2011). doi:10.1016/j.polymer.2010.11.042

Prabhu, P., Jawahar, P., Balasubramanian, M., Mohan, T.P.: Machinability study of hybrid nanoclay-glass fibre reinforced polyester composites. Int. J. Polym. Sci. (2013). doi:10.1155/2013/416483

Prasad, A.V.R., Rao, K.B., Rao, K.M., Ramanaiah, K., Gudapati, S.P.K.: Influence of nanoclay on the mechanical performance of wild cane grass fiber-reinforced polyester nanocomposites. Int. J. Polym. Anal. Charact. **20**, 541–556 (2015). doi:10.1080/1023666X.2015.1053335

Prasanth, R., Shubha, N., Hng, H.H., Srinivasan, M.: Effect of nano-clay on ionic conductivity and electrochemical properties of poly(vinylidene fluoride) based nanocomposite porous polymer membranes and their application as polymer electrolyte in lithium ion batteries. Eur. Polym. J. **49**, 307–318 (2013). doi:10.1016/j.eurpolymj.2012.10.033

Raghavan, P., Lim, D.-H., Ahn, J.-H., Nah, C., Sherrington, D.C., Ryu, H.-S., Ahn, H.J.: Electrospun polymer nanofibers: the booming cutting edge technology. React. Funct. Polym. **72**, 915–930 (2012). doi:10.1016/j.reactfunctpolym.2012.08.018

Rahman, N.A., Hassan, A., Yahya, R., Lafia-Araga, R.A.: Glass fiber and nano-clay reinforced polypropylene composites: morphological, thermal and mechanical properties. Sains Malasyana **42**, 537–546 (2013)

Rajini, N., Jappes, J.T.W., Jeyaraj, P., Rajakarunakaran, S., Bennet, C.: Effect of montmorillonite nanoclay on temperature dependence mechanical properties of naturally woven coconut sheath/polyester composite. J. Reinf. Plast. Compos. **32**, 811–822 (2013). doi:10.1177/0731684413475721

Raman, N., Sudharsan, S., Pothiraj, K.: Synthesis and structural reactivity of inorganic-organic hybrid nanocomposites—a review. J. Saudi Chem. Soc. **16**, 339–352 (2012). doi:10.1016/j.jscs.2011.01.012

Rawtani, D., Agrawal, Y.K.: Multifarious applications of halloysite nanotubes: a review. Rev. Adv. Mater. Sci. **30**, 282–295 (2012)

Ray, S.S., Okamoto, M.: Polymer/layered silicate nanocomposites: a review from preparation to processing. Prog. Polym. Sci. **28**, 1539–1641 (2003)

Nafchi, H.R., Abdouss, M., Najafi, S.K., Gargari, R.M., Mazhar, M.: Effects of nano-clay particles and oxidized polypropylene polymers on improvement of the practical properties of wood-polypropylene composite. Adv. Compos. Mater **24**, 239–248 (2015). doi:10.1080/09243046.2014.891341

Saba, N., Tahir, P., Jawaid, M.: A review on potentiality of nano filler/natural fiber filled polymer hybrid composites. Polymers (Basel) **6**, 2247–2273 (2014). doi:10.3390/polym6082247

Saba, N., Paridah, M.T., Abdan, K., Ibrahim, N.A.: Preparation and characterization of fire retardant nano-filler from oil palm empty fruit bunch fibers. Bio Resour. **10**, 4530–4543 (2015a)

Saba, N., Jawaid, M., Hakeem, K.R., et al.: Potential of bioenergy production from industrial kenaf (Hibiscus cannabinus L.) based on Malaysian perspective. Renew. Sustain. Energy Rev. **42**, 446–459 (2015b)

Saba, N., Paridah, M.T., Jawaid, M.: Mechanical properties of Kenaf fibre reinforced polymer composite: a review. Constr. Build. Mater. **76**, 87–96 (2015c)

Sandler, J.K.W., Pegel, S., Cadek, M., Gojny, F., van Es, M., Lohmar, J., Blau, W.J., Schulte, K., Windle, A.H., Shaffer, M.S.P., et al.: A comparative study of melt spun polyamide-12 fibres reinforced with carbon nanotubes and nanofibres. Polymer (Guildf) **45**, 2001–2015 (2004). doi:10.1016/j.polymer.2004.01.023

Sedaghat, S.: Synthesis of clay-CNTs nanocomposite. J. Nanostruct. Chem. **3**, 3–6 (2013)

Shalwan, A., Yousif, B.F.: Influence of date palm fibre and graphite filler on mechanical and wear characteristics of epoxy composites. Mater. Des. **59**, 264–273 (2014). doi:10.1016/j.matdes.2014.02.066

Shirini, F., Mamaghani, M., Atghia, S.V.: Sulfonic acid functionalized ordered nanoporous Na^+-Montmorillonite (SANM) as an efficient and recyclable catalyst for the chemoselective methoxymethylation of alcohols. J. Nanostruct. Chem. **3**, 1–5 (2012). doi:10.1186/2193-8865-3-2

Singh, A.P., Sharma, M., Singh, I.: A review of modeling and control during drilling of fiber reinforced plastic composites. Compos. Part B Eng. **47**, 118–125 (2013). doi:10.1016/j.compositesb.2012.10.038

Sothornvit, R., Rhim, J.W., Hong, S.I.: Effect of nano-clay type on the physical and antimicrobial properties of whey protein isolate/clay composite films. J. Food Eng. **91**, 468–473 (2009). doi:10.1016/j.jfoodeng.2008.09.026

Srasra, E., Bergaya, F., Fripiat, J.J.: Infrared spectroscopy study of tetrahedral and octahedral substitutions in an interstratified illite-smectite clay. Clays Clay Miner. F Full J. Title Clays Clay Miner. **42**, 237–241 (1994). doi:10.1346/CCMN.1994.0420301

Subramaniyan, A.K., Sun, C.T.: Interlaminar fracture behavior of nanoclay reinforced glass fiber composites. J. Compos. Mater. **42**, 2111–2122 (2008). doi:10.1177/0021998308094550

Tabari, H.Z., Nourbakhsh, A., Ashori, A.: Effects of nanoclay and coupling agent on the physico-mechanical, morphological, and thermal properties of wood flour/polypropylene composites. Polym. Eng. Sci. **51**, 272–277 (2011). doi:10.1002/pen.21823

Teacă, C.A., Bodîrlău, R., Spiridon, I.: Effect of cellulose reinforcement on the properties of organic acid modified starch microparticles/plasticized starch bio-composite films. Carbohydr. Polym. **93**, 307–315 (2013). doi:10.1016/j.carbpol.2012.10.020

Thakur, M.K., Gupta, R.K., Thakur, V.K.: Surface modification of cellulose using silane coupling agent. Carbohydr. Polym. **111**, 849–855 (2014). doi:10.1016/j.carbpol.2014.05.041

Thakur, V.K., Thakur, M.K.: Processing and characterization of natural cellulose fibers/thermoset polymer composites. Carbohydr. Polym. **109**, 102–117 (2014). doi:10.1016/j.carbpol.2014.03.039

Thakur, V.K., Singha, A.S., Thakur, M.K.: Surface modification of natural polymers to impart low water absorbency. Int. J. Polym. Anal. Charact. **17**, 133–143 (2012). doi:10.1080/1023666X.2012.640455

Thakur, V.K., Thakur, M.K., Gupta, R.K.: Graft copolymers from cellulose: synthesis, characterization and evaluation. Carbohydr. Polym. **97**, 18–25 (2013). doi:10.1016/j.carbpol.2013.04.069

Turku, I., Kärki, T.: The effect of carbon fibers, glass fibers and nanoclay on wood flour-polypropylene composite properties. Eur. J. Wood Wood Prod. **72**, 73–79 (2014). doi:10.1007/s00107-013-0754-8

Uddin, F.: Studies in finishing effects of clay mineral in polymers and synthetic fibers. Adv. Mater. Sci. Eng. **2013**, 1–13 (2013). doi:10.1155/2013/243515

Uddin, F.: Clays, nanoclays, and montmorillonite minerals. Metall. Mater. Trans. A **39**, 2804–2814 (2008)

Ursache, O., Rodrigues, S.: Combined effects of dam removal and past sediment mining on a relatively large lowland sandy gravel bed river (Vienne River, France). EGU Gen. **16**, 3870 (2014)

Valença, S.L., Griza, S., de Oliveira, V.G., Sussuchi, E.M., de Cunha, F.G.C.: Evaluation of the mechanical behavior of epoxy composite reinforced with Kevlar plain fabric and glass/Kevlar hybrid fabric. Compos. Part B Eng. **70**, 1–8 (2015). doi:10.1016/j.compositesb.2014.09.040

Wei, J., Meyer, C.: Sisal fiber reinforced cement composite with Portland cement substitution by a combination of metakaolin and nanoclay. J. Mater. Sci. **49**, 7604–7619 (2014). doi:10.1007/s10853-014-8469-8

Wilson, M.J.: Clay mineralogical and related characteristics of geophagic materials. J. Chem. Ecol. **29**, 1525–1547 (2003). doi:10.1023/A:1024262411676

Yan, L., Chouw, N., Jayaraman, K.: Flax fibre and its composites—a review. Compos. Part B Eng. **56**, 296–317 (2014). doi:10.1016/j.compositesb.2013.08.014

Yuan, P., Tan, D., Annabi-Bergaya, F.: Properties and applications of halloysite nanotubes: recent research advances and future prospects. Appl. Clay Sci. **112–113**, 75–93 (2015). doi:10.1016/j.clay.2015.05.001

Yunsheng, Z., Wei, S., Zongjin, L., Xiangming, Z., Eddie, Chungkong C.: Impact properties of geopolymer based extrudates incorporated with fly ash and PVA short fiber. Constr. Build. Mater. **22**, 370–383 (2008). doi:10.1016/j.conbuildmat.2006.08.006

Yusoh, K.B.: Mechanical and physical properties of wood-plastic composites made of polypropylene, wood flour and nanoclay. Proc. Int. Agric. Plant ISBN 978-9, 1–10 (2015)

Zahedi, M., Khanjanzadeh, H., Pirayesh, H., Saadatnia, M.A.: Utilization of natural montmorillonite modified with dimethyl, dehydrogenated tallow quaternary ammonium salt as reinforcement in almond shell flour–polypropylene bio-nanocomposites. Compos. Part B Eng. **71**, 143–151 (2015). doi:10.1016/j.compositesb.2014.11.009

Zhang, Y., Tang, A., Yang, H., Ouyang, J.: Applications and interfaces of halloysite nanocomposites. Appl. Clay Sci. (2015). doi:10.1016/j.clay.2015.06.034

Zotti, A., Borriello, A., Ricciardi, M., Antonucci, V., Giordano, M., Zarrelli, M.: Effects of sepiolite clay on degradation and fire behaviour of a bisphenol A-based epoxy. Compos. Part B **73**, 139–148 (2015)

Nanoclay and Natural Fibers Based Hybrid Composites: Mechanical, Morphological, Thermal and Rheological Properties

Hamid Essabir, Marya Raji, Rachid Bouhfid and Abou El Kacem Qaiss

Abstract The majority of the developing countries are extremely rich in natural resources such as natural fibers and clays. All these natural resources have exceptional properties and could be exploited effectively in the manufacturing of composite materials for a diversity of industrial applications. Combining two kinds of fillers is a potential way to improve the essential properties of natural filler-reinforced hybrid polymer composites. The hybridization is an efficient technique for making sustainable use of natural fillers. Natural fiber-reinforced polymers associated with particulate fillers as clay provide noteworthy advantages gains in terms of mechanical properties strength and stiffness, as well as improve the mechanical response these structural elements. Composites produced from natural resources are experiencing an increase in interest due to their high demand in the market for manufacturing, in addition to environmental and sustainability issues. This chapter focuses on the potential use of natural fibers and clay in hybrid composite materials, their availability, low cost, physical properties, structural properties and their friendless properties. The natural fiber/clay-reinforced hybrid polymer composites were prepared from Alfa fiber, Pine cone fiber, Sugarcane bagasse fiber and Coir fiber, polypropylene, and Moroccan clay named E1 using a extrusion and injection technique. Before being used, the natural fibers were alkali-treated to reduce their hydrophilicity, and the clay was purified. The chapter covers the advantages of natural fillers reinforcement in hybrid composites and future trends and applications.

H. Essabir · M. Raji · R. Bouhfid · A.E.K. Qaiss (✉)
Laboratory of Polymer Processing, Rabat Design Center,
Moroccan Foundation for Advanced Science, Innovation and Research (MAScIR),
Institute of Nanomaterial and Nanotechnology (NANOTECH), Rue Mohamed El Jazouli,
Madinat Al Irfane, 10100 Rabat, Morocco
e-mail: a.qaiss@mascir.com

M. Raji
Laboratoire de Chimie Organique Hétérocyclique, Université Mohammed V de Rabat,
Faculté des Sciences, Av. Ibn Battouta, BP 1014 Rabat, Morocco

M. Jawaid et al. (eds.), *Nanoclay Reinforced Polymer Composites*,
Engineering Materials, DOI 10.1007/978-981-10-0950-1_2

Keywords Natural fiber · Nanoclay · Hybrid composites · Rheological properties · Mechanical properties

1 Introduction

In the last decade, natural fiber composites have become extremely important materials. They were used in numerous industries such as aerospace, automotive, construction, and the construction and furniture industry (Elkhaoulani et al. 2013; Boujmal et al. 2014; Malha et al. 2013). Natural fiber composites are emerging as an alternatives to synthetic fiber-reinforced composites not only in food industries, but also and especially in non-structural applications (Joshi et al. 2004; El Mechtali et al. 2015). So far, a certain number of automotive components previously made with glass fiber composites are now being manufactured using environmentally friendly composites (Pracella et al. 2006; El Mechtali et al. 2015). But, their applications are limited mostly to semi- and nonstructural components due to their inherently inferior mechanical properties such as strength and rigidity, which are typically very dominant in the conventional fiber composites. Due to their low cost and their friendless environmental character, an increasing effort has emerged on the research, development, and application of composites based polymer and natural fibers (Qaiss et al. 2014, 2015a, b; Malha et al. 2013; Nekhlaoui et al. 2014). Hybrid composites reinforced by more than one type of filler (Verma et al. 2015; Jawaid et al. 2011a, b; Kakou et al. 2015), is convenient way to achieving suitable material properties and to create a multifunctional materials. Several researchers are focusing their attention on this type of composites mainly designed and processed by the combination of natural fiber with clay fillers in the same polymer matrix, due to their easy availability (Majeed et al. 2013), low density (Pracella et al. 2006), their low cost effective (Ku et al. 2011) and also to enhance the physical and mechanical composites properties, next to that the natural fibers are recyclable and biodegradable (Taj et al. 2007). Furthermore the hybridization is a technique that can avoid the drawbacks of the composite materials reinforced either by natural fiber or clay particles, such as low durability, poor resistance of natural fibers towards water absorption which make them inequitable with non-polar polymer matrix (Saba et al. 2014) and non-degradable character of plastic materials based on clay particles (Liu et al. 2012). Many studies reported that the most famous hybrid composite applications is as food packaging materials; beer bottles, carbonated drinks, juice bottles and thermoformed containers for industrial purposes, moreover in pharmaceutical packaging applications (Majeed et al. 2013; Jawaid et al. 2011a, b; Kakou et al. 2015). The hybrid composites based on natural fillers are expected to give less health risk and environmental hazard problems for people producing the composites as compared to composite based on synthetic fibers such as glass, aramid and carbon (Ku et al. 2011), which causing an environmental damage because they are characterized by higher pollutant emissions and higher green house gas emissions and their application domain has met a real decline with the appearance of the newest wave of green materials.

Besides the advantages of natural hybrid composites, natural fibres and clay fillers as renewable natural resources poses are al challenge caused by their hydrophilic characters (Xie et al. 2010) which generate a poor interfacial adhesion with the hydrophobic matrix (Oksman et al. 2003). This poor reinforcing ability of particles create the difficulty in achieving a homogeneous dispersion into the polymer matrix. Hence, it is necessary to modify the particle surface using an organic treatment before compounding under the appropriate manufacturing process, for ensuring a good adhesion and good dispersion/distribution into polymer matrix (Fook and Yatim 2015). In general, the natural fibers surface can be modified by using a chemical modification (Cao et al. 2007), which lead to remove lignin and other non-cellulosic compounds in the fiber surface. It is also expected to disrupt hydrogen bonds in the fiber surface, thus displaying extended surface roughness, which led to the better incorporation of fiber with the matrix, that may helped to significantly enhance the fiber–matrix interfacial adhesion which then reduce the possibility of the formation of strongly bonded fillers aggregates during the composite preparation, this later can cause the lack of filler dispersion within the polymer matrix. In other hand, the most commonly used inorganic fine powder as reinforcement in several industrial products such as paper, paints, rubbers and polymers applications, is calcium carbonate, talc and clay, they are used to improve the mechanical characteristics and the dimensional stability of the end product. The clay fillers currently dominate the thermoplastic industry, as an alternative filler of $CaCO_3$ and Talc (Šupová et al. 2011).

As previously mentioned, the manufacturing process plays a critical role in ensuring a good dispersion–distribution and an indiscriminate arrangement of the filler into the matrix (Ben Azouz et al. 2012). The most popular manufacturing process employed in thermoplastic polymer composites is melt compounding (Ben Azouz et al. 2012). This method is capable of facilitating chemical reactions between the components during the manufacturing process, is also regarded as more economically and flexibly for the formulation and it is compatible with the frequently used industrial process, environmentally benign as well (Hietala et al. 2012). The various natural fibers types; alfa, pine cone, coir, and sugarcane bagasse are selected to prepare a hybrid composites, thus the selected fibers were ground, treated and then melt-compounded with the natural clay fillers at various concentrations using the polypropylene (PP) as matrix, in each case the filler (fibers and particles) content ranging from 0 to 30 wt%. The chose of this types of fibers is based on their availability and their performance as reinforcement agent, in previous work these types of fibers was used with the thermoplastic matrices to improve the mechanical, thermal and structural properties of the thermoplastics polymer (Arrakhiz et al. 2012a, b, c, 2013a, b, c). The extruded materials were there after ground and then injected to preparing different test samples of natural hybrid composite based on natural fibers with clay fillers. In this book chapter, we try to studying the natural fibers effects and that of clay fillers on the hybrid composite properties, which can help to develop and promote the hybrid composites. The natural fibers, and clay particles properties and also that of their composite will be characterized using the following tests; scanning electron microscopy (SEM) to obtain some information on composites morphology. Fourier transforms infrared spectroscopy (FT-IR) to evaluate their structural properties, their thermal stability using TGA test, to complete the mechanical properties taken in tensile, and torsional test.

2 Disadvantages of Natural Fillers as Reinforcement for Polymer Matrix

The hydrophilic nature of the clay particles and of the natural fibers is a responsible of the major's problem in polymer matrix composites, which was blamed on poor adhesion between components and also the surface incompatibility between them. Which may, however negatively affect the physical and mechanical properties of the resulting composite. As results, the particles content tend to agglomerate and stick to each other. Consequently, bad dispersion-distribution of the charge in the matrix. Therefore, the composites exhibited worst properties, which often reduce the potential of natural fibers and the clay particle to act as fillers and thus limits their effective application.

As previously mentioned, the hydrophilic nature of the natural fillers influences the overall composites properties, for this reason, it is necessary to perform a chemical modification of the natural fibers in order to enhance the interfacial adhesion between them and the polymer matrix. There is much type of chemical treatments, but the Alkali treatment or delignification include NaOH is one of the simplest and the effective surface modification techniques. This treatment can firstly increase the surface roughness which leads to form new hydrogen bonds between the cellulose molecular chains and secondly this treatment may increase the amount of cellulose exposed on the fiber surface. The treatment process rests on the elimination of the amphioxus substances on the fiber surface. This can help for better interlocking between matrix and fiber. Alkali treatment is a common fiber chemical method that is extensively used by researchers to enhance the fiber–matrix interfacial adhesion and to improved fiber dispersion within the polymer matrix.

$$\text{Fiber—OH} + \text{NaOH} \rightarrow \text{Fiber —O}^-, \text{Na}^+ + \text{H}_2\text{O} + \text{impurities} \quad (1)$$

Various natural fillers from mineral resources such as calcium carbonate, clay, talc, graphite were tested as reinforcement in polymer composites. Moreover the inorganic fillers such as natural talc, calcium carbonate, clay, and graphite are now a day's present trend of research to enhance the mechanical, thermal, rheological and electrical properties of the polymer composite. Clay is considered a suitable alternative to $CaCO_3$ and talc fillers as reinforcement in the plastic industries. Before using, the raw clay has been purified and chemically modified. In this work, the clay used is from Morocco and it was made by mixing 10 g of clay E1 sample with 500 ml of deionized water in a 1 L beaker for 24 h. The supernatant was scrapped and the solids were combined and then suspended twice in 500 ml of demineralized water. To get an particle size of less than 15 μm; the suspension was centrifuged with optimized requirement in terms of speed (6000 rpm) and time (10 min).

3 Physicochemical Characteristics of the Used Filler

In recent years, more attention has gone to green materials that are compatible with the environment, as well as the increment of depletion of petroleum resources, the focus of the researcher is shifted to reinforcing the polymer matrix using the natural fiber with the application areas remaining the same. Many attempts has tried to replace the conventional fiber such as glass, aramid and carbon in some application domain, even they have high strength and modulus, by the natural fiber, Due to their many advantages. To give an example, low cost, renewable, does not cause irritation when handled or they are non-abrasive and are biodegradable under certain conditions. In addition, the natural fibers have a low density compared to glass fibers which may serve to reduce significantly the density of their composites. To compare the natural fibers properties and their composites, it is necessary to understand their physicochemical characteristics.

The chemical makeup of natural fibers is highly dependent on and nature of the fibers. The major components of natural fibers are; cellulose, hemicelluloses, lignin, and other non cellulosic components. The amount of major chemical constituents of the fiber varied from plant to plant, and with in different parts of the same plant. It has been already proven by many researchers that the fibers mechanical properties and their thermal properties depend largely on the fiber chemical composition which furthermore, has a vital role on the variation of overall composites properties. In order to evaluate the Moroccan resources, the Alfa, Pine cone, Sugarcane bagasse and Coir fibers were used. Alfa fibers (*Stipatenacissima L.*) with 25 cm in average length were collected from various rural areas of Morocco, the pine cone fibers were gathered from sources situated in the north of Morocco. The last Moroccan fiber used in this work, called sugarcane bagasse, obtained as a by-product of the Moroccan Sugar industry. In the other hand the coir fibers with a length of 20 cm were brought from Ivory Coast.

As the major components of the lignocellulosic fibers are cellulose, hemicelluloses, pectin and lignin, the obtained FTIR spectral bands was attributed primarily to these components. Figure 1 shows the FTIR spectra of the raw used fibers (Alfa, Pine cone, Sugarcane bagasse and Coir). The spectra shows the representing bands of the hydrogen bond and stretching of OH groups in the region of 3400 cm^{-1}, the absorptions can be assigned principally to the carbohydrates, including CO–C and CO stretching and bonds belonging to the glucoside bond and eventually the lignin from the lignin hydroxyphenyl, guaiacyl and syringyl groups, which are aromatic OH compounds (Qaiss et al. 2014, 2015a, b). The bands in the region of 2900 cm^{-1}

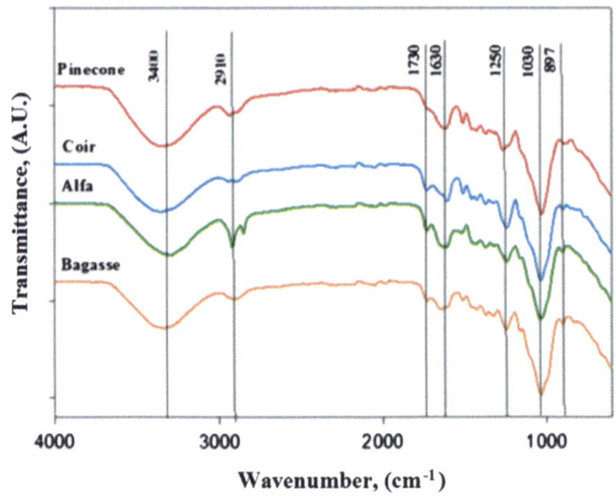

Fig. 1 FT-IR spectrum of the used natural fibers: Bagasse, Alfa, coir and pine cone

are due to the aliphatic saturated C–H stretching vibration in lignin/polysaccharides complex (Qaiss et al. 2014, 2015a, b). The bands in the region at 1730 cm^{-1} in all fibers are allocated to either the acetyl and uronic ester groups of the hemicelluloses or the ester linkage of carboxylic group of lignin and/or hemicelluloses (Qaiss et al. 2014, 2015a, b). The band at 1630 cm^{-1} may be due to the presence of water in the fibers. The bands in the 1250 cm^{-1} region include vibration C–O Stretching of primary and secondary aliphatic alcohols of the cellulose, hemicelluloses, lignin and extractives (Qaiss et al. 2014, 2015a, b). The peak at 897 cm^{-1} is due to β-glycosidic connections of glucose ring of cellulose (Qaiss et al. 2014, 2015a, b).

The quantitation of the main components of the understudy of fibers was carried out by conventional methods and the results have been included in Table 1. These analyses show that the Sugarcane Bagasse fibers have the superior content in cellulose (50 %). On the other hand, Alfa and Pine cone present the smaller content in lignin (23 %) and the Coir has the high content in lignin (45 %).

Cellulose the most important component of all fibers in the study may be found in the crystalline or amorphous and can be viewed by X-ray diffraction patterns. X-ray diagrams of Alfa fibers such example is shown in Fig. 2. The appearance of

Table 1 Physico-chemical properties of selected bio-filler (Paiva et al. 2007; Defoirdt et al. 2010; Geethamma et al. 1998; Bledzki and Gassan 1999; Ben and Ben 2007)

	Chemical constituents (%)			Physical properties		
	Cellulose	Lignin	Waxes	Densilty (g/cm^3)	Tensile strength (MPa)	Tensile modulus (GPa)
Alfa	45	23	5	1.2	134–220	13–17.8
Coir	43	45	–	1.15–1.25	120–304	4–6
Pine cone	42–49	23	–	–	–	–
Sugarcane bagasse	50	25	–	–	96.24	6.42

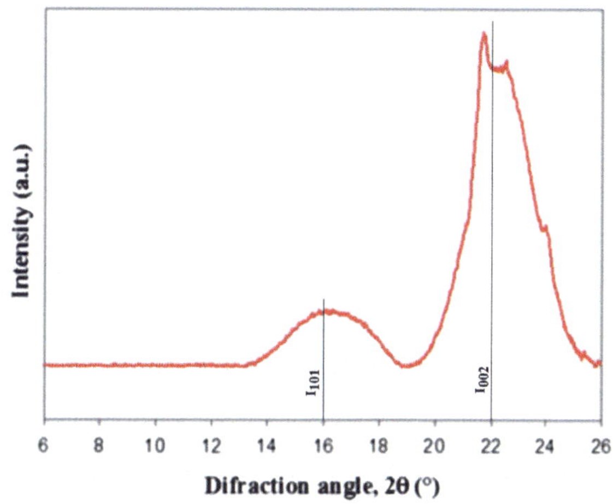

Fig. 2 X-ray diffractogram of Alfa fibers

two most intense peaks, closer to the from 2 values θ = 16° and 2 θ = 22° representing cellulose crystallographic planes I_{101} and I_{002}, respectively, can be observed. These peaks can be assigned to the crystal dissemination and diffuse background associated with disordered regions.

To improve fibers/matrix interaction, the full fibers were subjected to a chemical treatment of lignin, wax and oils that covers the external surface of fibers cell wall. Based on previous studies, the alkali treatment Removes non-cellulosic components, mining, waxes, oil, etc., from fiber surfaces. The extraction of then on cellulosic components Can be observed in FTIR spectrum, exemplified by the spectra of Coir and pine cone fibers before and alkaline after-treatment (Fig. 3).

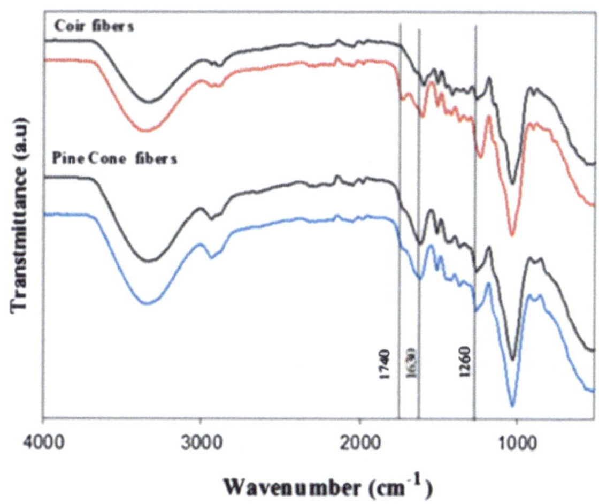

Fig. 3 FT-IR spectrum of the raw and alkali treated fibers: Pine cone and and Coir fibers

Based on previous studies, the alkali treatment eliminates noncellulosic components, mining, waxes, oil, etc., from the fiber surfaces. The extraction of the non-cellulosic ingredients can be seen in the FTIR spectrum, exemplified by the spectra of Coir and pine cone fibers before and after alkaline treatment (Fig. 3). It follows from this figure the disappearance of the band in the region at 1740 cm^{-1}, in the raw fibers assigned to the hemicelluloses and lignin, indicating a complete cleavage of these ester bonds. The intensity of the peaks is attributed to C=C of aromatic skeletal vibrations in lignin, in the region of 1500 cm^{-1}, has decreased in chemically modified fibers indicating partial removal of lignin. Also, the intensity reductions of the peak around 1260 cm^{-1} after alkalization reflects the preferential removal of hemicellulose materials instead of than lignin. Although, this chemical change in the composition of the fibers, the alkalization of lignocellulosic plant fibers changes the surface topography and their crystallographic structure.

Much research effort has explored the improvement in thermoplastic composites performance. At the Toyota research centre in Japan, the first group who worked with clay as a reinforcing phase dispersed in an engineering polymer matrix. The clay particle showed a good performance, which afford remarkable property enhancements of the polymer matrices. There are many types of clay depend on their structure. Although all clay minerals contains two types of leaves (Sheets), tetrahedron (T) and octahedron (O). In general the clay structure and the particle size of clay had an obvious effect on the composite mechanical and thermal properties. The used clay as fillers is a local variety noted E_1 (density of 1.63 g/cm^3) with diameter average of 12 μm (Fig. 4), from North of Morocco (Nador). This type of clay was more effective than talc to enhance the mechanical and thermal properties of the polymer matrix (Nekhlaoui et al. 2014). The chemical composition of the clay was determined by X-ray fluorescence (XRF) spectrometry using Axios XRF Wavelength Dispersive Spectrometer (PANalytical) (Table 2).

The recognition of the chemical components and crystallographic structure of clay noted E1 was achieved by X-ray diffraction (XRD) analysis. The diffraction patterns were analyzed using X'Pert High Score software. The X-ray diffraction pattern of the clay E_1 (Fig. 5) presents the characteristic peaks for the Quartz (SiO_2) and other associated minerals of clay such as Dolomite ($MgCa(CO_3)_2$), Illite (($K,H_3O)(Al,Mg,Fe)_2(Si,Al)_4O_{10}[(OH)_2, (H_2O)]$), and Calcite ($CaCO_3$) (Table 3).

To investigate the structural properties of the chemical nature of the clay the FTIR study was done and presented in figure. The FTIR spectra of the used clay named E_1 show intensive vibrations in the bands of 3620, 1407, 1007, 881, 874, and 778 cm^{-1} (Fig. 6). A wide absorption band at 3620 cm^{-1} matching Al-O-H stretching vibration, of attached water and Al_2OH on illite (Martinez et al. 2010). The peaks at 1407 cm^{-1} corresponds to the C–O stretching vibration in calcite (Wang et al. 2010); whilst the absorbance at 1007 cm^{-1} is typical of Si–O–Si and Si–O–Al lattice vibrations (Echeverria et al. 2005). The absorption band at 778 cm^{-1} verify the presence of a Si-translation (Martinez et al. 2010).

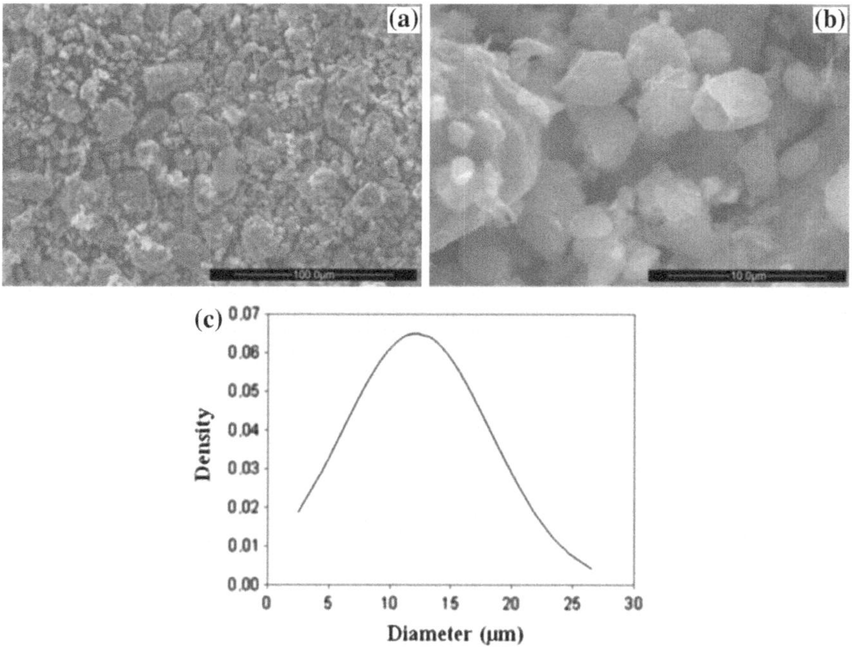

Fig. 4 **a**, **b** SEM image of the used clay (E1), **b** Diameter distribution of clay E1

4 Processing Techniques

Different techniques have been employed in order to transform the materials to the final shape without causing any defect of product, as like in situ polymerization, emulsion polymerization, sol-gel templating and melt compounding (Fornes et al. 2001), the effective manufacturing process utilized, in producing reinforced thermoplastic hybrids is melt compounding processing methods. From the industrial point of view, this technique has great advantages related to the absence of organic solvents, environmentally unfriendly, is also considered more techno-economically viable, because of its compatibility with the frequently used industrial process, giving end-use producers of numerous degrees of freedom concerning the specifications end product (e.g. choice of polymer grade, fillers types, reinforcement level, etc.) (Qaiss et al. 2015a, b). In other words, melt processing of composites allowed to be prepared right at using conventional mixing devices like as extruders or mixers. Meanwhile, melt processing is environmentally friendly as no solvents are needed; and it enhances the specificity for the polymer intercalation, eliminating solvent-host and concurrent polymer-solvent interactions. Albeit, from a chemical point of view, one can distinguish between several ways to incorporate the particles loading into polymers matrix is their homogeneity of dispersion/distribution, depending on the degree of the interfacial strength between filler and polymer,

Table 2 Chemical composition of clay E_1

Chemical composition	SiO_2	CaO	Al_2O_3	MgO	Fe_2O_3	Na_2O	K_2O	TiO_2	Pr_2O_3	P_2O_5	V_2O_5	SO_3	Cl	ZnO	MnO_2	SrO	ZrO_2	LOI
Weight (%)	48.2	11.4	10.9	3.48	2.01	0.866	0.729	0.269	0.218	0.157	0.154	0.127	0.126	0.0503	0.0255	0.0242	0.0124	21.2516

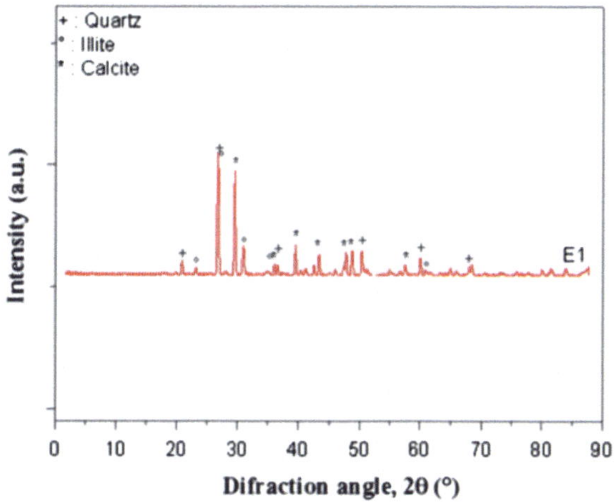

Fig. 5 X-ray Diffractogram of clay E1

Table 3 X-ray diffraction 'd-spacing' values of E_1

d-spacing (Å)	Mineral
4.2689, 3.3515, 2.4604, 2.2855	Quartz
3.8627, 3.0388, 2.4960, 2.2855	Calcite
4.4961, 3.5133, 3.3515, 3.1835	Illite

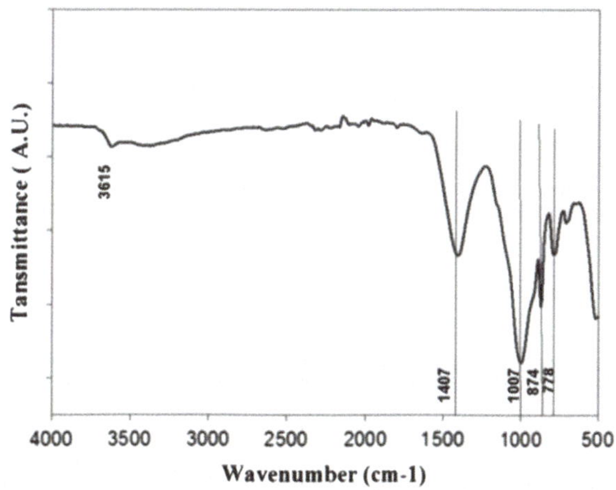

Fig. 6 FT-IR spectrum of clay E1

because they affect overall properties of the final composite (Qaiss et al. 2015b). In this direction, numerous studies have demonstrate that the melt compounding produce better composites mechanical properties, include less agglomerates, and contain fewer defects. The hybrid composites containing a natural clay fillers and the natural fibers at various concentrations using the polypropylene (PP) as polymer matrix, with the particles loading ranging from 0 to 30 wt%, were blended using a Leistritz ZSE-18 Twin Screw Extruder (Leistritz Extrusions technik GmbH,

Germany) equipped with seven heating zones, the temperature profile in the various zones was set at 200, 200, 200, 200, 180, 180, 180 °C and the die temperature was 180 °C. Screw speed for melting was using a speed of 125 rpm for the main screws and a 40 rpm side-feeder screw speed for particles feeding. The melt extruded composites were cooled in a water bath and then pelletized into granules of 2–3 mm in length. Hybrid composites granules are injection molded using an Engel e-Victory injection molding machine, to produce mainly the specimens for mechanical and rheological tests. The temperature was set at 200, 200, 200 and 180 from the barrel to the nozzle while the temperature mold was set at 45 °C (Essabir et al. 2013a, b, c).

5 Properties of Hybrid Composite

5.1 Morphological Properties

The mechanical properties of material composites depends strongly on the fillers morphology, fillers content, dispersion/distribution of fillers into matrix, and fillers/matrix adhesion. A excellent fillers dispersion/distribution on the polymer matrix can be obtained by effective mixing of the various components and by a suitable mixing method (Nekhlaoui et al. 2015; Essabir et al. 2013a, b; Qaiss et al. 2015b). Regardless of fillers type (fibers or clay), the fractured surfaces of the hybrid composites (15:15) Coir fibers as exampling clearly showed a excellent fillers dispersion/distribution into polymer matrix is an indication of the relatively homogeneous nature of these hybrid composites as seen in SEM (Fig. 7). Furthermore, it is observed that the smooth surfaces of the breakage with a homogeneous matrix, as well as excellent adhesion between fillers and thermoplastic matrix, which play an important role in enhancing mechanical performance of the composite as described earlier. The fillers/matrix interface is obscure, and expenses were fully covered by the matrix. Fillers were fractured, indicating that the

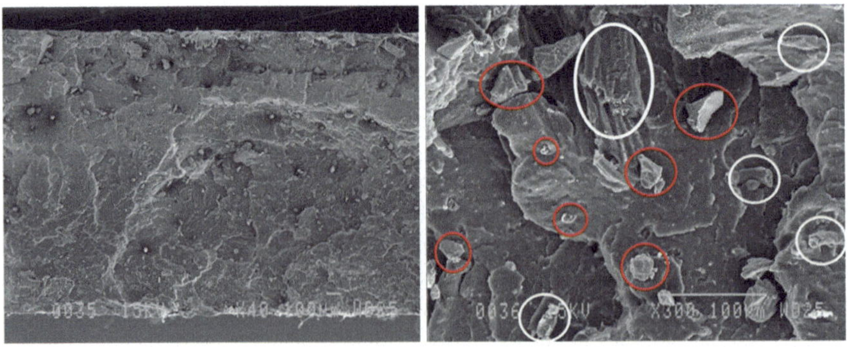

Fig. 7 SEM image of the hybrid composites 15:15 (Alfa fibers/E1)

stress was effectively transferred across the interface to the reinforcing fillers. This may be due to the interfacial interactions between the fillers and the matrix.

5.2 *Thermal Properties*

The thermal stability of the hybrid composites were studied using the Thermogravimetric (TGA) by determining their mass loss during heating, Thermogravimetric analysis was performed using Q500 instrument from TA Instruments. Roughly 20 mg of each sample was placed in a platinum pan and heated under air in the range of 30–600 °C at a heating rate of 10 °C/min to yield the onset temperature of decomposition (Fig. 8).

The initial weight loss in the range 100–200 °C may be due to the evaporation of water contained in the natural fiber. The thermal degradation of 30:0 composite presented in Table 4, its notice that the thermal decomposition of the polymer matrix, usually located between 320 and 370 °C (Arrakhiz et al. 2012a, b), is shifted to 386 °C with the incorporation of clay particles, more thermally stable. For the hybrid composite reinforced by the natural fibers, is located that, the thermal decomposition of the polymer composite were shifted to lower temperature raison of the reduction of the thermal stability of the melt by the incorporation of natural fibers which are lower

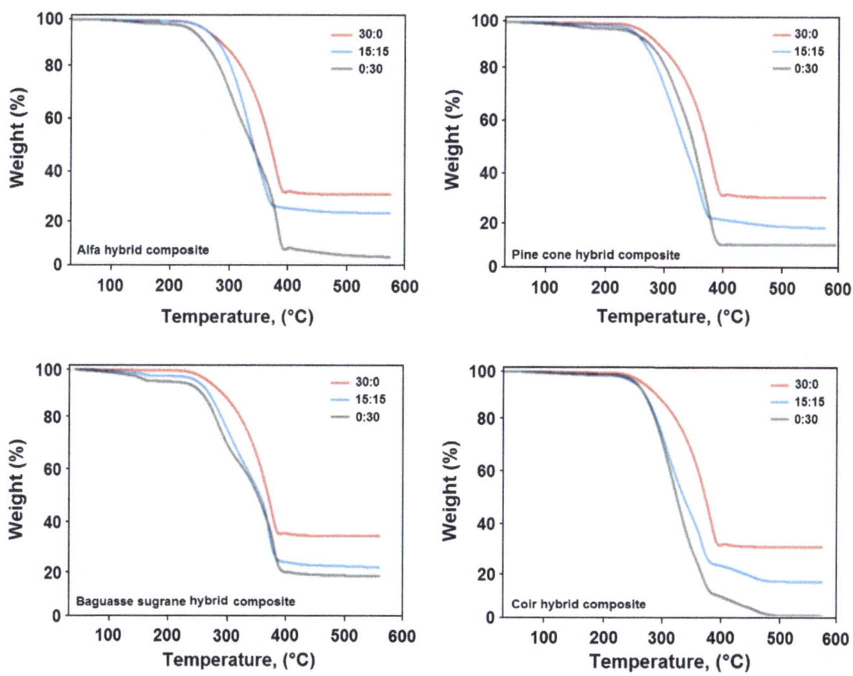

Fig. 8 TGA curves of hybrid composite: (**a**) alfa composite, (**b**) pine cone composites, (**c**) baguasse sugrane, (**d**) coir

Table 4 Thermal degradation of different hybrid composite

	T_{MAX} (°C) Alfa hybrid composite				T_{MAX} (°C) Pine cone hybrid composite				T_{MAX} (°C) baguassesugrane hybrid composite				T_{MAX} (°C) coir hybrid composite			
30:0	–	–	386	–	–	–	386	–	–	–	386		–	–	386	–
15:15	111	342	368	440	142	327	364	430	157	297	380	422	156	310	371	455
0:30	156	307	386	456	156	361	381	445	155	286	389	426	156	320	372	460

thermally stable than clay. Excluding for baguasse sugrane hybrid composite, it was found that the polymer decomposition temperature is higher, due to the high cellulose content in the baguasse sugrane fibers. The thermal decomposition of all fibers consisted of three decomposition steps; the first stage of decomposition is due to the hemicellulose degradation in the temperature range 180–310 °C. The second takes place in the range of 320–340 °C, this is due to the decomposition of cellulose, which is more thermally stable because of its crystalline nature, The third stage corresponds to the thermal degradation of lignin which occurs between 440 and 480 °C, characterized by their complex structure. The variation of the thermal degradation peak temperature is due to the variation of the natural fibers composition.

5.3 Mechanical Properties

The mechanical test is used to investigate the positive or negative hybrid effect on the composite properties. The mechanical properties of the manufactured composites were first evaluated in tension: Young's modulus, tensile strength, and strain at yield (Fig. 9). For this reason three specimens for each type of composites were tested and the average value reported. The experiment was performed under quasi-static tensile loading for all the specimens according to ISO 527-1:2012, using a universal testing machine INSTRON 8821S (Instron, USA), with a cross-head speed of 3 mm/min using a 5 KN load cell. Tensile properties such as Young's modulus, tensile strength, and strain at yield of the composites were obtained from the stress-strain curves (Fig. 9).

Figure 9a shows the variation of Young's modulus values as function of particles loading for different natural fibers. It's clear that the Young's modulus of all hybrid composite increase with the clay loading up to maximum for 30:0 at 3167 MPa compared to the neat PP (1034 MPa), due to the rigid inorganic clay particles (Essabir et al. 2015a, b; Arrakhiz et al. 2013b). This is a common behavior when rigid fillers are incorporated into softer polymer matrix. These results strongly align with the results of Jawaid et al. (2011a, b), Kakou et al. (2015) and Essabir et al. (2014) which investigates that the natural lignocellulosic fillers have been reported to have a higher elastic modulus than neat polymer. Therefore, the incorporation of rigid fillers such as clay particles to a polymer matrix can improve composite stiffness because the stiffness of the inorganic particles is commonly higher than that of organic fillers such natural fibers. Some authors have also related the increase in composites rigidity with the reduction of polymer chains mobility in the presence of fillers. In the other hand, Fig. 9b shows the evolution of the tensile strength of the hybrid composite as a function of fillers content species. The tensile strength of hybrid composite exhibit an optimum at 15:15 loading (34.52 MPa) compared to that of the neat PP (33 MPa). This improvement related to the physical anchoring between polymer chains and fibers, the tensile strength presents also a high value for alfa fibers hybrid composite due to their low percentage of lignin which means by the presence of more pendant hydroxyl on the alfa surface fibers,

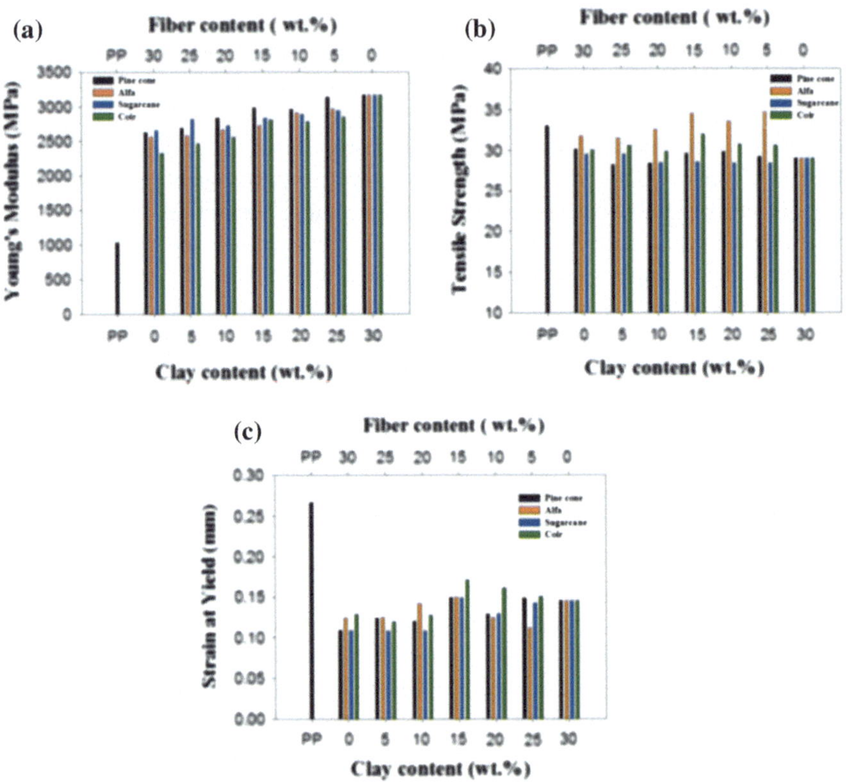

Fig. 9 **a** Young modulus of hybrid composite, **b** Tensile strength of hybrid composites, **c** strain at Yield of hybrid composite at various loading content and with different types of fibers

could lead to make the strong ester bonds between fibers and polymer matrix. This is based on the efficiency of the stress transfer between the matrix and fillers. Figure 9c present the strain at yield of the hybrid composite at various loading content and with different types of fibers. From the Fig. 9, it is clear that neat PP characterized by high strain at yield value compared to the hybrid composite made with either by alfa, coir, pine cone and sugarcane bagasse due to the rigid character of fillers content, which create a bad adhesion between fibers and PP. This debonding generates a stress concentration which accelerates the break of the sample. It's also observed that the elongation at yield values of the hybrid composites have improved until an optimum at 15:15 and then decreases with increasing amount of clay fillers result of the loss of ductility of the material with the addition of rigid particles in a ductile matrix. The increased strain at yield performance can be attributed to the good adhesion between the fibers and the matrix (Essabir et al. 2013b; 2014). The strain at yield maximum value presented for the coir hybrid composite due to their high elongation (Table 1) values which make it more flexible and easily deform.

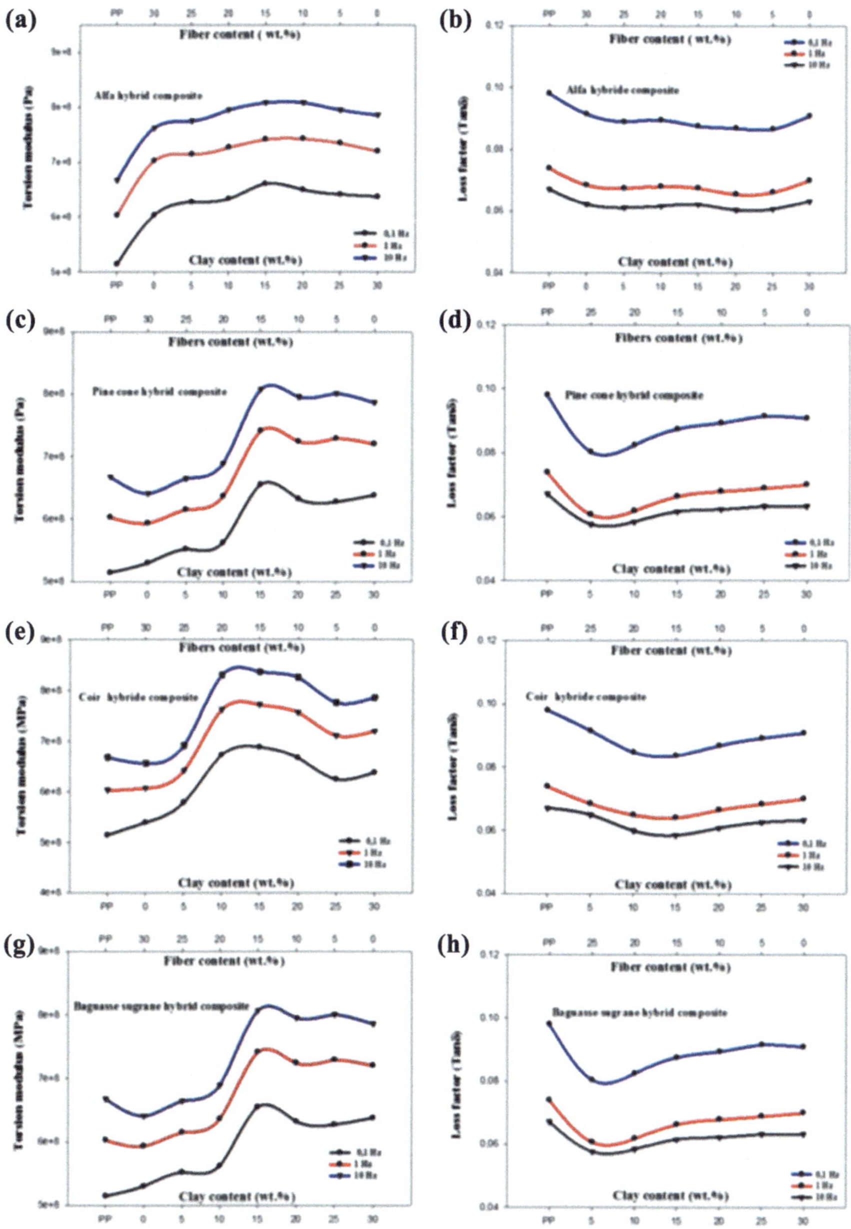

Fig. 10 Torsional properties of hybrid composite at various loading content and with different types of fibers: (**a**, **c**, **e**, **g**) torsional modulus; and (**b**, **d**, **f**, **h**) damping factor

In the other hand, torsional testing is used to describe the hybrid composites response to shear stress. Figure 10 shows the torsion modulus and the damping factor for the hybrid composite at various loading content and with different types of fibers, for all composites, the torsion modulus increase with clay particles loading (Kakou et al. 2015; Essabir et al. 2016a, b, c). On the basis of filler loading, hybrid composite with 15:15 clay/fiber content reinforced hybrid composites had the optimum set of torsional modulus among all composites manufactured. Incorporation of stiff filler as clay and oil palm fiber into the PP matrix considerably increased the torsional modulus of the manufactured composites compared to the neat PP. The variation of the torsional modulus is also marked by the variation of frequency from 0.1 to 10 Hz, leading to conclude that the hybrid composites response is like an elastic solid (Essabir et al. 2016a, b, c). This has been reported before where the elastic character of the material at higher frequency prevails over a viscous behavior. The polymer molecular time response is responsible of this enhancement in complex modulus. At higher frequency, the polymer chains do not have enough time to relax. These results agree with those observed in our previous works (Essabir et al. 2015a, b; Qaiss et al. 2015a, b) in term of the variation of rheological properties of composites as function of frequency compared to those of neat polymer matrix.

To investigate the elastic and viscous behaviours of hybrid composites, the evolution of tan δ versus frequencies and particles loading is illustrate in Fig. 10b for all composites tan δ decrease with increasing frequencies. Noteworthy is the curves in each composite have the same shape with increasing frequency from 0.1 to 10 Hz. The elastic character of the material prevails over a viscous behaviour.

Therefore, the hybrid composites have intermediate values between the composites reinforced with oil palm fibers or clay particles alone. Similar results have been reported by Jawaid et al. (2011a, b), Arrakhiz et al. (2013b), and Kakou et al. (2015). According to this result it can be concluded that hybridization has a direct effect on the mechanical properties of the composites. The hybridization of natural fibers and E_1 as clay has a positive effect on the mechanical properties of resulted hybrid composite.

6 Conclusion

The natural fibers and natural clay hybrid polymer composites have been fabricated using extrusion process followed by injection moulding technique. The prepared samples were characterized and results are reported. From this chapter, we may conclude that natural fibers (Coir, bagasse, Pine cone and Alfa) can be combinated with the natural clay as a filler in a hybrid composite material, which ultimately contributes to resolve environmental problems usually arises from natural resources. The chapter explored the potential of combination of natural fibers and clay in composites specifically hybrid polymer composites and also studied and showed the structural, morphological, mechanical and rheological properties of developed hybrid composites. This chapter also discussed the surface modification (chemical modification) of fibers which is extremely important and provides good interfacial adhesion between the fiber and matrix. Results of this study can be summarized with the following key conclusions:

Wide variety of application of natural fillers (fibers and clay) composites can also be found like packaging, doors, furniture, etc. The following application of the hybrid polymer composites can be observed, based on the achieved results, in different areas: Building component, transport sector (railway coach and Vehicle), seat and backrest and Furniture (Tables, Chairs, Kitchen, cabinet, etc.).

References

Arrakhiz, F.Z., El Achaby, M., Kakou, A.C., Vaudreuil, S., Benmoussa, K., Bouhfid, R., Fassi-Fehri, O., Qaiss, A.: Mechanical properties of high density polyethylene reinforced with chemically modified coir fibers: impact of chemical treatments. Materials Design **37**, 379 (2012a)

Arrakhiz, F.Z., Elachaby, M., Bouhfid, R., Vaudreuil, S., Essassi, M., Qaiss, A.: Mechanical and thermal properties of polypropylene reinforced with alfa fiber under different chemical treatment. Materials Design **35**, 318 (2012b)

Arrakhiz, F.Z., El Achaby, M., Benmoussa, K., Bouhfid, R., Essassi, E.M., Qaiss, A.: Evaluation of mechanical and thermal properties of Pine cone fibers reinforced compatibilized polypropylene. Materials Design **40**, 528 (2012c)

Arrakhiz, F.Z., Malha, M., Bouhfid, R., Benmoussa, K., Qaiss, A.: Tensile, flexural and torsional properties of chemically treated alfa, coir and bagasse reinforced polypropylene. Composites Part B Eng. **47**, 35 (2013a)

Arrakhiz, F.Z., Benmoussa, K., Bouhfid, R., Qaiss, A.: Pine cone fiber/clay hybrid composite: mechanical and thermal properties. Materials Design **50**, 376 (2013b)

Arrakhiz, F.Z., El Achaby, M., Malha, M., Bensalah, M.O., Fassi-Fehri, O., Bouhfid, R., Benmoussa, K., Qaiss, A.: Mechanical and thermal properties of natural fibers reinforced polymer composites: doum/low density polyethylene. Materials Design **43**, 200 (2013c)

Ben Azouz, K., et al.: Simple method for the melt extrusion of a cellulose nanocrystal reinforced hydrophobic polymer. ACS Macro Lett. **1**(1), 236–240 (2012)

Ben, Brahim S., Ben, Cheikh R.: Influence of fibre orientation and volume fraction on the tensile properties of unidirectional Alfa-polyester composite. Compos. Sci. Technol. **67**, 140 (2007)

Bledzki, A.K., Gassan, J.: Composites reinforced with cellulose based fibres. Prog. Polym. Sci. **24**, 221 (1999)

Boujmal, R., Essabir, H., Nekhlaoui, S., Bensalah, M.O., Bouhfid, R., Qaiss, A.: Composite from polypropylene and henna fiber: structural, mechanical and thermal properties. J. Biobased Mater. Bioenergy **8**, 246 (2014)

Cao, Y., Sakamoto, S., Goda, K.: Effects of heat and alkali treatments on mechanical properties of kenaf fibers. In: 16th International Conference on Composite Materials, pp. 1–4 (2007)

Defoirdt, N., Biswas, S., De Vriese, L., Ngoc Tran, L., Van Acker, J., Ahsan, Q., Gorbatikh, L., Van Vuure, A., Verpoest, I.: Assessment of the tensile properties of coir, bamboo and jute fibre. Composites: Part A **41**, 588 (2010)

Echeverrıa, J.C., Zarranz, I., Estella, J., Garrido, J.J.: Simultaneous effect of pH, temperature, ionic strength, and initial concentration on the retention of lead on illite. Appl. Clay Sci. **30**, 103 (2005)

Elkhaoulani, A., Arrakhiz, F.Z., Benmoussa, K., Bouhfid, R., Qaiss, A.: Mechanical and thermal properties of polymer composite based on natural fibers: Moroccan hemp fibers/polypropylene. Materials Design 49, 203 (2013)

El Mechtali, F.Z., Essabir, H., Nekhlaoui, S., Bensalah, M.O., Jawaid, M., Bouhfid, R., Qaiss, A.: Mechanical and thermal properties of polypropylene reinforced with almond shells particles: impact of chemical treatments. J. Bionic Eng. **12**(3), 483 (2015)

Essabir, H., Hilali, E., Elgharad, A., El Minor, H., Imad, A., Elamraoui, A., Al Gaoudi, O.: Mechanical and thermal properties of bio-composites based on polypropylene reinforced with nut-shells of Arganparticles. Mater. Des. **49**, 442 (2013a)

Essabir, H., Elkhaoulani, A., Benmoussa, K., Bouhfid, R., Arrakhiz, F.Z., Qaiss, A.: Dynamic mechanical thermal behavior analysis of doum fibers reinforced polypropylene composites. Mater. Des. **51**, 780 (2013b)

Essabir, H., Nekhlaoui, S., Malha, M., Bensalah, M.O., Arrakhiz, F.Z., Qaiss, A., Bouhfid, R.: Bio-composites based on polypropylene reinforced with almond shells particles: mechanical and thermal properties. Mater. Des. **51**, 225 (2013c)

Essabir, H., Bensalah, M.O., Bouhfid, R., Qaiss, A.: Fabrication and characterization of apricot shells particles reinforced high density polyethylene based bio-composites: mechanical and thermal properties. J. Biobased Mater. Bioenergy **8**(3), 344 (2014)

Essabir, H., Hilali, E., El Minor, H., Bensalah, M.O., Bouhfid, R., Qaiss, A.: Mechanical and thermal properties of polymer composite based on natural fibers: Moroccan luffa sponge/high density polyethylene. J. Biobased Mater. Bioenergy **9**(3), 350 (2015a)

Essabir, H., Achaby, M., Hilali, E.M., Bouhfid, R., Qaiss, A.: Morphological, structural, thermal and tensile properties of high density polyethylene composites reinforced with treated argan nut shell particles. J. Bionic Eng. **12**(1), 129 (2015b)

Essabir, H., Bensalah, M.O., Rodrigue, D., Bouhfid, R., Qaiss, A.: Structural, mechanical and thermal properties of bio-based hybrid composites from waste coir residues: Fibers and shell particles. Mech. Mater. **93**, 134–144 (2016a)

Essabir, H., Rodrigue, D., Bouhfid, R., Qaiss, A.: Effect of nylon 6 (PA6) addition on the properties of glass fiber reinforced acrylonitrile-butadiene-styrene. Polymer (2016b). doi:10.1002/pc.23895

Essabir, H., Bensalah, M. O., Denis, R., Bouhfid, R., Qaiss, A.: Biocomposites based on Argan nut shell and a polymer matrix: effect of filler content and coupling agent. Carbohydrate Polym. (2016c). doi:10.1016/j.carbpol.2016.02.002

Fook, L.T., Yatim, J.M.: An experimental study on the effect of alkali treatment on properties of kenaf fiber for reinforced concrete elements. Int. J. Res. Eng. Technol. **4**(08), 37–40 (2015)

Fornes, T., et al.: Erratum to "Nylon 6 nanocomposites: the effect of matrix molecular weight". Polymer **42**(7), 9929–9940 (2001)

Geethamma, V.G., Thomas, Mathew K., Lakshminarayanan, R., Thomas, S.: Composite of short coir fibres and natural rubber: effect of chemical modification, loading and orientation offibre. Polymer **39**, 1483 (1998)

Hietala, M., Mathew, A.P., Oksman, K.: Bionanocomposites of thermoplastic starch and cellulose nanofibers manufactured using twin-screw extrusion. Eur. Polym. J. **1** (2012)

Jawaid, M., Abdul Khalil, H.P.S., Abu, Bakar A.: Chemical resistance, void contents and tensile properties of oil palm/jute fibre reinforced polymer hybrid composites. Mater. Des. **2011**(32), 1014 (2011a)

Jawaid, M., Abdul Khalil, H.P.S., Abu Bakar, A.: Hybrid composite made from oil palm empty fruit bunches/jute fibres: Water absorption, thickness swelling and density behaviour. J. Polym. Environ. **19**, 106 (2011b)

Joshi, S.V., et al.: Are natural fiber composites environmentally superior to glass fiber reinforced composites? Compos. A Appl. Sci. Manuf. **35**(3), 371–376 (2004)

Kakou, C.A., Essabir, H., Bensalah, M.-O., Bouhfid, R., Rodrigue, D., Qaiss, A.: Hybrid composites based on polyethylene and coir/oil palm fibers. J. Reinf. Plast. Compos. **34**(20), 1684 (2015)

Ku, H., et al.: A review on the tensile properties of natural fiber reinforced polymer composites. Compos. B Eng. **42**(4), 856–873 (2011)

Liu, M., Pu, M., Ma, H.: Preparation, structure and thermal properties of polylactide/sepiolitenanocomposites with and without organic modifiers. Compos. Sci. Technol. **72**(13), 1508–1514 (2012)

Majeed, K., et al.: Potential materials for food packaging from nanoclay/natural fibres filled hybrid composites. Mater. Des. **46**, 391–410 (2013)

Malha, M., Nekhlaoui, S., Essabir, H., Benmoussa, K., Bensalah, M.-O., Arrakhiz, F.-E., Bouhfid, R., Qaiss, A.: Mechanical and thermal properties of compatibilized polypropylene reinforced by woven doum. J. Appl. Polym. Sci. **130**(6), 4347 (2013)

Martinez, R.E., Sharma, P., Kappler, A.: Surface binding site analysis of Ca2+-homoionized clay–humic acid complexes. J. Colloid Interface Sci. **352**, 526 (2010)

Nekhlaoui, S., Essabir, H., Bensalah, M.O., Fassi-Fehri, O., Qaiss, A., Bouhfid, R.: Fracture study of the composite using essential work of fracture method: PP-SEBS-*g*-MA/E1 clay. Mater. Des. **53**, 741 (2014)

Nekhlaoui, S., Essabir, H., Kunal, D., Sonakshi, M., Bensalah, M.O., Bouhfid, R., Qaiss, A.: Comparative study for the talc and two kinds of moroccan clay as reinforcements in polypropylene-SEBS-*g*-MA matrix. Polym. Compos. **36**(4), 675 (2015)

Oksman, K., Skrifvars, M., Selin, J.F.: Natural fibres as reinforcement in polylactic acid (PLA) composites. Compos. Sci. Technol. **63**(9), 1317–1324 (2003)

Paiva, M.C., Ammar, I., Campos, A.R., Cheikh, R.B., Cunha, A.M.: Alfa fibres: mechanical, morphological and interfacial characterization. Compos. Sci. Technol. **67**, 1132 (2007)

Pracella, M., et al.: Functionalization, compatibilization and properties of polypropylene composites with Hemp fibres. Compos. Sci. Technol. **66**(13), 2218 (2006)

Qaiss, A.E.K., Bouhfid, R., Essabir, H.: Natural fibers reinforced polymeric matrix: thermal, mechanical and interfacial properties. In: Biomass and Bioenergy Processing and Properties. Springer International Publishing Switzerland, pp. 225–244 (2014)

Qaiss, A.E.K., Bouhfid, R., Essabir, H.: Characterization and use of coir, almond, apricot, argan, shells, and wood as reinforcement in the polymeric matrix in order to valorize these products. In: Biomass and Bioenergy Agricultural Biomass Based Potential Materials. Springer International Publishing, Switzerland, pp. 305–339 (2015a)

Qaiss, A.E.K., Bouhfid, R., Essabir, H.: Effect of processing conditions on the mechanical and morphological properties of composites reinforced by natural fibres. In: Manufacturing of Natural Fibre Reinforced Polymer Composites. Springer International Publishing Switzerland, pp. 177–197 (2015b)

Saba, N., Tahir, P., Jawaid, M.: A review on potentiality of nano filler/natural fiber filled polymer hybrid composites. Polymers **6**, 2247–2273 (2014)

Šupová, M., Martynková, G.S., Barabaszová, K.: Effect of nanofillers dispersion in polymer matrices: a review. Sci. Adv. Mater. **3**(1), 1–25 (2011)

Taj, S., Munawar, M.A., Khan, S.: Natural fiber-reinforced polymer composites. Proc. Pakistan Acad. Sci. **44**(2), 129–144 (2007)

Verma, D., Varanasi, B.H.U., Gope, P.C.: The use of coir/coconut fibers as reinforcements in composites. In: Biofiber Reinforcements in Composite Materials, pp. 285 (2015)

Wang, C., Piao, C., Zhai, X., Hickman, F.N., Li, J.: Synthesis and characterization of hydrophobic calcium carbonate particles via a dodecanoic acid inducing process. Powder Technol. **198**, 131 (2010)

Xie, Y., et al.: Silane coupling agents used for natural fiber/polymer composites: A review. Compos. Appl. Sci. Manuf. **41**(7), 806–819 (2010)

Hybrid Composite Foams Based on Nanoclays and Natural Fibres

Hajar Yousefian and Denis Rodrigue

Abstract In this chapter, a brief introduction on natural fibers (types, properties and advantages) is presented first to show their application in various industries. Then, nanoclay as one the most widely used fillers is presented. The structures, properties and applications of this filler are described with examples related to the nanocomposite industries. Then, the combination of nanoclay with synthetic and natural fibers to produce hybrid composite systems is reviewed showing recent achievements and activities in this field. Foamed materials are then introduced. The basic concepts of foaming are discussed followed by the effect of adding particles on structures and properties. Finally, composites and hybrid foams based on nanoclay and natural fiber is reviewed followed by a conclusion and openings for future works.

Keywords Nanoclay · Natural fiber · Hybrid composites · Foam

1 Natural Fibers

Interests of replacing synthetic fibers with natural fibers to produce fiber reinforced polymers increased substantially in recent years according to the many advantages provided by these eco-friendly fibers. Natural origin, bio-degradability, low cost, relatively good mechanical properties, and high specific strength make fibers such as jute, flax, hemp, kenaf, cotton, sisal, and ramie good replacements for glass, aramid, and other non-natural fillers (Ku et al. 2011; Koronis et al. 2013; Yousefian and Rodrigue 2014).

H. Yousefian · D. Rodrigue (✉)
Department of Chemical Engineering, Université Laval,
Quebec, QC G1V 0A6, Canada
e-mail: Denis.Rodrigue@gch.ulaval.ca

H. Yousefian
e-mail: Seyedeh-hajar.yousefiandivkoalei.1@ulaval.ca

M. Jawaid et al. (eds.), *Nanoclay Reinforced Polymer Composites*,
Engineering Materials, DOI 10.1007/978-981-10-0950-1_3

The ultimate challenge while using these natural fibers is their variability and wide range of characteristics and properties depending on their source, type, processing methods, and any modifications affecting the final properties of a composite (Faruk et al. 2012).

One of the most used natural filler is wood (flour and fiber). It can be easily and economically prepared (wood waste of different processing like pulp and paper, building and construction, furniture, etc.) and used after a simple sieving. The main drawback of applying natural fibers in polymer matrices is their poor compatibility, water absorption and thermal stability. Nevertheless, these limitations can be improved by using physical (stretching, calendaring, etc.) or chemical (alkylation, esterification, etc.) modifications. Other natural fillers of potential interest are residues of the food and agriculture industry such as nutshells, oil palm empty fruit bunch fibers, corn plant extracted fibers, etc. Nevertheless, these particles are less studied for polymers reinforcement since they provide limited mechanical properties improvements (La Mantia and Morreale 2011).

2 Nanoclays

Until the late 1970s, minerals (talc, calcium carbonate, mica, etc.) were widely used as fillers in polymers. The main goal was economic as those materials were less expensive than polymers. Nowadays, with the developments associated with nanotechnology, one of the most used mineral fillers are nanoclays (Anadão 2013).

Nanoclays or naturally formed alumina silicate (generally referred to as phyllosilicates) are well-known for their platelet-like shape. These multifunctional materials are environmentally friendly and also cost effective (Majeed et al. 2013). Nanoclays were extensively studied and applied in agriculture and medicine. These minerals are fundamentally made of tetrahedral and octahedral sheets. Depending on the arrangement and/or number of these sheets, they are categorized in three main classes:

a. 1:1 clays containing one tetrahedral and one octahedral sheet.
b. 2:1 clays having one octahedral sheet sandwiched between two tetrahedral sheets per layer.
c. 2:1:1 clays containing one octahedral sheet next to a 2:1 layer.

2.1 *Structure*

Layered silicates are very small crystalline substances from the 2:1 phyllosilicates group. They could be either natural (montmorillonite, hectorite and saponite) or synthetic (fluorohactorite, laponite, and magatiide) minerals (Majeed et al. 2013). They mainly contain an octahedral sheet of alumina or magnesia and two tetrahedral

Table 1 Formula and characteristics of three natural clay minerals [7]

Name	Formula	CEC (Mequiv/100 g)	Length of particle (nm)
Montmorillonite	$M_x(Al_{4-x}Mg_x)Si_8O_{20}(OH)_4$	110	100–150
Hectorite	$M_x(Mg_{6-x}Li_x)Si_8O_{20}(OH)_4$	120	200–300
Saponite	$M_xMg_6(Si_{8-x}Al_x)$ $Si_8O_{20}(OH)_4$	86.6	50–60

M monovalent cation; *x* degree of isomorphous substitution (between 0.5 and 1.3), *CEC* cation exchange capacity

Fig. 1 Structure of Na-MMT

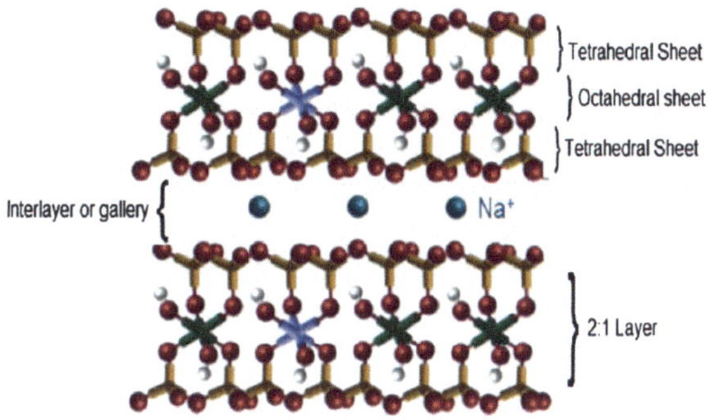

sheets of silica. The side dimensions of these layers range from 3000 nm to several microns with thicknesses of about 1 nm. A weak electrostatic force keep these layers together, so various agents such as water, organic cations or polar organic liquids can penetrate into the space between them. Table 1 summarizes some characteristics of three natural mineral clays.

Among these nanoclays, montmorillonite (MMT) is the most common filler in nanocomposite production due to its large aspect ratio (50–1000). A typical structure of MMT is presented in Fig. 1 where an octahedral layer (thickness = 1 nm) of alumina or magnesia located between two silica tetrahedral layers. Besides, metal cations such as Na^+ and Ca^{+2}, placed between the sheets, usually balance the overall negative charge caused by isomorphic exchange of Al^{3+} in the octahedral layer by Mg^{2+} or Fe^{2+}, and Mg^{2+} by Li^+.

2.2 *Polymer Clay Nanocomposite*

To date, the use of nanoparticles in composites has been an efficient way to improve the mechanical (toughness, stiffness and heat distortion), thermal, electrical, and barrier properties of composites (Ali and Ahmad 2012). But the effectiveness of nanofillers as reinforcement strongly depends on their concentration, shape, size, surface nature, dispersion, and interaction with the polymer matrix. As nanotechnology progresses, nano-size becomes of interest as it was reported that these particles, due to their small sizes, high surface areas and high aspect ratios, could bring substantial improvement to the polymer world compared to their micron-size counterparts (Ali and Ahmad 2012).

Among the different mineral fillers, nanoclays attracted much attention, especially for packaging industries who recently concentrated on nanoclays because they are low cost and available (Majeed et al. 2013). Compared to commonly use fine particles such as talc, they are much smaller in size (1400 times smaller).

Nanocomposites based on clay were primarily produced in 1950, but the turning point of their development occurred in late the 1980s (Anadão 2013). In particular, Toyota used organophilic modified clays in polyamide 6 (Nylon) for the timing belt covers of their cars (collaborating with UBE Industries, a Japanese provider of polyamide 6). The nanocomposites produced contained only 5 % wt. clay but this substantially improved the mechanical properties of polyamide 6: 60 % in tensile modulus, 126 % in flexural modulus, 40 % in stress at break and a 65–152 °C improvement in the heat distortion temperature.

Moreover, a major advantage of working with nanoclay is they can be easily used with most polymers via standard processing techniques like extrusion, injection, compression, and solution casting to produce various products such as films, membranes, automotive and electronics parts, packaging materials, etc. (Anadão 2013).

According to their difference in surface energy, interfacial bonding between polymers and clays can be thermodynamically classified as (Ray 2013a):

a. Intercalated nanocomposites (Fig. 2 left) in which the polymer chains regularly penetrate inside clay layers.
b. Flocculated nanocomposites (Fig. 2 middle) are similar to the intercalated state, with the exception for some silicate layers which are flocculated because of hydroxyl group interactions.
c. Exfoliated nanocomposites (Fig. 2 right) are randomly dispersed clay layers in the polymer matrix. Commonly, the clay content is much lower than the other two groups.

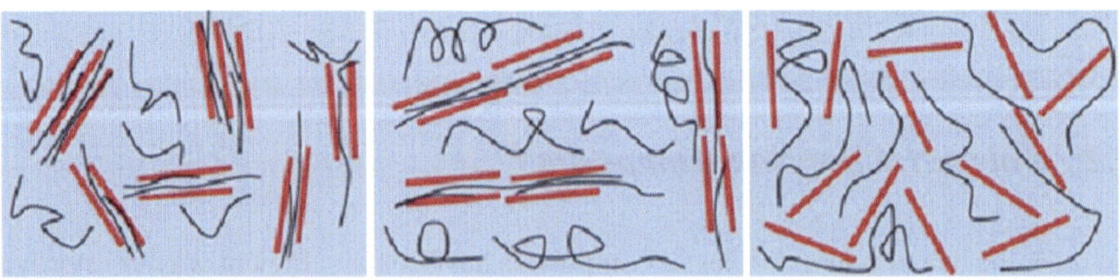

Fig. 2 Three thermodynamically different polymer clay composites: (*left*) intercalated, (*middle*) flocculated, and (*right*) exfoliated

2.3 *Properties*

Nanoclay filled thermoplastics received a great deal of attention due to the improved performances they bring to nanocomposite industries. In particular, exfoliated nanocomposites have high modulus, enhanced strength and heat resistance, as well as reduced gas permeability and flammability compared to neat polymers, along with increased biodegradability for biodegradable polymers (Majeed et al. 2013). The three most influenced properties in polymer clay nanocomposite are discussed next.

2.3.1 Mechanical Properties

Nanoclays, modified or even unmodified, have significant effects on the mechanical properties of polymer matrices. It was observed that nanoscale interactions between the matrix and nanoclay particles lead to new and unique properties in the final nanocomposite which is not like traditional composites (Ray 2013b). It has been shown that nanoclay, compared to conventional fillers such as glass fibers, can increase the elastic modulus of Nylon 6 at very low content (less than 10 % wt.) as shown in Fig. 3.

It can be seen that only 6.5 % of nanoclay will double the modulus, while 20 % of glass fiber is needed to get the same increase. Lower content usually leads to economic and processing advantages (Ray 2013b). With lower filler content, raw material costs can decrease and lighter composites (lower overall density since fillers are usually denser than polymer matrices) with higher performances are produced leading to higher specific properties (property per unit weight) which are highly desirable for packaging and automotive applications.

In another study, the effect of nanoclay on the mechanical properties of high density polyethylene (HDPE) with and without compatibilizer (polyethylene grafted maleic anhydride: MAPE) were evaluated and the results are reported in Table 2

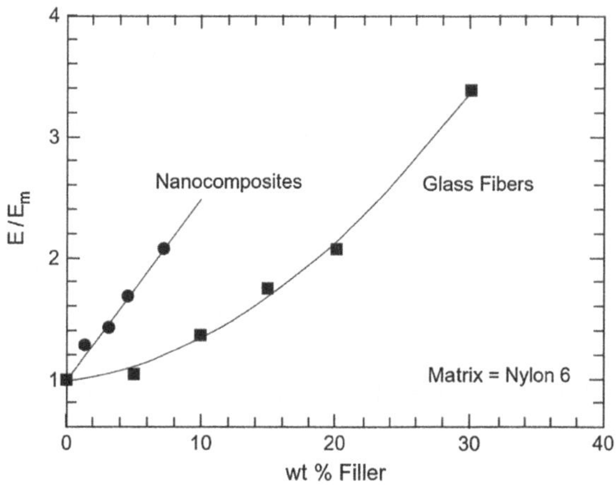

Fig. 3 Comparison between nanoclay and glass fiber on the relative elastic modulus of Nylon 6

Table 2 Tensile properties of HDPE-clay composites (Rezanavaz and Aghjeh 2012)

Sample	Young modulus (MPa)	Strain at yield (%)	Elongation at break (%)
HDPE	512	24.7	1790
HDPE-(15 %) MAPE	545	24.1	1090
HDPE-(3 %) clay	576	17.5	1200
HDPE-(15 %) MAPE-(3 %) clay	651	18.6	1250

All samples contain very small amount of antioxidant (~0.1 % wt.) and all percentages are based of weight

(Rezanavaz and Aghjeh 2012). It can be seen that with the addition of only 3 % wt. clay, the tensile modulus increased (from 512 to 576 MPa) and this increase was more noticeable when adding MAPE (from 512 to 651 MPa) to enhance the interfacial bonding between HDPE and clay. Elongation at break also changed and significantly decreased with clay addition, but was improved with MAPE (Kumar and Singh 2013). For polypropylene (PP), similar trends were observed for mechanical properties improvement: a 76 % increase in tensile modulus when 7.7 % wt. clay was added to PP-(5 % wt.) MAPP (Tabuani and Camino 2011).

Overall, nanoclays in polymers can alter the mechanical properties of the composites in either of these three possible ways (Ray 2013b):

- Clay particles bring their inherent toughness and strength to the matrix and improve the mechanical properties of the composites.
- Clay sheets can improve composite failure by stopping crack propagation.
- Polymer chains close to the nanoclay surface go through structural variations changing the overall mechanical properties.

Generally, it is believed that both degree of dispersion and interfacial bonding have important effects on the mechanical properties of final nanocomposites as described next.

2.3.2 Barrier Properties

In each material, barrier properties determine how substances such as gases and liquids permeate from a high concentration side to a low concentration side (Ray 2013c). This process consist of four steps:

- Adsorption
- Dissolution
- Diffusion
- Desorption

Gases or liquids enter the substance on the high concentration side, then diffuse into the material and finally desorb on the other side.

Clay particles, according to their specific shape, are assumed to provide a tortuous pathway in the polymer composite consequently improving the barrier properties as depicted in Fig. 4. Another important advantage of this longer diffusion path is higher thermal insulation. Nanoclay based composites, even at low contents, were shown to be highly impermeable to small size molecules such as O_2, H_2O, CO_2, etc. For instance, 2 % wt. of nanoclay decreases the permeability of polyamide by a factor of around 10. Comparing different nanoclays according to their shapes, it was observed that clays having higher length to diameter ratio (aspect ratio) produced highly impermeable composites (Ray 2013c).

A study on polyethylene terephthalate (PET) showed that incorporation of nanoclay significantly increased the oxygen barrier properties of virgin PET. Composites containing 1–4 % wt. of clay were shown, through TEM and XRD, to be intercalated composites and the oxygen barrier results are presented in Fig. 5. It is clear that even low nanoclay content can substantially reduce the oxygen permeability of PET-nanoclay; 3 % wt. decreased the permeability of the neat matrix by 50 %. This information is important for applications like packaging.

As dispersion is a key factor in polymer nanocomposites, an interesting study on using CO_2 in polystyrene (PS) was presented to improve clay dispersion

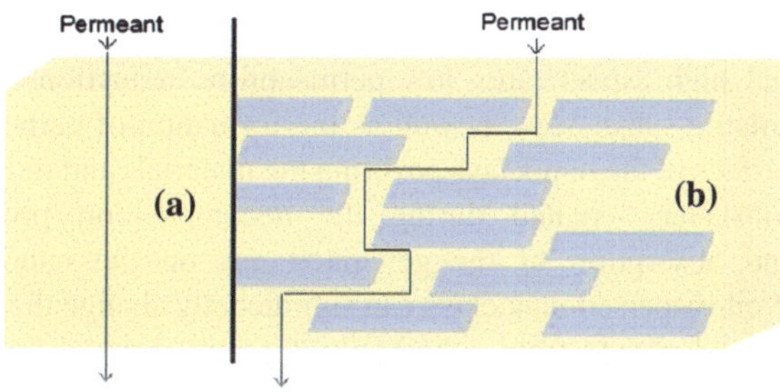

Fig. 4 Schematic representation of the origin of improved barrier properties (higher tortuosity) of polymers filled with nanoclays

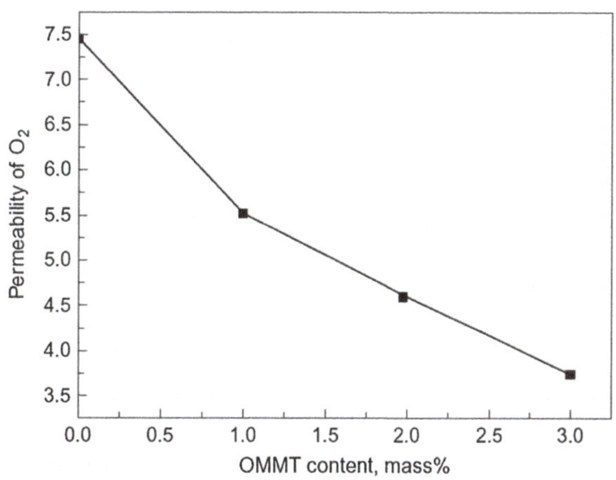

Fig. 5 Effect of clay content on PET oxygen permeability

(Yang et al. 2014). Morphological results showed that this methodology help clay particles to lose their layered structure and produce an intercalated/exfoliated structure. In this case, O_2 permeability was substantially reduced (around 49 %) using only 1.09 % wt. of clay.

Barrier properties of polylactic acid (PLA)/nanoclays were also studied (Picard et al. 2011). Organo-modified MMT was shown to have a significant effect on PLA crystallisation and oxygen permeability of annealed PLA samples was reduced (13 %) while using 4 % wt. of MMT.

The effect of chain extenders (multifunctional epoxy based: Joncryl®) was also studied on PET-organo-modified clay systems (Ghanbari et al. 2013). It was found that the chain extender react with PET end groups and consequently increased the viscoelastic properties of the matrix, leading also to improved clay exfoliation level. This is why the samples with chain extenders exhibited higher barrier properties. In particular, the O_2 permeability of PET decreased by 46 % for a composite containing 4 % wt. clay and 1 % wt. Joncryl.

2.3.3 Thermal Stability

Generally, the presence of nanoclay improves the thermal stability of polymers which can be associated to various effects like silicate content, platelet-like geometry and high aspect ratio, low permeability, reduction in the rate of development of volatile substances, as well as the formation of carbonaceous silicate chars on the surface of the nanoclays covering the materials and reducing the discharge of volatile substances created during the decomposition process (oxidation/combustion) and absorption of the generated gas on the nanoclay platelets (Ray 2013d). Well-dispersed nanoclays can significantly change the decomposition mechanism of the polymer to from a thermally different material than the virgin matrix.

During the past years, several works were done on the effect of nanoclays on the thermal stability of polymers such as polystyrene (PS) and ethylene-vinyl acetate (EVA) (Ray 2013d). In the case of PS, thermal stability of clay containing nanocomposites showed a significant improvement with clay content. Like most of the nanocomposites, degree of dispersion has a direct influence on the thermal stability of PS/nanoclay composite systems. Different kinds of modified clays have been used in PS to study their effect on the thermal stability of the composite and in most samples, the thermal stability was reasonably improved compared to virgin PS (up to 50 °C improvement) (Ray 2013d). Similar trends were observed for low density polyethylene (LDPE). Addition of nanoclay, improved the thermal stability of LDPE and in most cases, 4 % wt. is the optimum value (see Fig. 6) (Ray 2013d).

PP clay composites were also shown to behave like LDPE. High improvement in thermal stability level were related to polymer chains entrapped between clay platelets. It is also possible that the production of carbonaceous silicate char lead to a surface barrier (Ray 2013d).

The thermal stability of PLA also was investigated whit different clay (N. Najafi et al. 2012). The rheological results also showed that PLA degraded faster with the

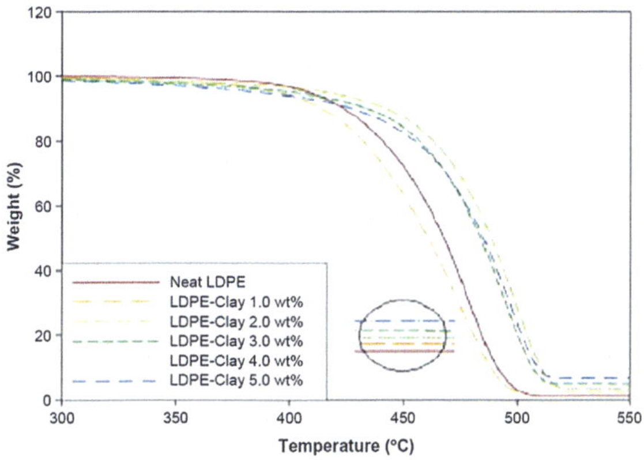

Fig. 6 Effect of clay content on LDPE thermal stability

addition of organo-modified clays. However, using a chain extender such as polycarbodiimidein (PCDI) at its optimum value (2 % wt.) could help controlling the rate of polymer degradation and viscosity change.

3 Hybrid Composites

The idea of reducing synthetic fibers use along with minimizing production costs and environmental issues led to the development of new composite systems called hybrid composites. With hybridization, the weaknesses of one component will be balanced by the strengths of another. In these systems, a combination of synthetic and natural fibers is generally used to achieve the goals of hybrid systems. Nanoclay-natural fiber system is one of the most interesting hybrid systems in which the strength and rigidity of the nanoclay balance the flexibility and specific stiffness of natural fibers (Ali and Ahmad 2012). Natural fiber/nanofiller hybridization in polymers lead to lower water uptake and improved mechanical properties. Several investigations were devoted to hybrid systems producing improved mechanical and thermal properties. For example, rice husk flour/HDPE composite with the addition of small amount of nanoclay were produced (Kord 2011). Different polymers, such as polypropylene (PP), epoxy, polyester and polyurethane (PU), have been used to produce hybrid systems (Saba et al. 2014).

3.1 Clay Based Hybrid Composites

One of the most traditional and widely used fillers in automotive and aerospace industries is glass fiber (GF). Due to its inherent high strength, GF has been used to improve the mechanical properties of polymers such as PP (Rahman et al. 2012). However, because of high stress concentration and week GF-matrix bonding, it was observed that in PP-GF composites the tensile modulus increases while tensile strength and impact strength decrease. On the other hand, nanoparticles were believed to significantly improve the physical and mechanical properties of the composites. The most important effect, while being used in semi-crystalline thermoplastics, is their nucleating agent role. As crystallinity of a polymer such as PP increases, this modifies the mechanical properties of the composites: increased modulus while decreasing strength and deformability. So in such systems (PP-GF), adding nanoclays as a third component to produce hybrid systems leads to a balance of the overall properties. In a work on PP-GF-nanoclay systems, the morphological and mechanical analyses showed that the presence of nanoclay in PP-GF composites improved the overall properties of composite system (Rahman et al. 2012). DSC results showed that the melting temperature of the system shifted to higher temperature when nanoclay was added to PP-GF system (from 161.6 to 165.5 °C while adding 6 % wt. of clay) and the degree of crystallinity was strongly affected by both nanoclay and GF in the system (14 % increase when 15 % wt. GF and 6 % wt. nanoclay were added to neat PP). As seen in Fig. 7, the authors reported some evidence of ductile deformation in Fig. 7a due to fiber debonding and pull-out. As nanoclay was added (Fig. 7b–d), ductility decreased which is clear with smaller deformation pattern. These observations were also in agreement with impact results.

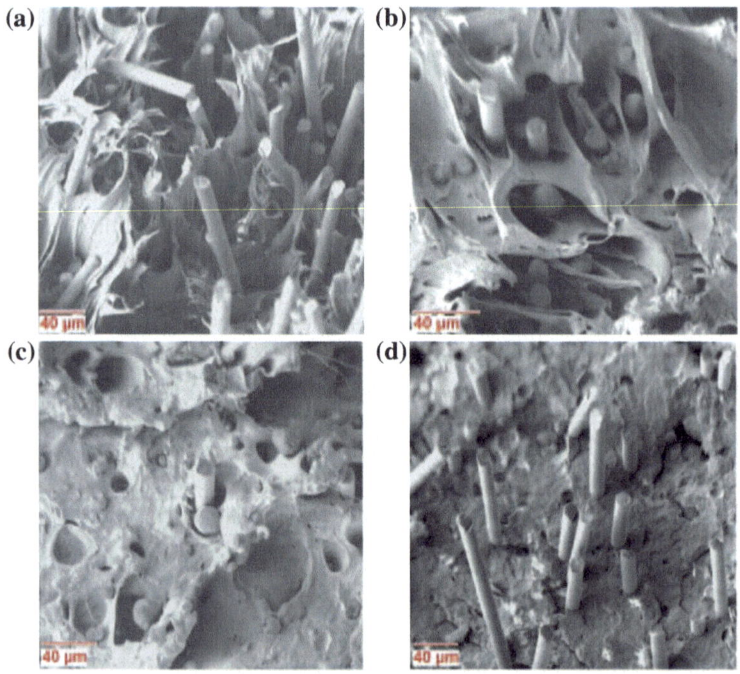

Fig. 7 SEM images of **a** PP-GF (15 % wt.), **b** PP-GF (15 % wt.)-nanoclay (3 % wt.), **c** PP-GF (15 % wt.)-nanoclay (6 % wt.), and **d** PP-GF (15 % wt.)-nanoclay (9 % wt.)

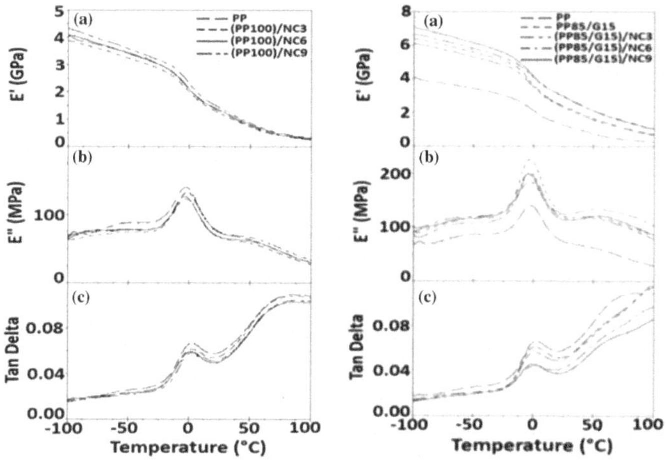

Fig. 8 DMA results of PP-nanoclay (*left*) and PP-GF-nanoclay hybrid system (*right*)

Dynamic mechanical analysis of both composites and hybrid systems also showed that the storage modulus (E′) increased with the addition of nanoclay to PP-GF and PP resulting in higher stiffness. Tan δ obtained via dynamic mechanical analysis (DMA) was also shown to be significantly influenced from clay and GF content in the composites as presented in Fig. 8.

3.2 *Clay-Natural Fiber Based Hybrid Composites*

Recently, the combination of nanoclay/natural fibers became of high interest to produce hybrid systems. This is related to reduced water absorption and dimensional swelling with improvement in tensile properties of PP-reed flour composites with the addition of nanoclay (Figs. 9 and 10). The effect of coupling agent on the

Fig. 9 Effect of nanoclay content on water absorption of PP-reed flour composites

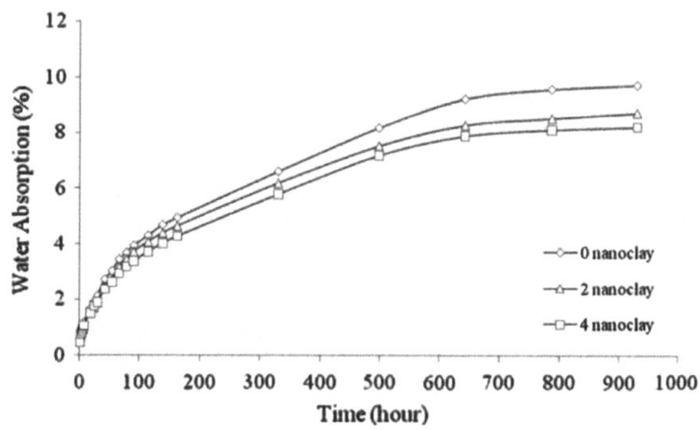

Fig. 10 Effect of nanoclay content on thickness swelling of PP-reed flour composites

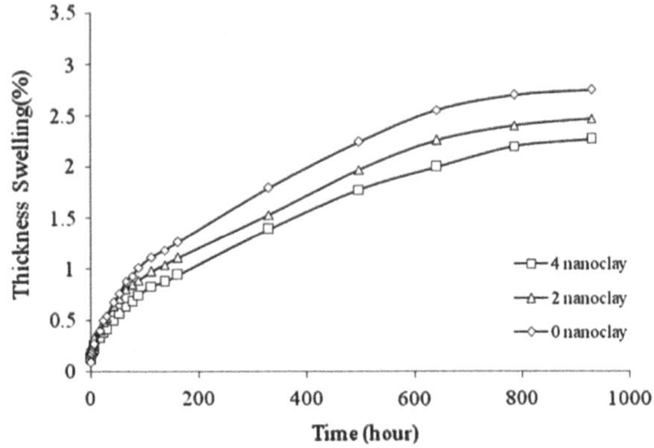

mechanical and physical properties of this system was also studied (A. Najafi et al. 2012). It was observed that the addition of 4 phc (parts per hundred contents) nanoclay with polypropylene grafted with maleic anhydride (MAPP) increased the tensile strength and modulus by around 96 and 89 %, respectively.

In another work, the positive effect of adding nanoclay and rice husk on the mechanical properties of HDPE was reported (Kord 2011). It was observed that the composite having the best structure (intercalated) contained 2 phc of nanoclay. Moreover, nanoclay had an effect on the crystallization behavior of rice husk filled HDPE by increasing the crystallization temperature, enthalpy, and crystallinity. To obtain an exfoliated structure, the authors suggested to increase the compatibilizer (MAPE) content (2 phc).

A combination of nanoclay and hemp was also found to have synergistic effects on the flexibility and stiffness of PP as reported in Fig. 11 (Kord 2012). A noticeable reduction in water absorption and thickness swelling (Fig. 12) was also observed with increasing nanoparticles content for the range studied (0–5 % wt.). Dynamic mechanical behavior as well as fire retardancy and dimensional stability were also improved with the presence of nanoparticles (Figs. 13 and 14).

In another work on microcrystalline cellulose (MCC) filled ethylene-propylene (EP) copolymer, addition of 5 % wt. of nanoclay was found to have a significant

Fig. 11 Effect of nanoclay content on the tensile strength and modulus of PP-hemp composites

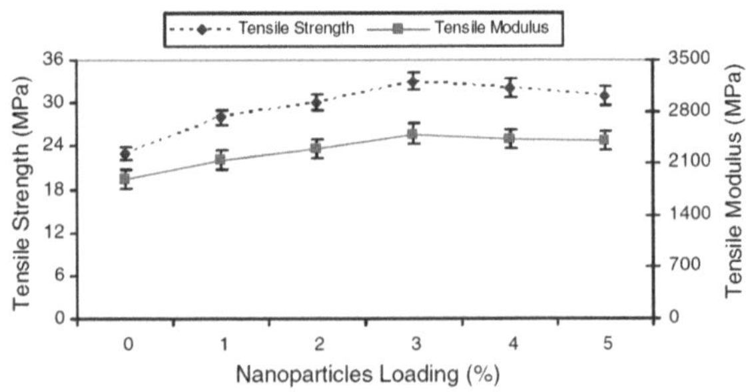

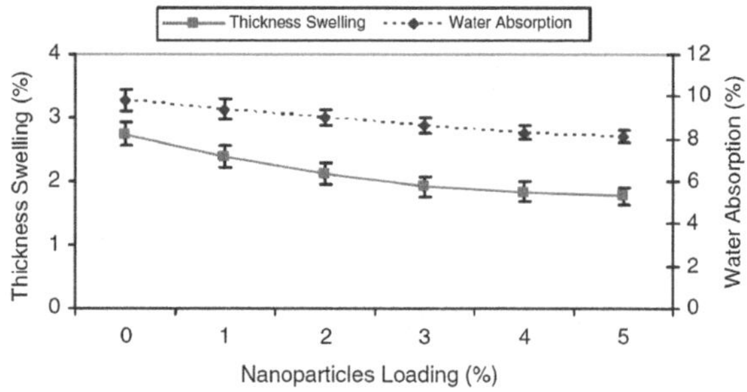

Fig. 12 Effect of nanoclay content on water absorption and thickness swelling of PP-hemp composites

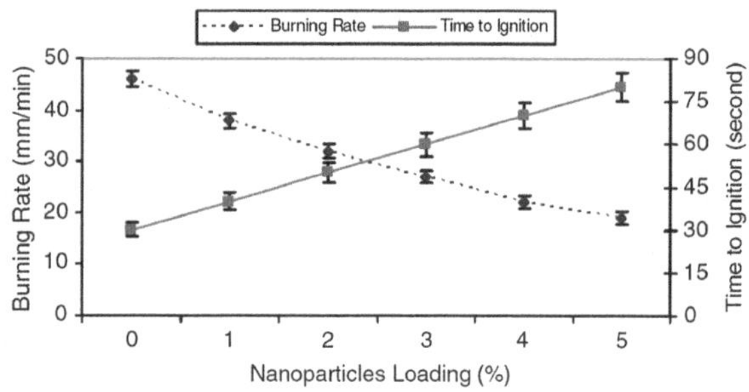

Fig. 13 Effect of nanoclay on the burning rate and time to ignition of PP-hemp composites

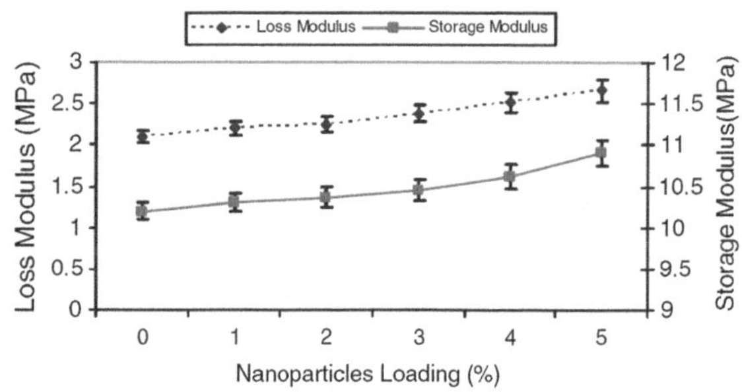

Fig. 14 Effect of nanoclay content on the loss and storage modulus of PP-hemp composites

effect on the elastic modulus of the composites (increase from 1.04 to 1.24 GPa) (Saba et al. 2014).

The effect of nanoclay on the mechanical, thermal and physical properties of hybrid systems such as PP-pine cone fiber (Arrakhiz et al. 2013) and PP-GF-wood flour (Kord and Kiakojouri 2011) was also studied. Nanoclay and MAPP were added to PP-GF-wood flour systems and the tensile properties (modulus and strength) were found to be optimum for a nanoclay content of 4 phc (see Figs. 15 and 16). The positive effect of nanoclay addition on water absorption properties is also presented in Fig. 17.

As all the studies showed, well-dispersed nanoparticles in the polymer matrix can have an effective role as to improve the mechanical, barrier, and fire resistance properties of natural fiber polymer composites (Johansson et al. 2012).

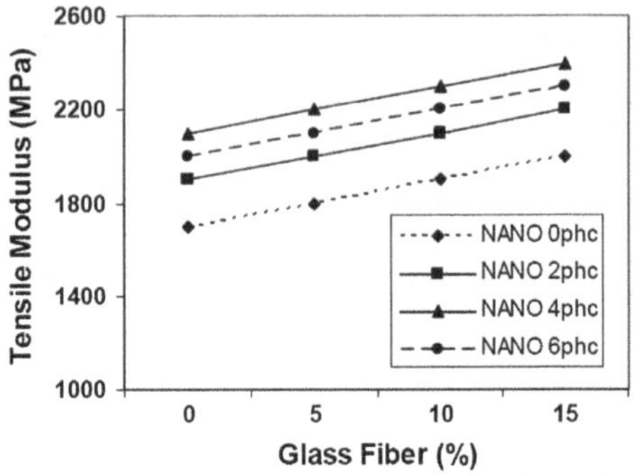

Fig. 15 Tensile modulus as a function of GF and nanoclay contents in PP-wood flour composites

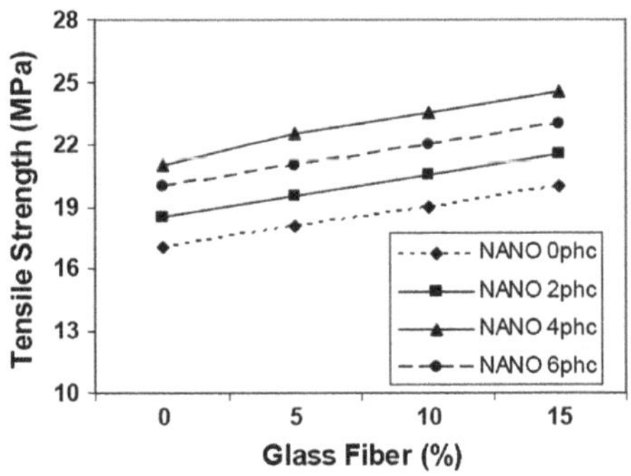

Fig. 16 Tensile strength as a function of GF and nanoclay contents in PP-wood flour composites

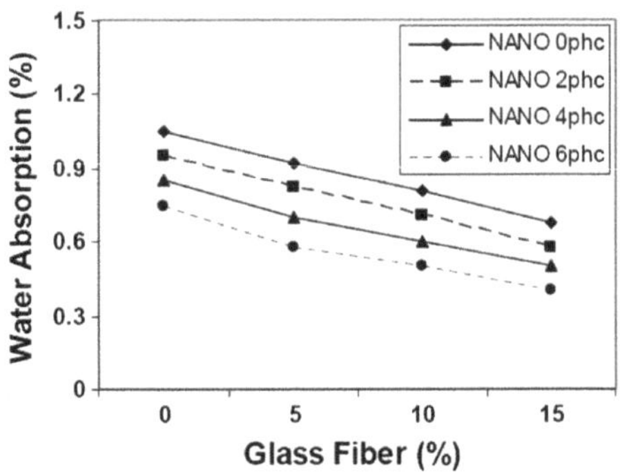

Fig. 17 Water absorption as a function of GF and nanoclay contents in PP-wood flour composites

4 Foams

Foams are porous materials consisting of numerous gas bubbles which were produced in the polymer melt and then stabilized during solidification (Anadão 2013). Foams with different morphologies are one solutions to use less amount with satisfactory mechanical properties. Many industries, especially construction, automotive, and packaging, are interested to use foams because of their light-weight, high specific strength, as well as high thermal and sound insulation properties. Besides cell geometry, two parameters are important to characterize a foam morphology: cell size and cell density (number of cells per unit volume). According to cell size, foams can be divided into 4 categories (Anadão 2013):

1. Macrocellular have cell sizes above 100 μm. This group is attractive for sound or heat insulating properties and furniture industries. Generally and like other foams they are cost effective compared to their unfoamed counterparts.
2. Microcellular have cell sizes in the range of 1–100 μm. For over three decades now, this group has been developed due to their good mechanical properties such as impact strength and toughness compared to traditional foams. They also have better acoustic and thermal insulation properties and this is why there is still high interest to develop these foams for industrial applications like construction and automotive.
3. Ultramicrocellular have cell sizes in the range of 0.1–1 μm.
4. Nanocellular foams have cell sizes in the range of 0.1–100 nm.

Other types of foam classification are possible. Depending on their rigidity, they can be classified as rigid and flexible foams. Finally, depend on cell connections, they can be classified as open (in which cells are connected to each other and the wall between cells are broken) and closed (cells are surrounded by polymer matrix and isolated from each other) cell foams (Chen et al. 2013). Relative density (foam density divided by un-foamed density) is also an important property as it mainly controls the overall mechanical and thermal properties of the foams.

4.1 Nanocomposite Foams

The mechanical properties of polymer foams are generally lower than their unfoamed counterparts because less material is available to sustain the applied stresses. One way to overcome this limitation is to use fillers inside the foamed structure (Yousefian and Rodrigue 2015). But high filler content can bring processing issues such as dispersion and aggregation problems, as well as higher viscosity. So, nanofillers instead of larger size fillers have been used in foam structure. Besides their lower content, they can provide improved mechanical, barrier and flame retardancy properties to the foam. They also improve nucleation (crystallization and foam cell) by acting as nucleating agents leading to higher cell densities and lower cell sizes. In fact, all these improvement are directly related to the nanofiller dispersion and content, as well as interfacial bonding (Sorrentino and Iannace 2013).

4.2 Clay Based Nanocomposite Foams

Nanoclays (especially MMT) are one of the most widely used nanofillers in polymer composite foams (Ahmadzadeh et al. 2015). Indeed, they provide better conditions to control the foam morphology. Besides excellent effect on nucleation due to their flat surface and specific surface area, well-dispersed clay particles were shown to produce excellent foam structures (uniform and small cell size with higher cell density) and improved mechanical properties (Chen et al. 2013). There are several works on clay based nanocomposite foams and the most recent ones are presented next.

In a study done by Zandi et al. (2011), the effect of organophilic MMT (oMMT) on the morphological, thermal, and mechanical properties of uncrosslinked LDPE via extrusion was studied. Their results showed that the presence of nanoclay changed the foam morphology by decreasing cell size and increasing cell density (2 times increase in cell density while nanoclay increased from 0 to 0.6 % wt.). Probably due to this effect, these nanocomposite foams were shown to have lower thermal conductivity than neat LDPE (7 % reduction when clay increased from 0 to 0.6 % wt.). The mechanical results also showed that for an optimum clay content (0.3 % wt.), Young's modulus increased (31 % compared to neat LDPE) and due to presence of nanoclay, these nanocomposite foams had better fire retardancy properties (Fig. 18) related to char formation on the burning face.

In a study by Chaudhary and Jayaraman (2011), several PP-clay nanocomposites were foamed by chemical foaming agent via extrusion. The authors found that based on different melt extensional strain hardening and crystallization, different foam structures were developed. In the samples that did not stain harden, a slower crystallization rate led to the production of foams with a closed cell structure with smaller cell sizes and higher cell densities. However, in samples with noticeable strain hardening effect, the foam cells were the highest in number and smallest in size with much less cell coalescence. Overall, the best foam structure was obtained with a linear PP containing 3 % wt. clay with a compatibilizer (MAPP) to increase melt strength (see Table 3 and Fig. 19).

Another study was done by Yeh et al. (2013) on five different matrices: PET, amorphous PET (APET), thermoplastic polyurethane (TPU), thermoplastic polyether ester elastomer (TPEE), and PP which are widely used in textile applications.

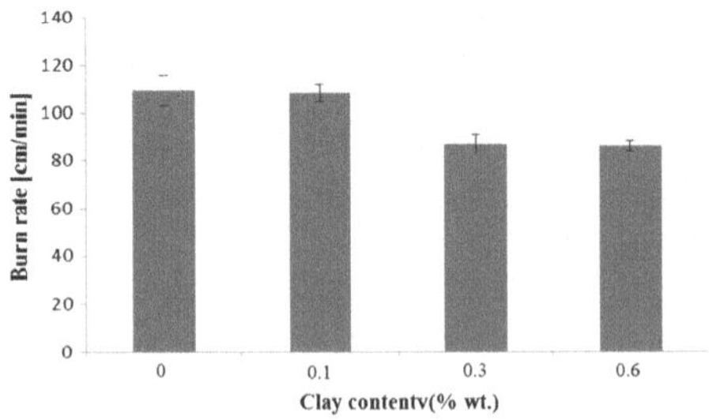

Fig. 18 Effect of clay content on the burning rate of LDPE foam

Table 3 Coding of PP-clay nanocomposites (Chaudhary and Jayaraman 2011)

Sample	PP	MAPP (% wt.)	Organoclay
PPNC-N1	68	24	8
PPNC-S1	40	52.8	7.2
PPNC-S2	76	21	3

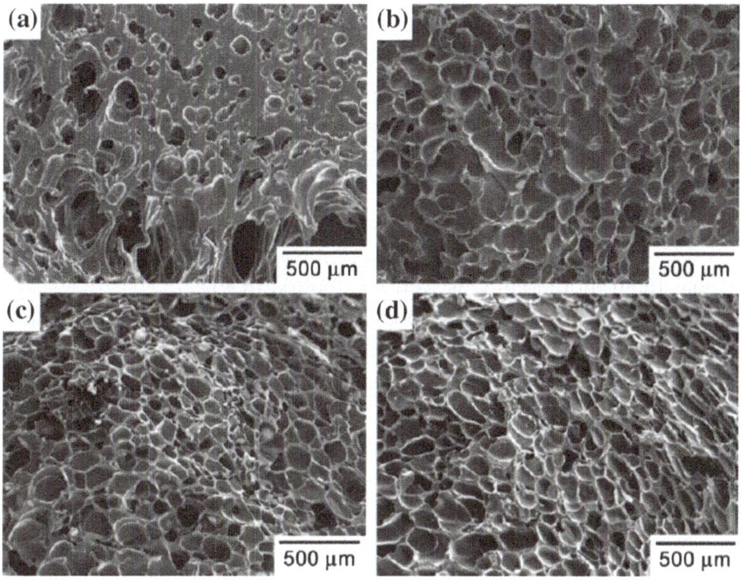

Fig. 19 SEM images of **a** linear PP foam, **b** PPNC-N1 foam, **c** PPNC-S1 foam, and **d** PPNC-S2 foam

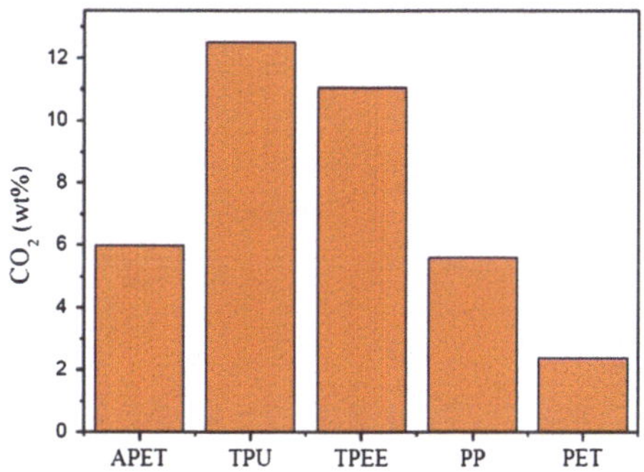

Fig. 20 Solubility of CO_2 in different polymers at 13.6 MPa and 40 °C for 6 h

These polymers were foamed through a batch process using CO_2 as the foaming agent. According to CO_2 solubility results in polymers (Fig. 20), PU was expected to be a good candidate for foaming and this was confirmed with experiments. This sample had the best foam structure with the highest cell density and lowest cell size.

Moreover, nanoclay was added to PU foams to improve even more on the foam structure. The authors mentioned that from their foaming results, they obtained an intercalated/exfoliated clay structure in the polymer matrix, especially at low clay content (1 % wt.). With short foaming time (5 s) at 150 °C, it was also possible to get nanocomposite foams with 1 μm in cell size and 10^{11} cells/cm^3 in cell density.

In another study by Keshtkar et al. (2014), the effect of nanoclay on the foamability of PLA in continuous extrusion foaming with supercritical CO_2 was explored. Clay content in PLA samples varied from 0 to 5 % wt. and CO_2 from 5 to 9 % wt. A morphological analysis showed that the clays were highly exfoliated in the PLA matrix. The results on foamed samples revealed that the expansion rate and cell density were highly promoted by clay addition and dissolved carbon dioxide. The authors also showed that, by changing the temperature profile along the extruder, faster crystallization rate was responsible for higher cell density and expansion ratio. The effect of die temperature on cell size and expansion ratio can be seen in Fig. 21. This also changed the strain hardening behavior of the materials and the final foam morphology as well as reported in Fig. 22.

In another work, the effect of nanoclay surface chemistry and content on the cellular structure of poly(D,L lactic acid) (PDLLA) was studied (Tsimpliaraki et al. 2011). Three sets of composite were prepared with PDLLA and nanoclay. In the first set, extrusion was used to prepare composites with different clay contents (0, 1, 3, 5, 9, and 15 % wt.). In the second set, MMT modified with various type of alkylammonium surfactants were used. In the last set, the organic cation content (surfactants) was changed. Then, composite foam of each series were prepared through isothermal pressure quench with supercritical CO_2. Morphological analysis revealed that the foam structures were highly dependent on clay content and chemistry of the clay surface (type and content of the organic cation used for the surfactant for modification of MMT). Increasing modified MMT into the polymer matrix improved heterogeneous nucleation leader to a higher number of smaller cells, as well as increased foam density. The authors also mentioned that the foam structure can be controlled by CO_2-clay interactions and content. In each case,

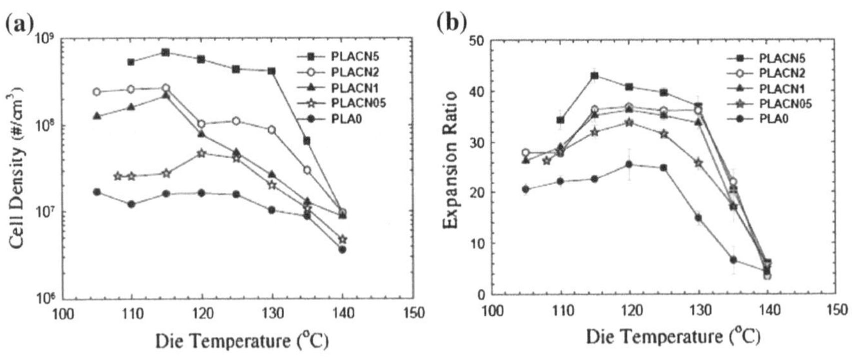

Fig. 21 Morphology of PLA and PLA-nanoclay foams obtained at various die temperatures. **a** Cell density and **b** expansion ratio

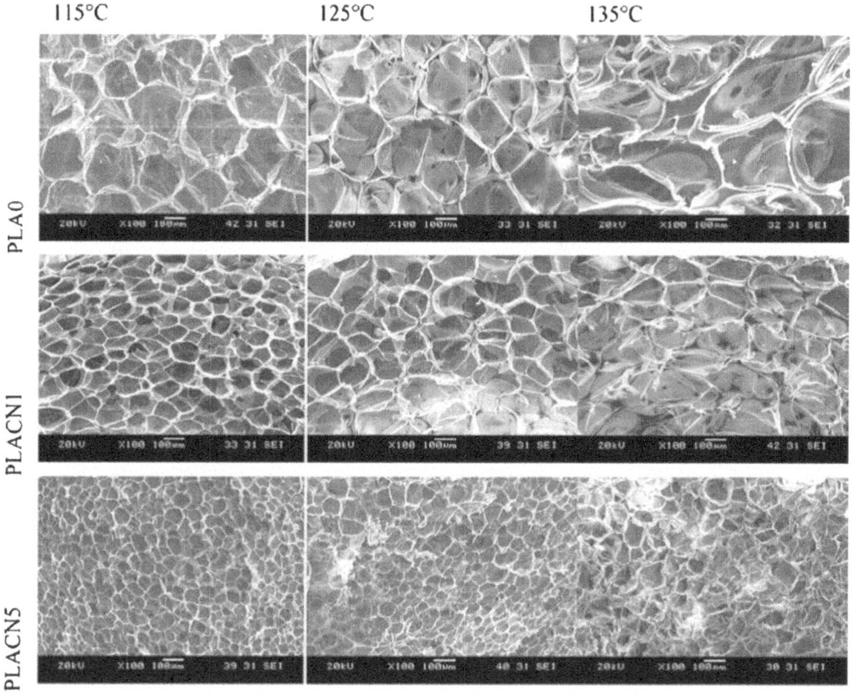

Fig. 22 Cell morphologies of neat PLA and PLA-nanoclay foams obtained at different die temperatures

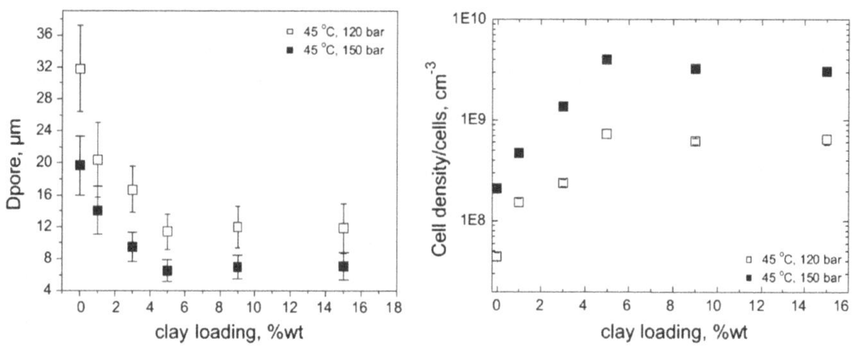

Fig. 23 Effect of clay content on cell size (*left*) and cell density (*right*) of PDLLA foams

the factors studied had a strong effect on the cell structure: higher clay content, longer organic tail of the surfactant, and higher modification level led to higher cell density and smaller cell sizes (Figs. 23 and 24). Moreover, CO_2 was believed to have a significant effect on dispersion besides its inherent role of a foaming agent.

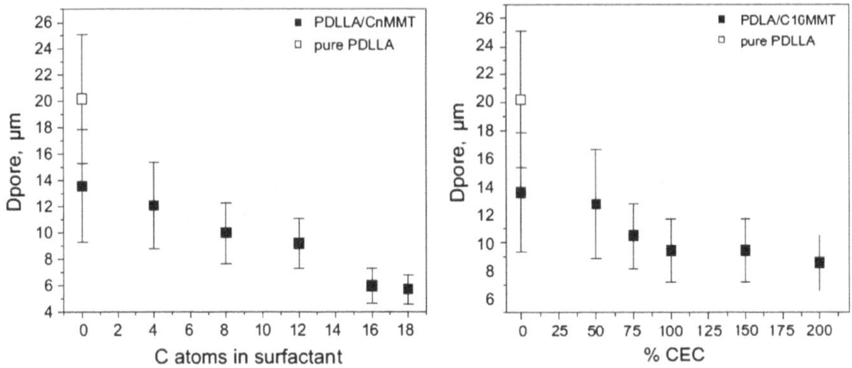

Fig. 24 Effect of surfactant type (*left*) and clay cation exchange capacity, CEC, (*right*) on the average cell diameter of PDLLA foams

Nanoclay effect was also investigated on biopolymers such as poly (3-hydroxybutyrate-co-3-hydroxyvalerate: PHBV) foaming through extrusion with supercritical CO_2 (Le Moigne et al. 2014). Two methods were used to produce foam samples. The first one was a one-step extrusion process of PHBV and nanoclay with CO_2, while the second one was a two-step extrusion method: compounding a PHBV-nanoclay composites which was foamed in a second extrusion run. The samples were analyzed in terms of physical and morphological properties. In the two-step method, compounding (masterbatch) was shown to have a positive effect on clay dispersion and limited PHBV degradation. From Fig. 25, it can be observed that mostly aggregates in the 20 % wt. clay masterbatch were obtained (Fig. 25c), while mostly intercalated and probably exfoliated structures were obtained in Fig. 25b, d where 2.5 % wt. clay was used in the matrix via the one step extrusion and the two-step with dilution, respectively. The authors also mentioned that by controlling CO_2 content, well dispersed clay was assumed to play the role of nucleating agent reducing cell coalescence and resulting in a good and homogeneous foam morphology: low cell size and high cell density by up to 50 % as seen in Fig. 26.

4.3 *Clay-Natural Fiber Hybrid Composite Foams*

Nanoclays have also been used with various natural fibers in different matrices to produce hybrid foam systems. Nanoclay hybridization with empty fruit bunch (EFB) was studied in foamed PU (Ali and Ahmad 2012). Composites of PU and EFB, silane treated EFB (sEFB) and oMMT were compared with hybrid systems of EFB/oMMT and sEFB/oMMT at different ratios (1/4, 1/2, and 3/4). The authors observed that hybridization improved the barrier properties of PU. As seen in Table 4, the thermal stability was influenced by both nanoclay and EFB. Higher

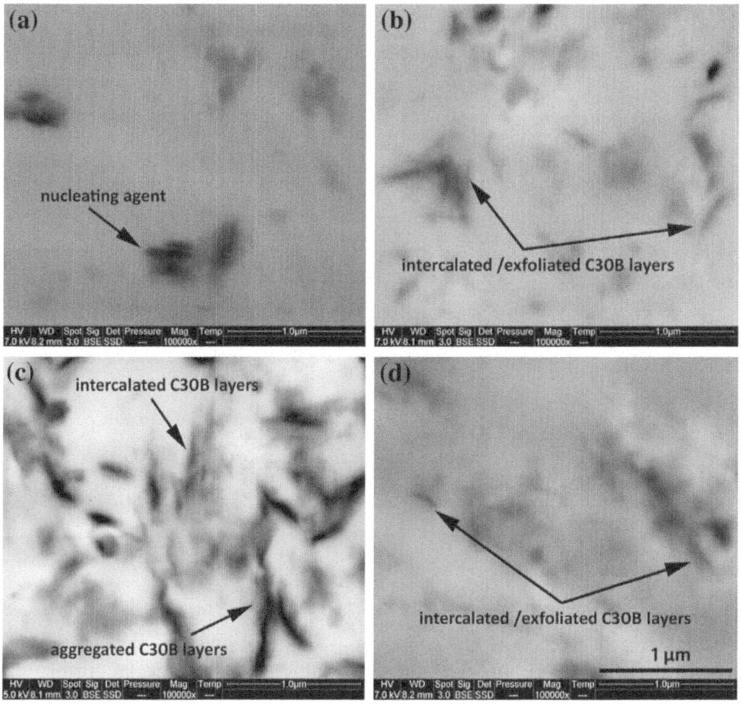

Fig. 25 SEM images of **a** extruded PHBV, **b** PHBV-clay (2.5 % wt., extruded), **c** PHBV-clay (20 % wt. extruded) masterbatch, and **d** PHBV-clay (2.5 % wt. dilution of 20 %, extruded) nanocomposite

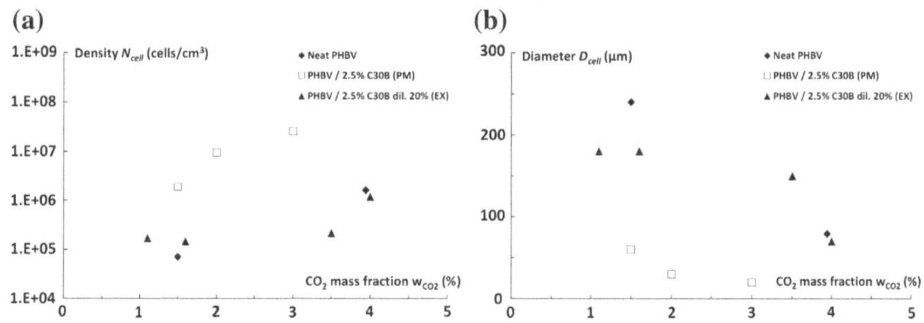

Fig. 26 Foam morphology as a function of CO_2 content for foams based on virgin PHBV, PHBV-clay (2.5 % wt., PM), and PHBV-clay (2.5 % wt., dilution of 20 %, extruded). **a** Cell density and **b** cell size

degradation temperature of the hybrid systems compared to the biocomposite shows the effect of nanoclay to improve the thermal stability of the system which is probably related to limited mobility of PU chains between the intercalated clay structures.

Table 4 Thermogravimetric analysis results of different PU composites (Ali and Ahmad 2012).

Sample	Degradation temperature (°C)	Weight loss (%)
Bio-composite	440	93.6
Hybrid bio-nano-composite	450	92.0
Nano-composite	460	93.4

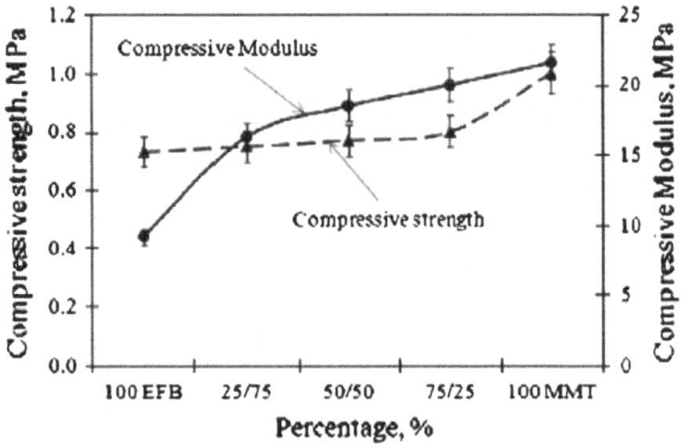

Fig. 27 Effect of filler content on the compressive modulus and strength of PU

The hybridization also affected the mechanical properties by improving the compression modulus and strength (Fig. 27). Since the mechanical properties of composites mostly depend on the properties and volume fraction of each component, Fig. 27 clearly shows that with increasing nanoclay content, both modulus and strength of the hybrid systems increased (120 and 4 % respectively, at 75 % of nanoclay and 25 % of EFB). With nanoclay addition in the composite, the silicate layers can well disperse in the system and increase the available surface between the matrix and the fillers. So the mechanical properties are improved.

Nanoclay effect was also investigated for HDPE-wheat straw flour (WSF) foam system according to their mechanical, heat resistance, permeability, fire retardancy, and physical properties (Babaei et al. 2014). The samples were fixed at 40 phr WSF and the nanoclay content was selected 0, 2, and 5 phr (parts per hundred resin). An internal batch mixer was used for compounding while foaming was conducted using an injection molding machine using azodicarbonamide as a chemical foaming agent. The foams were shown to have a microcellular closed cell structure and it was observed that by using nanoclay up to 5 phr, cell density (Fig. 28) increased while cell size (Fig. 29) decreased.

It was also observed that increasing foaming agent from 0 to 4 phr increased water absorption (Fig. 30) and swelling thickness (Fig. 31) by around 19 and 26 % respectively, while for a fixed foaming agent content, due to the nucleating agent role of the nanofillers, the addition of nanoclay decreased both water absorption and thickness swelling. However, increasing foaming agent content from 2 to 4 phr decreased the overall density (Fig. 32) by around 21 %, but nanoclay addition (5 % wt.) increased foam density (around 10 %).

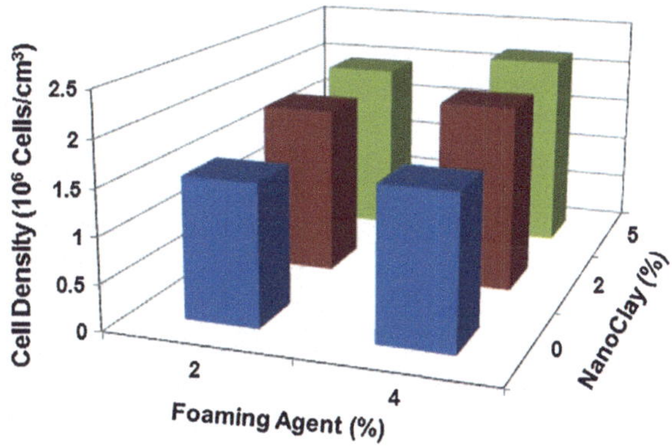

Fig. 28 Cell density of HDPE-WSF composites as a function of foaming agent and clay content

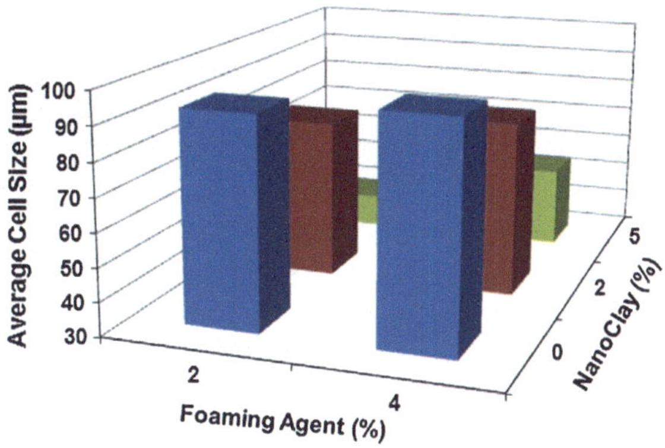

Fig. 29 Average cell size of HDPE-WSF composites as a function of foaming agent and clay content

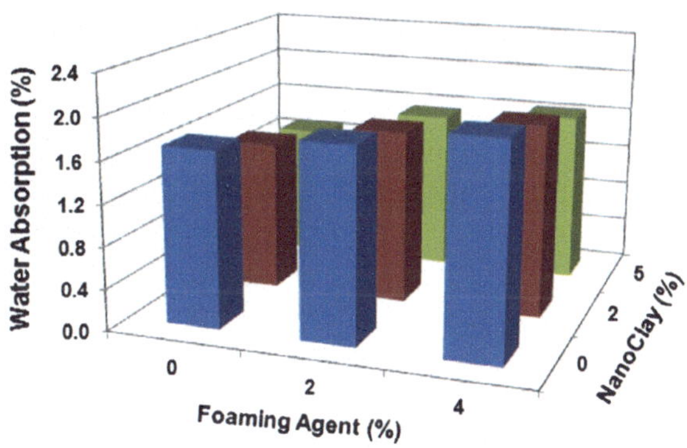

Fig. 30 Water absorption of HDPE-WSF composites as a function of foaming agent and clay content

The mechanical analysis also showed that tensile modulus (Fig. 33) and strength (Fig. 34) of the foams increased with nanoclay addition up to 2 %. However, foaming and nanoclay incorporation had negative effect on impact strength (Fig. 35) as confirmed by SEM.

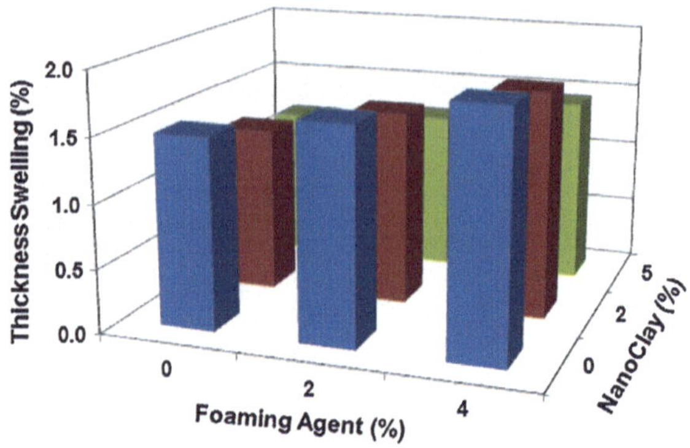

Fig. 31 Thickness swelling of HDPE-WSF composites as a function of foaming agent and clay content

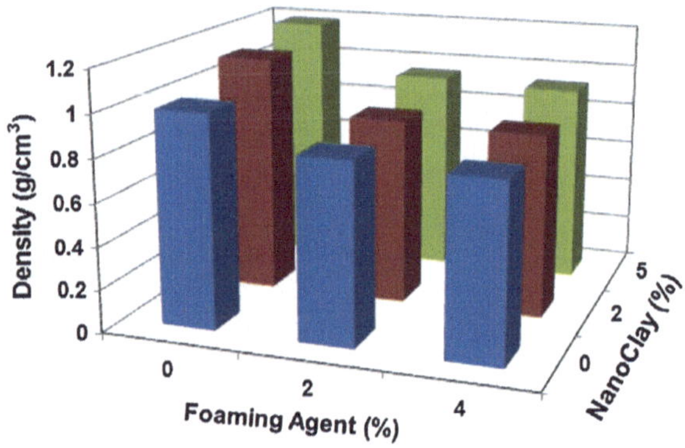

Fig. 32 Density of HDPE-WSF composites as a function of foaming agent and clay content

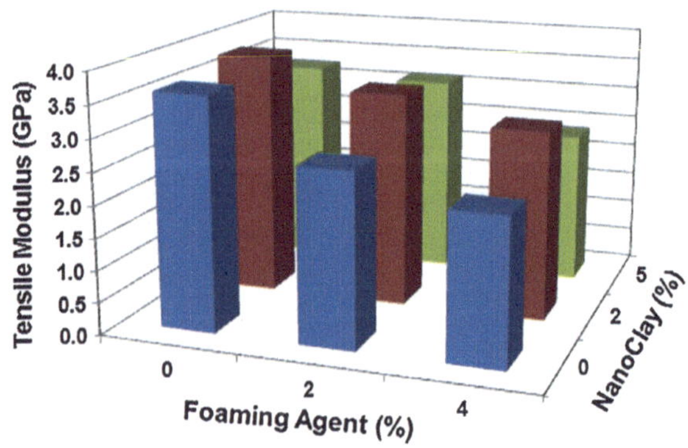

Fig. 33 Tensile modulus of HDPE-WSF composites as a function of foaming agent and clay content

In another study on hybrid foam systems based on clay and natural fibers, cellulose was used to produce foam trays for applications in food packaging (Ahmadzadeh et al. 2015). The process used was dissolution/regeneration, then drying by lyophilization. Different analyses on the composite foams showed that nanoclay presence brought more homogeneity to the structure of cellulose foams.

Fig. 34 Tensile strength of HDPE-WSF composites as a function of foaming agent and clay content

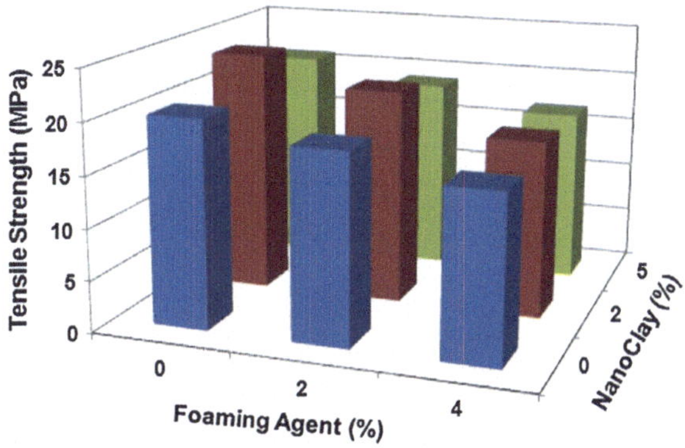

Fig. 35 Impact strength of HDPE-WSF composites as a function of foaming agent and clay content

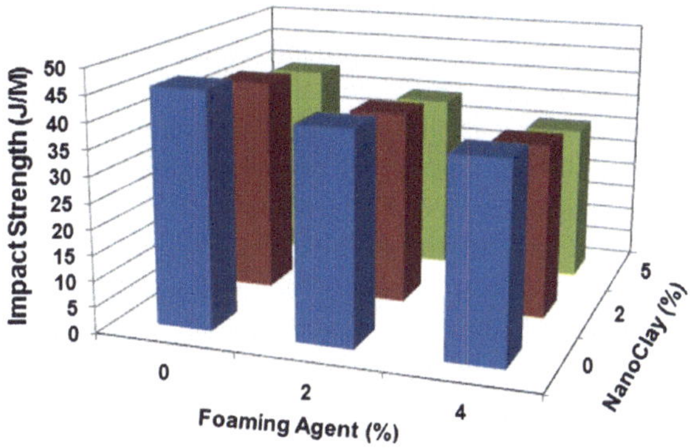

Fig. 36 Typical stress-strain curve for a cellulose foam

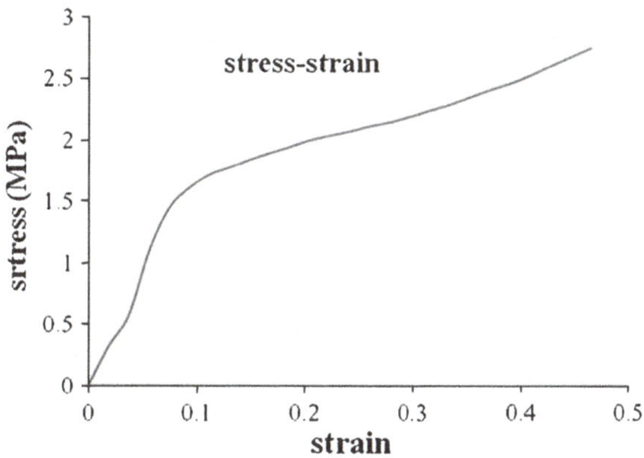

It also decreased the average cell size which mainly affected foam density, porosity, and mechanical properties. A typical stress-strain behavior of this composite foam can be seen in Fig. 36. According to the authors, the morphological results confirmed that intercalation of cellulose into clay layers improved both compressive

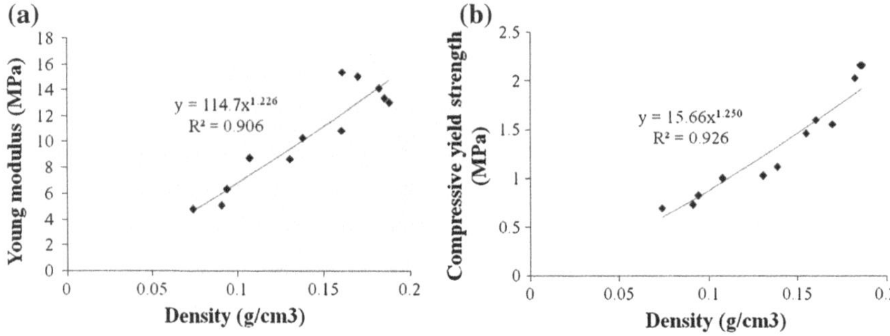

Fig. 37 **a** Young's modulus and **b** compressive yield strength as a function of foam density

modulus and strength (Fig. 37). Although more works are needed on this subject, it seems that nanocomposite foams based on cellulose and nanoclay could be good replacements for petroleum-based materials in food packaging industries.

Some works using only natural polymers instead of petroleum derived polymers were published, especially Vercelheze et al. (2012) using cassava starch, sugarcane bagasse fiber, and Na-MMT to produce nanocomposite trays. They were able to reach well-formed trays where the addition of natural fibers and clay resulted in lower density and rigidity. With well mixed nanoclay in the matrix, they also claimed to obtain an exfoliated structure resulting in good barrier and mechanical properties. Well dispersed clay in the starch matrix can be obtained by first making the nanocomposite in diluted aqueous solution and then plasticization (Mali et al. 2013). Vercelheze et al. (2012) and Matsuda et al. (2013) used the same method to produce starch/fiber/nanoclay nanocomposite foams and confirmed that the samples contained exfoliated MMT. The main issue while using starch base biodegradable foams is cost and performance when compared with conventional polymers. The use of natural fibers and nanoparticles such as nanoclay can help to improves on these challenges. Material performance is directly related to the type of starch, processing parameters, filler content, and foam formulations. Typical SEM images of starch foams containing different nanoclays are presented in Fig. 38.

Fig. 38 SEM of foam trays cross-sections: **a** cassava starch, **b** 95 % wt. cassava starch with 5 % wt. Cloisite 10A, and **c** 95 % wt. cassava starch with 5 % wt. Cloisite 30B

Although several works have been presented here, there is still potential to develop new materials, formulations, and processes to combine clay and natural fibers in polymer matrices to produce hybrid composites foams with higher specific properties.

5 Conclusion and Prospective Works

Nanoclays, due to their specific structure (small size, high aspect ratio, and high specific area), are known to provide improved mechanical, barrier, and fire retardancy properties to polymers which is desirable for many industries, especially packaging and membrane technologies. Dispersion degree according to intercalation/exfoliation state is one of the key factors determining their reinforcing role while being used in polymer matrices.

Following numerous studies done on the incorporation of nanoclays in polymer matrices, the use of nanoclay in hybrid composite systems (with natural and synthetic fillers) was also investigated in many works as this material has the potential of being easily processed in different polymer matrices through different processing methods (extrusion, injection, compression, and solution based processes).

In the future, nanoclay combined with natural fibers (jute, hemp, flax, cellulose, etc.) to produce hybrid composites and foam systems will be further developed. The effect of different parameters such as type and amount of compatibilizer, fiber type, size, and modification as well as process conditions also need more investigation. Finally, these systems are expected to have important advantages such as: light-weight, high specific mechanical properties, high barrier properties, flexibility, well defined and controlled structure, as well as being ecofriendly, especially when using biopolymers as matrices.

References

Ahmadzadeh, S., Keramat, J., Nasirpour, A., Hamdami, N., Behzad, T., Aranda, L., Vilasi, M., Desobry, S.: Structural and mechanical properties of clay nanocomposite foams based on cellulose for the food-packaging industry. J. Appl. Polym. Sci. (2015)

Ali, E.S., Ahmad, S.: Bionanocomposite hybrid polyurethane foam reinforced with empty fruit bunch and nanoclay. Compos. B Eng. **43**, 2813–2816 (2012)

Anadão, P.: The use of montmorillonite clay in polymer nanocomposite foams. In: Mittal, V. (ed.) Polymer Nanocomposite Foams, pp. 149–168 (2013)

Arrakhiz, F.Z., Benmoussa, K., Bouhfid, R.: Pine cone fiber/clay hybrid composite: mechanical and thermal properties. Mater. Des. **50**, 376–381 (2013)

Babaei, I., Madanipour, M., Farsi, M., Farajpoor, A.: Physical and mechanical properties of foamed HDPE/wheat straw flour/nanoclay hybrid composite. Compos. B Eng. **56**, 163–170 (2014)

Chaudhary, A.K., Jayaraman, K.: Extrusion of linear polypropylene–clay nanocomposite foams. Polym. Eng. Sci. **51**, 1749–1756 (2011)

Chen, L., Rende, D., Schadler, L.S., Ozisik, R.: Polymer nanocomposite foams. J. Mater. Chem. A **1**, 3837–3850 (2013)

Faruk, O., Bledzki, A.K., Fink, H.P., Sain, M.: Biocomposites reinforced with natural fibers: 2000–2010. Prog. Polym. Sci. **37**, 1552–1596 (2012)

Ghanbari, A., Heuzey, M.C., Carreau, P.J., Ton-That, M.T.: A novel approach to control thermal degradation of PET/organoclay nanocomposites and improve clay exfoliation. Polymers **54**, 1361–1369 (2013)

Johansson, C., Bras, J., Mondragon, I., Nechita, P., Plackett, D., Šimon, P., Svetec, D.G., Virtanen, S., Baschetti, M.G., Breen, C., Clegg, F., Aucejo, S.: Renewable fibers and bio-based materials for packaging applications—a review of recent developements. BioResources **7**, 2506–2552 (2012)

Keshtkar, M., Nofar, M., Park, C.B., Carreau, P.J.: Extruded PLA/clay nanocomposite foams blown with supercritical CO_2. Polymers **55**, 4077–4090 (2014)

Kord, B.: Nanofiller reinforcement effects on the thermal, dynamic mechanical, and morphological behavior of HDPE/rice husk flour composites. BioResources **6**, 1351–1358 (2011)

Kord, B.: Effect of nanoparticles loading on properties of polymeric composite based on hemp fiber/polypropylene. J. Thermoplast. Compos. Mater. **25**, 793–806 (2012)

Kord, B., Kiakojouri, S.M.H.: Effect of nanoclay dispersion on physical and mechanical properties of wood flour/polypropylene/glass fibre hybrid composites. BioResources **6**, 1741–1751 (2011)

Koronis, G., Silva, A., Fontul, M.: Green composites: a review of adequate materials for automotive applications. Compos. B Eng. **44**, 120–127 (2013)

Ku, H., Wang, H., Pattarachaiyakoop, N., Trada, M.: A review on the tensile properties of natural fiber reinforced polymer composites. Compos. B Eng. **42**, 856–873 (2011)

Kumar, V., Singh, A.: Polypropylene clay nanocomposites. Rev. Chem. Eng. **29**, 439–448 (2013)

La Mantia, F.P., Morreale, M.: Green composites: a brief review. Compos. A **42**, 579–588 (2011)

Le Moigne, N., Sauceau, M., Benyakhlef, M., Jemai, R., Benezet, J.C., Rodier, E., Lopez-Cuesta, J.M., Fages, J.: Foaming of poly(3-Hydroxybutyrate-Co-3-Hydroxyvalerate)/organo-clays nano-biocomposites by a continuous supercritical CO_2 assisted extrusion process. Eur. Polymer J. **61**, 157–171 (2014)

Majeed, K., Jawaid, M., Hassan, A., Abu Bakar, A., Abdul Khalil, H.P.S., Salema, A.A., Inuwa, I.: Potential materials for food packaging from nanoclay/natural fibres filled hybrid composites. Mater. Des. **46**, 391–410 (2013)

Mali, S., Yamashita, F., Grossmann, M.V.E.: Foams based on starch, bagasse fibers, and montmorillonite. In: Mittal, V. (ed.) Polymer Nanocomposite Foams, pp. 79–92 (2013)

Matsuda, D.K.M., Verceheze, A.E.S., Carvalho, G.M., Yamashita, F., Mali, S.: Baked foams of Cassava starch and organically modified nanoclays. Ind. Crops Prod. **44**, 705–711 (2013)

Najafi, A., Kord, B., Abdi, A., Ranaee, S.: The impact of the nature of nanoclay on physical and mechanical properties of polypropylene/reed flour nanocomposites. J. Thermoplast. Compos. Mater. **25**, 717–727 (2012a)

Najafi, N., Heuzey, M.C., Carreau, P.J., Wood-Adams, P.M.: Control of thermal degradation of polylactide (PLA)-clay nanocomposites using chain extenders. Polym. Degrad. Stab. **97**, 554–565 (2012b)

Picard, E., Espuche, E., Fulchiron, R.: Effect of an organo-modified montmorillonite on PLA crystallization and gas barrier properties. Appl. Clay Sci. **53**, 58–65 (2011)

Rahman, N.A., Hassan, A., Yahya, R., Lafia-Araga, R., Hornsby, P.: Polypropylene/glass fiber/nanoclay hybrid composites: Morphological, thermal, dynamic mechanical and impact behaviors. J. Reinf. Plast. Compos. **31**, 1247–1257 (2012)

Ray, S.S.: Thermodynamics, molecular modeling, and kinetics of nanocomposite formation. In: Clay-Containing Polymer Nanocomposites: From Fundamentals to Real Applications, pp. 25–37 (2013a)

Ray, S.S.: Mechanical properties. In: Clay-Containing Polymer Nanocomposites: From Fundamentals to Real Applications, pp. 171–226 (2013b)

Ray, S.S.: Barrier properties. In: Clay-Containing Polymer Nanocomposites: From Fundamentals to Real Applications, pp. 227–241 (2013c)

Ray, S.S.: Thermal stability. In: Clay-Containing Polymer Nanocomposites: From Fundamentals to Real Applications, pp. 243–261 (2013d)

Rezanavaz, R., Aghjeh, M.K.R.: Relationship between the morphology and physico-mechanical properties of polyethylene/clay nanocomposites. In: IOP Conference Series: Materials Science and Engineering, vol. 40, pp. 1–10 (2012)

Saba, N., Tahir, P., Jawaid, M.: A review on potentiality of nano filler/natural fiber filled polymer hybrid composites. Polymers **6**, 2247–2273 (2014)

Sorrentino, L., Iannace, S.: Nanocomposite foams from high-performance thermoplastics. In: Mittal, V. (ed.) Polymer Nanocomposite Foams, pp. 189–231 (2013)

Tabuani, D.S.C., Camino, G.: Polypropylene nanocomposites, study of the influence of the nanofiller nature on morphology and material properties. In: Macromolecules Symposium, pp. 114–127 (2011)

Tsimpliaraki, A.I.T., Marras, S.I., Zuburtikudis, I., Panayiotou, C.: The effect of surface chemistry and nanoclay loading on the microcellular structure of porous Poly(d, l Lactic Acid) nanocomposites. J. Supercrit. Fluids **57**, 278–287 (2011)

Vercelheze, A.E.S., Fakhouri, F.M., Dall'Antônia, L.H., Urbano, A., Youssef, E.Y., Yamashita, F., Mali, S.: Properties of baked foams based on Cassava starch, sugarcane bagasse fibers and montmorillonite. Carbohydr. Polym. **87**, 1302–1310 (2012)

Yang, F., Manitiu, M., Kriegel, R., Kannan, R.M.: Structure, permeability and rheology of supercritical CO_2 dispersed polystyrene-clay nanocomposites. Polymers **55**, 3915–3924 (2014)

Yeh, S.-K., Liu, Y.-C., Wu, W.-Z., Chang, K.-C., Guo, W.-J., Wang, S.-F.: Thermoplastic polyurethane/clay nanocomposite foam made by batch foaming. J. Cell. Plast. **49**, 119–130 (2013)

Yousefian, H., Rodrigue, D.: Effect of nano-crystalline cellulose on morphological, thermal and mechanical properties of Nylon 6 composites. Polym. Compos. (2014)

Yousefian, H., Rodrigue, D.: Effect of nano-crystalline cellulose, chemical blowing agent and mold temperature on the morphological, physical and mechanical properties of polypropylene. J. Appl. Polym. Sci. (2015)

Zandi, F., Rezaei, M., Kasiri, A.: Effect of nanoclay on the physical-mechanical and thermal properties and microstructure of extruded noncross-linked LDPE nanocomposite foams. Key Eng. Mater. **471–472**, 751–756 (2011)

Nanoclay Based Natural Fibre Reinforced Polymer Composites: Mechanical and Thermal Properties

Carlo Santulli

Abstract This chapter discussed the introduction of nanoclay into natural fibre reinforced composites (in practice, reinforced with lignocellulosic fibres from plants, therefore referred here as plant fibre composites, abbreviated as PFC) in terms of improvement of mechanical and thermal properties. A number of issues are to be accounted for, which include the specificity and hierarchical structure of the different lignocellulosic fibres, therefore their significant variability in mechanical terms, which results also in a substantial dimensional variability when introduced in the composite. More specifically, the main effects encountered by this introduction, in amounts normally variable between 1 and 5 wt %, would concern the reduction of viscoelastic behaviour of the composite and achieving improved dimensional control, provided a sufficient interaction of the nanoclay in the composite is obtained. This is particularly significant on one side aiming at an effective production of injection moulded PFCs and on the other side in connection with the use of biodegradable matrices, focusing on the production of a fully sustainable composite.

Keywords Nanoclay · Lignocellulosic fibres · Mechanical properties · Thermal properties

1 Introduction

When dealing with plastics, the addition of nanofillers can be useful to modify their properties, with notable effects on heat distortion, hardness, toughness, stiffness and mould shrinkage (Barmar et al. 2006). In the particular case of nanoclay, its use as a filler has been proposed in polymer composites, where at the expense of some increase in weight, limited by the small amount of nanoclay introduced, advantages in mechanical terms and as far as processability is concerned, were encountered.

C. Santulli (✉)
School of Architecture and Design (SAAD), viale della rimembranza,
63100 Ascoli Piceno, Italy
e-mail: carlo.santulli@unicam.it

M. Jawaid et al. (eds.), *Nanoclay Reinforced Polymer Composites*,
Engineering Materials, DOI 10.1007/978-981-10-0950-1_4

In principle, different ways to introduce nanoclay in composites can be proposed, in particular directly in the matrix, which is known as the melt blending process or else its addition when producing the composite: the two methods were e.g., compared in the case of high density polyethylene (HDPE) wood flour composites (Faruk and Matuana 2008). In addition, the introduction of fillers of agricultural origin, but not strictly speaking of fibrous geometry, was also proposed in combination with nanoclay: this was the case for example for almond shells, a typical waste from the confectionery industry, introduced in maleic anhydride grafted polypropylene matrix with the main limitation being a reduced impact strength (Lashgari et al. 2013). A similar introduction of empty fruit bunches (EFB) in a polyurethane foam matrix together with organic montmorillonite (o-MMT) improved by the synergy of the two fillers the stiffness of the composites, although EFB did not prove particularly effective in combination with o-MMT as far as thermal stability is concerned (Ali and Ahmad 2012).

On the other side, the introduction of nanoclay proved also beneficial also in cement-based composites, where its synergy with fibres of vegetable origin, was studied. In this case, it was observed for example that hemp fabric in the amount of 2.4 wt% had a beneficial effect on the reduction of porosity and the consequent increase of density and particularly on flexural strength and toughness, when coupled to 1 wt% of nanoclay (Cloisite 30) in a cement matrix with a cement/binder ratio equal to 0.48. A larger amount of hemp fabric (up to 6.9 wt%) was also accommodated in the composite by treating the fibres by mercerization (Hakamy et al. 2015). The addition of Cloisite was not effective though when increasing the amount of nanoclay up to 3 wt%, in which case the presence of more pores, calcium hydroxide crystals, ettringite and microcracks weaken the structure was revealed (Hakamy et al. 2014a). However, in the amount of 1 wt%, nanoclay was demonstrated to act as an activator to promote pozzolanic reaction therefore modifying cement matrix and improving the hemp fabric-matrix adhesion, promoting also thermal stability of the composite (Hakamy et al. 2013, 2014b).

In other cases, even more complex situations have been envisaged, such as the combined introduction of multiwalled carbon nanotubes (MWNT) and sepiolite nanoclay in a plant fibre composite, especially in view of obtaining a superior flame retardancy effect, a result which has been the object of specific reviews, as regards nanoclays (Kiliaris and Papaspyrides 2010). As a matter of fact, the introduction of 30 wt% hemp, using a particularly coarse fabric (areal weight 800 g/m^2) in a ternary composite comprising polylactic acid (PLA), 10 wt% sepiolite + 2 wt% MWNT was measured to have a 25 % reduction in peak heat release (PHRR) under the cone calorimeter (Hapuarachchi and Peijs 2010).

Dealing with plant fibre composites (PFCs), the main synergistic effect of nanoclay with plant fibres, e.g. flax, apart the aforementioned mechanical advantages, was demonstrated in hindering the dissolution of lignin in the alkaline environment present in cement (Aly et al. 2011), an aspect that proved its usefulness also in the case of geopolymers (Assaedi et al. 2015). Also, in combination with flax fibres, different types of nanoclay indicated a significant and yet again in synergy with the natural fibres, improvement of the fire resistance in soy-based matrices (Bensadoun

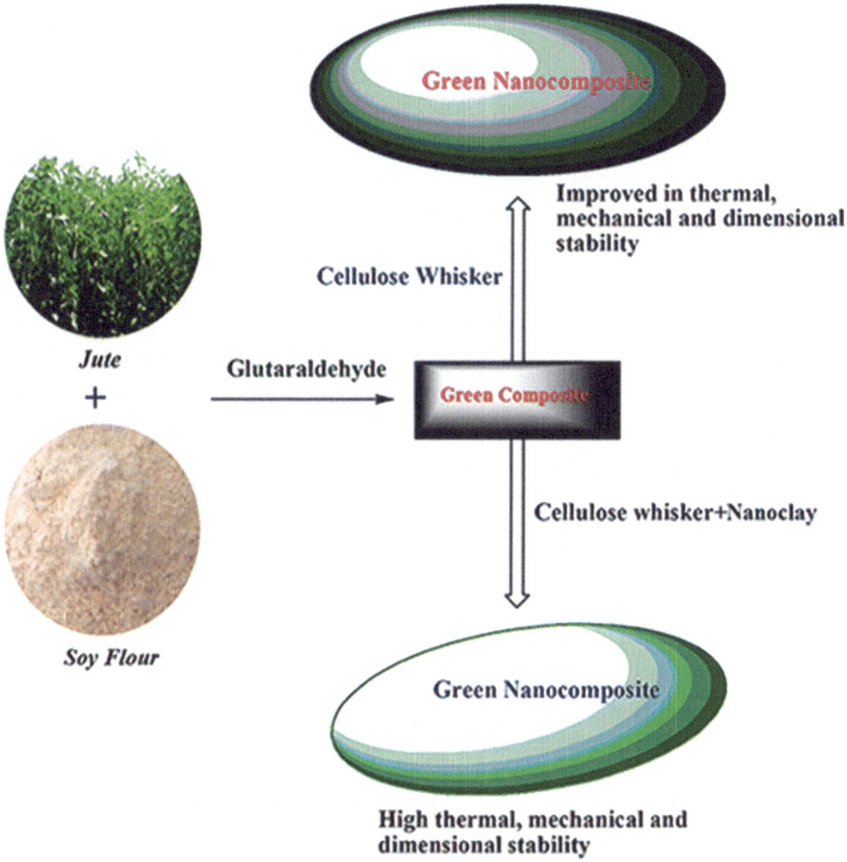

Fig. 1 System for creation of a green nanocomposite (Iman 2013)

et al. 2011). As a matter of fact, these attempts in producing nanoclay filled plant fibre composites often resort to structures of significant complexity, where a number of components need to effectively interact (the use of biomatrices in this sense can represent a further complication for the assessment of performance in the widest sense). An example suggestive of this intricate yet fascinating synergy can be offered by the diagram depicted in Fig. 1: to try and elucidate these issues, we need to refer back to the reasons for development of PFCs.

2 Plant Fibre Composites (PFCs): Perspectives and Issues

Plant fibre composites (PFCs) have been developed with the idea to replace traditional composites, in particular fibreglass, in view of their reduced carbon footprint: this is of specific interest in sectors such as automotive industry (Holbery and

Houston 2006). This drive towards improved environmental sustainability suggested that also intermediate approaches, such as the fabrication of hybrids, including both glass fibres and vegetable fibres, such as jute, flax, hemp, etc., would be of some interest for the purpose, allowing to take better control of the variables, which are essential for the success of the production and semi-structural use of PFCs, for example impact performance (Santulli 2007).

Significant problems still exist in the production and introduction into service of PFCs: on one side the importance is perceived of introducing plant fibres into bio-sourced, hence biodegradable and compostable matrices, some of which have become increasingly popular in fields such as packaging, in particular poly(lactic acid) (PLA). However, it needs to be pointed out that in other sense these materials do not always appear to be quite ready for these purposes. For example, whilst on one side PLA appears mostly suitable for packaging rigid objects, its use in soft films requires conversely the use of plasticizers, such as for example glycerol, sorbitol (Li and Huneault 2011) or limonene (Arrieta et al. 2013). This can not only ease the processing of the polymer, but also enhance some mechanical properties, particularly toughness. However, common issues encountered in biopolymers are not limited to mechanical and thermal properties. For example, also their permeability to gases and vapours and their generally scarce barrier properties are critical: amongst innovative strategies to improve these properties can be the addition of fillers such as nanoclay (Majeed et al. 2013). Of course, all fillers that are added to the polymer present a synergic effect on the properties of the composite, which is not easy to clarify. In particular, whenever nanoclay is added to PFCs, vegetable fibres are present, which present an inherent variability, due to a number of factors, which start from the ones depending on the plant species and the different agronomical parameters of its cropping, but are not limited to that, including the effects of extraction, treatment and fibre arrangement.

In particular, vegetable fibres are compounds combining different polysaccharides, in particular variable amounts of cellulose, hemicellulose and lignin, while pectin is normally removed during *fibre extraction* from the plant, as it is markedly non structural (Faruk et al. 2012). In particular, cellulose is a semicrystalline polysaccharide consisting of D-glucopyranose units, linked together by β-(1-4)-glycosidic bonds, therefore linearly arranged, such as those shown in Fig. 2. In contrast, hemicellulose, joined to cellulose fibrils by hydrogen bonds, has a branched structure, is fully amorphous and has lower molecular weight. Lignins are also amorphous and are mostly formed by aromatic phenylpropane polymers, whereas pectins are glue-like polysaccharide acid polymers. Passing from cellulose to

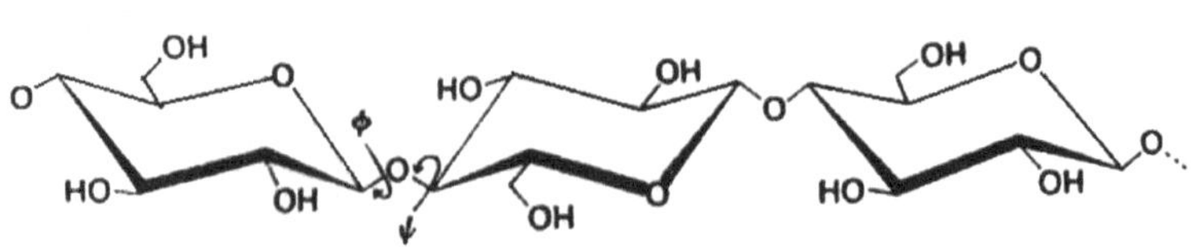

Fig. 2 Arrangement of β-1,4 glycosidic bonds

hemicellulose and lignin the hydrophilic character gradually decrease and in general terms hemicellulose, pectins and lignin are not thermally stable, tending to degrade at a temperature lower than 500 °C (Li et al. 2007).

After this, the influence of the species needs to be taken into account, often referred to as specificity (Klemm et al. 2005). Vegetable fibres can be derived from various parts of the plant, including leaf (e.g., sisal or pineapple), bast (e.g., flax, hemp and jute), seed (e.g., cotton) and fruit (e.g., coir). Once extracted the fibres, a process which mainly occurs by microbial retting, whereas only in some cases by action of enzymes, such as pectinase, these may be treated to further remove non structural parts and regularise their shape and geometry, reducing the extension of the internal voids, commonly referred to as lumens or lacunae: these are a characteristic feature of vegetable fibres and were thoroughly investigated for their effect on mechanical performance (Satyanarayana et al. 1981).

The treatment, as described above, is normally defined as primary and is essential for the manufacturing of textile products: most common primary treatments are mercerization by an alkali, such as typically sodium hydroxide (George et al. 2001), or bleaching by sodium hypochlorite (Saheb and Jog 1999), acetylation, benzoylation or potassium permanganate treatment, all of which are dealt with in La Mantia and Morreale (2011). Secondary treatments may also be applied, in come cases following a primary one, which tend to be more protective for the fibres, also to reduce moisture penetration (Kalia et al. 2009). A final factor, among the principal ones contributing to the variability, is the type of fibre architecture by which the vegetable fibres are introduced in the composite: they might be in the form of short random fibres, or else to be in other randomly oriented forms, such as mats. Oriented geometries include plain weave, twill weave and satin weave, together with other forms of more complex arrangements of multiple warps and wefts, such as leno, also referred to as cross-weave.

3 General Considerations on Nanoclay

Clays are defined as phyllosilicates (from the Greek word “φυλλον” = leaf), in that they have a layered structure formed by octahedral layers formed by aluminium or magnesium hydroxide and tetrahedral silicon dioxide structure layers. They can be divided in 1:1 phyllosilicates, such as kaolinite and serpentine, in which each octahedral layer is alternate to a tetrahedral one, or 2:1 phyllosilicates, such as montmorillonite and saponite, the latter being of much more frequent use in nanocomposites, in which every octahedral layer is sandwiched and edge-fused between two layers of tetrahedrally coordinated silicon atoms. The thickness of every layered sheet is normally in the order of 1 nm with a length going from tens of nanometres to over a micron. In octahedral layers it is often possible that aluminium is replaced by iron hydroxide or magnesium by lithium hydroxide with no significant change in the geometry of the clay.

Clays are hydrophilic, because between the layers, where as an effect of stacking Van der Waals gaps are created, usually referred to as interlayer or gallery, are normally present ions of alkali metals, such as Na+ and K+, or alkaline-earth, such as Ca2+. These counterbalance the negatively charged platelets and therefore allow more water to be chemically linked and allocated within the structure.

Normally though, to improve the affinity between the filler and the matrix, which will have a beneficial effect on material properties, the inorganic cations located in the galleries (Na+, Ca2+, etc.) are generally exchanged by ammonium or phosphonium cations bearing at least one long alkyl chain, and possibly other substituted groups. The resulting clays are referred to as organomodified layered silicates (OMLS) and, in the case of montmorillonite (MMT), are abbreviated OMMT. As a matter of fact, most nanoclays used in composites are MMTs, with a possible competitor in synthetic fluorine mica (SFM) (patented by Tateyama et al. 1993), which is produced using 10–35 wt% of an alkali silicofluoride (possibly in addition with some alkali fluoride) and the balance of talc. Some types of nanoclay are reported in Table 1, while the characteristics that are of interest for a specific one of these are offered in Table 2.

Table 1 Main types of nanoclay

Commercial (O)MMT and their characteristics						
Commercial clays Supplier/trade name/designation		Clay type	Organomodifier type	Modifier concentration (meq/100 g)	Δw^a (%)	*d*-spacing (Å)
Southern clay products (USA)						
Cloisite®Na	CNa	MMT	–	–	7	11.7
Cloisite®15A	C15A	MMT	$N^+(Me)_2(tallow)_2$	125	43	31.5
Cloisite®20A	C20A	MMT	$N^+(Me)_2(tallow)_2$	95	38	24.2
Cloisite®25A	C25A	MMT	$N^+(Me)_2(C_8)(tallow)$	95	34	18.6
Cloisite®93A	C93A	MMT	$NH^+(Me)(tallow)_2$	90	37.5	23.6
Cloisite®30B	C30B	MMT	$N^+(Me)(EtOH)_2(tallow)$	90	30	18.5
Sud-Chemie (Germany)						
Nanofil®804	N804	MMT	$N^+(Me)(EtOH)_2(tallow)$		21	18
Laviosa Chimica Mineraria (Italy)						
Dellite®LVF	LVF	MMT	–	105	4–6	9.8
Dellite®43B	D43B	MMT	$N^+(Me)_2(CH_2\text{-}\phi)$ (tallow)	95	32–35	18.6
CBC Co. (Japan)						
Somasif	MEE	SFM	$N^+(Me)(EtOH)_2$(coco alkyl)	120	28	
	MAE	SFM	$N^+(Me)_2(tallow)_2$	120	41	

Tallow: ~65 % C_{18}; ~30 % C_{16}; ~5 % C_{14}

[a]%Weight loss on ignition

From Bordes et al. (2009) (Table 3)

Table 2 Physical properties of nanoclay Cloisite 30B

Physical properties of the nanoclay (Cloisite 30B)	
Physical properties of Cloisite 30B	
Colour	Off white
Density (g/cm^3)	1.98
d-spacing (0 0 1) (nm)	1.85
Aspect ratio	200–1000
Surface area (m^2/g)	750
Mean particle size (μm)	6

From Alamri et al. (2012) (Table 1)

4 Objectives of the Introduction of Nanoclay in PFCs

The introduction of nanoclay in polymers does occur in two different configurations, depending on the degree of polymer polarity and Van der Waals interaction with the nanoclay, therefore the orientation of filler particles: these are respectively described as intercalated and exfoliated, as represented in Fig. 3. In any case, such an introduction produces the combination of two effects: it acts as filling the polymer with a particle-shaped reinforcement and restricts the movement of the polymer chains. While the former effect results in an increase of mechanical strength and rigidity, improving load transfer at matrix-fibres interface, reducing on the other side gas and (if relevant) moisture penetration and flammability, the latter has an outcome similar to crosslinking (Singh et al. 2009; Katschnig and Battisti 2015). To obtain these results, the uniform dispersion of nanoclay in the polymer is of crucial importance, for which mechanical stirring and sonication processes have also been experimented: of course, there is an optimal amount of nanoclay not to be exceeded to yield the best properties to the composites: this is often around 5 wt% (Chan et al. 2011).

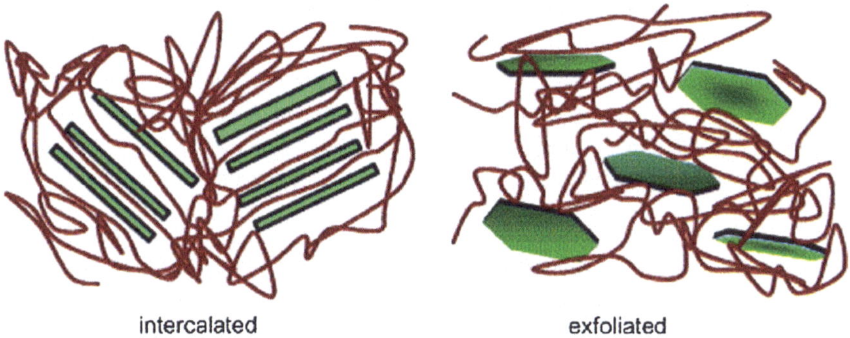

Fig. 3 Schematic illustration of two different types of thermodynamically achievable polymer/layered silicate nanocomposites (Ray and Bousmina 2005)

From what exposed above, it is clear that the introduction of nanoclay in plant fibre composites presents some additional issues, but on the other side also suggests possibilities: it has to be considered that vegetable fibres, being natural hierarchical structures, do not offer particular compactness and normally present internal gaps among the filaments, which are bound in a helical-like structure. Moreover, they have a considerable dimensional variability, which has an effect on properties. Therefore, introducing nanoclay in plant fibre composites, small quantities, such as 1 wt%, are likely to offer better stress transfer throughout the matrix, resulting typically in better impact and flexural properties: however, further increase would result in decreased properties, due to agglomeration, whose effect is worsened by the geometrical mismatches in the composites, owed to the presence of vegetable fibres (Rozman et al. 2010). In other words, optimisation of the amount and distribution of nanoclay in the composite is strongly needed. However, it is important to notice that another objective of introduction of nanoclays is the improvement of the behaviour at high temperature of plant fibre composites, which has notable effects on processing. In this sense, also studies of thermal and thermo-mechanical properties are highly required and it is important to evaluate the deviation from elastic behaviour, which tends to uncontrolled damping at high temperature. This is important, since damping behaviour is one of the great limits to the widespread application of plant fibre composites, especially as regards fatigue resistance, even for low number of cycles and low loading levels. Studies on that are limited, yet of considerable interest, in particular regarding the fact that fibre secondary treatments (Gassan and Bledzki 1997) or matrix grafting with maleic anhydride (Mohanty et al. 2006) may have a significant role in modifying this behaviour. In another sense, it is very likely that the introduction of nanoclay produces even larger effect on these properties, for this reason thermal properties of the obtained composites are of crucial importance.

5 Strategies Toward Improving the Mechanical Performance of Nanoclay-Filled PFCs

In general terms, several possibilities are open when adding nanoclay to plant fibre composites, a strategy which normally results in an increase in mechanical performance, in particular in tensile and flexural properties (Ku et al. 2011), although in some cases other characteristics might also be of crucial importance, such as for example impact performance and enhanced toughness.

A viable route is using matrices among the most diffuse for commodity uses, which are for example polyolefins, therefore practically mainly polypropylene and different grades of polyethylene, such as high density and low density. In this case, interest is concentrated on obtaining an effective moulding of these materials, despite the introduction of nanoclay essentially for mechanical purposes. This occurs particularly because thermoplastic matrices have been traditionally preferred, mainly since their processing is reasonably easy, in that manufacturing technologies such as injection moulding can be used, leading to high production rates with excellent dimensional tolerance. It is important to observe that the introduction of nanoclay has an important role in modifying the glass transition temperature of the polymer and reducing the deviation from the elastic behaviour for use at temperature higher than ambient one (Saw 2015).

Another possible strategy, which is significantly different as a concept, is using biodegradable matrices, such as starch-based ones. It is worth noting that there is a partial superposition with what described above, in the sense that the success of some biodegradable polymers, such as poly(lactic acid) (PLA) in their use as matrices, also notably in plant fibre composites, is also to be ascribed to their relatively easy moulding. However, in this case the addition of nanoclay is expected to modify their degradation profile, the preservation of which is ultimately of interest, together with the maximum possible flexibility in terms of manufacturing technologies (Lim et al. 2008). The largest majority of biodegradable matrices are thermoplastics: there is nonetheless a significant amount of studies aimed at the crosslinking of some of these matrices, in particular those based on vegetable oils, such as soy, castor, palm oil, towards the fabrication of thermosetting matrices to be processed at temperatures as close as possible to ambient temperature. In this case, an additional factor of possible modification of the properties is introduced, in the sense that moulding temperatures need to undergo only limited variations as the effect of the introduction of nanoclay. A comprehensive review of these attempts, in the view of reducing environmental impact of matrices used, while retaining the initial mechanical performance was offered by Mosiewicki and Aranguren (2013): these involved the use of fillers of different size and geometries, including also nanoclay, as received or vinyl modified.

5.1 Nanoclay-Filled PFCs with Oil-Based Matrices

In the case of traditional thermoplastic matrices, such as polyolefines, it can be suggested that nanoclay can in a sense replace those ceramics that exert some action of making the flow of melt polymer more regular during moulding such as for example talc. This would come together with some enhancement of properties, such as rigidity, deriving from a substantial hindrance to the motion of polymer chains, provided the distribution of nanoclay platelet is uniform enough (Chen et al. 2003). It is worth noting that in most cases polyolefin-based plastics are injection moulded. In this case, to possibly create the pellet to be injected, short random vegetable fibres are used with substantial difficulties in their dispersion due to the combination between non-polar matrix and highly polar lignocellulosic reinforcement (Herrera-Franco and Valadez-González 2005). This situation can be improved in the case of introduction of nanoclay: in a composite including kenaf fibres (KF) and coir fibres (CF) in a polypropylene matrix the interfacial properties were found to be substantially improved with a number of surface micro-voids significantly reduced as the result of MMT addition and fostering the achievement of good wettability

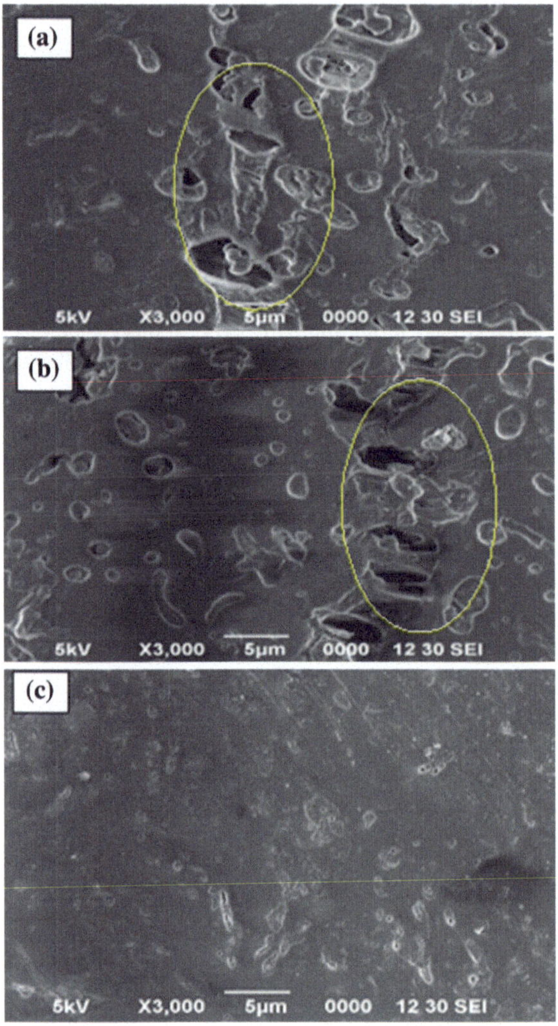

Fig. 4 Typical SEM micrographs of tensile fracture surfaces of **a** KF/PP composite, **b** KF/CF/PP composite, and **c** KF/CF/PP/MMT hybrid nanocomposites (Islam et al. 2012a, b)

properties (Islam et al. 2012a, b), as illustrated in Fig. 4. As a matter of fact, the introduction of nanoclay has been attempted in polypropylene matrix with different amounts of kenaf fibres (20–40 wt%), with the reinforcement with the higher amount of fibres resulting in a clear decrease of tensile and Izod impact properties, despite the simultaneous introduction of nanoclay in the amount of 6 wt% (Kamardin et al. 2015). The interaction of thermoplastic polymers with two different lignocellulosic fibres (Brown Rubber bark wood and coir) in a hybrid plant fibre polypropylene composite together with montmorillonite (MMT) resulted in some kind of synergic behaviour to be better investigated. For example, interestingly enough, the addition of MMT to the bare coir fibre composite did not substantially modify water absorption profile, whereas adding MMT to the composite including both fibres

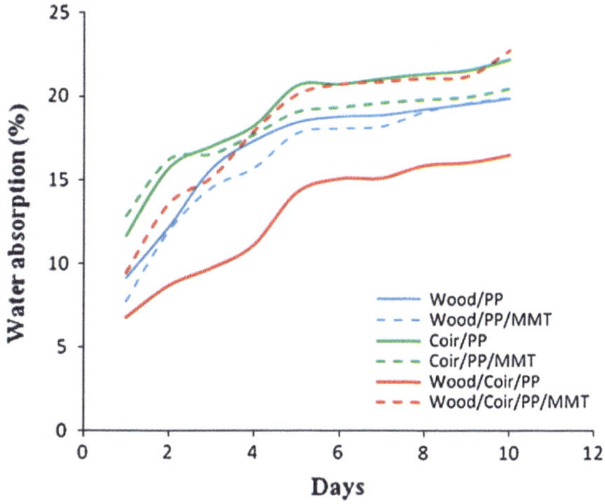

Fig. 5 Water absorption of different composites (Islam et al. 2015)

made the composite more vulnerable to water (Islam et al. 2015), as observed in Fig. 5. This observation can be a significant caveat on the modification of the effect of nanoclay depending on the very nature of the specific PFC it is introduced in.

The introduction of nanoclay (in the specific case Cloisite 30B in sisal/epoxy composites) proved on the other side effective to reduce the degradation of tensile properties after water immersion, since the amount of water absorbed at equilibrium is equally diminished. This has a significant interest, although at high amounts, such as 5 wt%, of nanoclay introduced, some contrasting effect from agglomeration and not fully effective mixing is also revealed (Mohan and Kanny 2011). Agglomeration, which has the consequence to increase the viscosity of resin-clay mixture, is represented in Fig. 6, is particularly experimented in the case of addition of Cloisite Na, hence with no modifiers. As far as the study of mechanical behaviour in temperature is concerned, it can be noticed that, despite normally the introduction of nanoclay reduces the viscoelasticity of the matrix, different evidences were found with typical treatments, such as alkali and silane, on coir sheath/polyester composites. In this case, a slight increase in tan δ values was noticed with the addition 3 wt% of clay, possibly due to non-uniform dispersion of clay and the agglomeration of residual coupling agent over the fibre surface (Rajini et al. 2013).

5.2 Nanoclay-Filled PFCs with Biomatrices

A comprehensive review on the use of nanoclay in biodegradable polymers is offered by Ray and Bousmina (2005). This review, which concentrates substantially on poly(lactic acid) (PLA) and polyhydroxybutyrate (PHB) suggests in particular

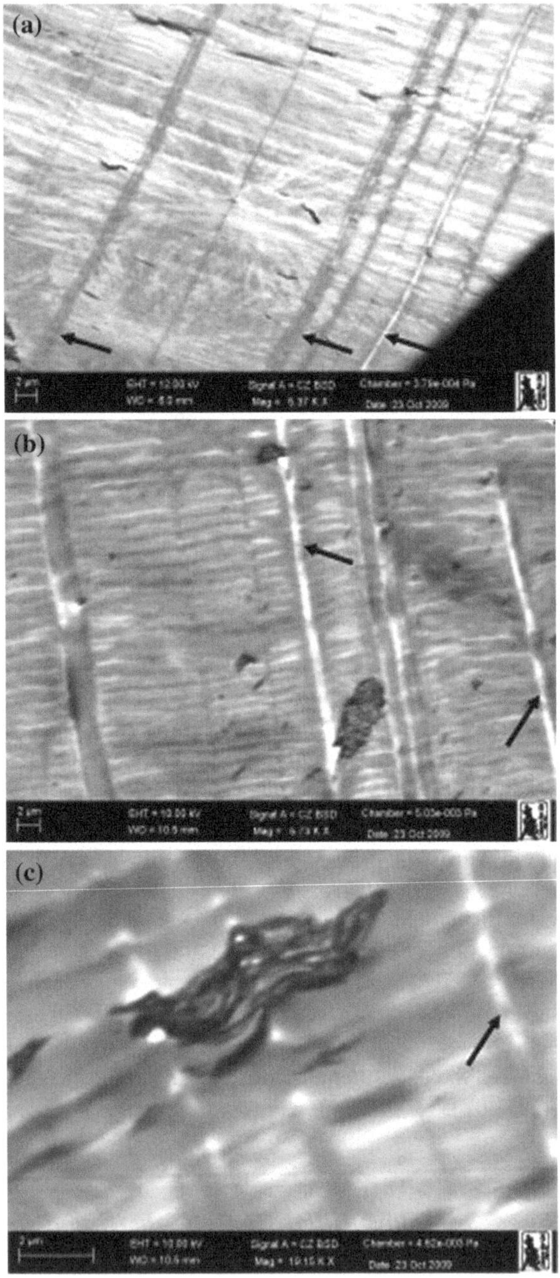

Fig. 6 TEM of sisal fibre composites with **a** 3 wt% nanoclay, **b** 5 wt% nanoclay and **c** 5 wt% Na+ microclay (Mohan and Kanny 2011)

that the incorporation of nanoclays does enhance compostability of the material, with judicious selection of the type on nanoclay. On the other side, it warns that the maximum viable amount of nanoclays for substantial increase of tensile properties on the polymer might not exceed 5 wt%. As regards PLA, which is the most frequently used biopolymer, an important consideration needs to be done on the fact that the addition of plasticizers increases the chain motion at the same time yielding crystallization, which at the same time reduces the glass transition temperature of the polymer (Pillin et al. 2006). It is largely suggested instead that the introduction of nanoclay has an opposite effect on PLA. In the case of use of these thermoplastic biomatrices for PFCs, some importance is assigned to the fibre treatments, whose effect cannot be neglected. In the case for example of the reinforcement of chemically treated (by alkali and acetic acid) 30 wt% jute fibres into a polyhydroxybutyrate-co-valerate (PHBV) matrix with the addition of 2 %, 3 %, and 4 wt% montmorillonite K10 nanoclay: the composite with the higher nanoclay content offered lower moisture absorption, better dynamic mechanical properties, and flexural properties. It can be noticed that the rougher surface of jute fibres due to treatment resulted in a substantially improved compaction of the composites under resin infusion (Hossain et al. 2011).

On the other side, adhesion to vegetable-oil derived matrices proved effective in the case for example of partially bio-based polyurethane derived from *Mesua ferrea* L. seed, where the tensile strength was increased by about 60 % for a maximum amount of clay introduced of 5 %, obtaining notable advantages also in terms of scratch hardness, thermostability, water vapour permeability, and adhesive strength, while on the other side there a slightly positive effect on impact resistance and bending, and a limited reduction of elongation at break (Deka and Karak 2009). The addition of 21–22 wt% of industrial hemp fibres, chopped at a length of 25 mm, together with 1.5 wt% of Cloisite 30B nanoclay in a blend of traditional unsaturated polyester and epoxidized methyl soyate (EMS) was also attempted in the view to tailor the properties of the obtained composite, especially as regards stiffness, ultimate tensile stress and toughness parameters, such as impact strength and ductility (Haq et al. 2008). In this regard, coupling of jute fibres with different fillers in a soy-based matrix, which was crosslinked to confer to it quasi-thermosetting properties, was often proposed (Iman and Maji 2012). These included nanoclay in combination with cellulose whiskers (Iman et al. 2013), or else with zinc oxide, offering in the latter case a good compromise between improved dimensional stability and effective biodegradation (Iman et al. 2015). Cloisite nanoclay was also added in an amount of 5 wt% directly to a soy protein concentrate (SPC) resin, modified by 15 wt% of glycerol, in order to improve its mechanical properties and its thermal stability. To offer a general idea of the level of enhancement obtained, this allowed introducing a higher tenor of flax fibres, arranged in unidirectional yarns, up to 48 wt%, and offered tensile strength as high as 62 and 82 MPa in the warp and weft directions, respectively (Huang and Netravali 2007). In Table 3 a number of studies that are particularly significant on the role of nanoclay in modifying PFCs are reported.

Table 3 Some studies on the introduction of nanoclay in PFCs

Matrix	Fibre type (and amount when available)	Nanoclay amount (%)	Improved variables by nanoclay addition	Reference
Unsaturated polyester	Coir 25 %	3	Tensile strength, Izod impact strength, compression and resin viscosity	Deepak et al. (2015a)
Unsaturated polyester	Jute 25 %	3	Tensile strength, Izod impact strength, compression and resin viscosity	Deepak et al. (2015b)
Unsaturated polyester	Jute 40 vol%	1–2	Flexural, compression, and interlaminar shear strength (ILSS) (maximum with 1 wt%)	Dewan et al. (2013)
Unsaturated polyester	Kenaf	1	Flexural and impact properties	Rozman et al. (2010)
Unsaturated polyester	Coir sheath	1	Specific wear rate (particularly with silane treatment)	Divya et al. (2014)
Unsaturated polyester	Coir sheath	1–2–3–5	Alkali and silane treated	Rajini et al. (2013)
Epoxy	Sisal	1–3–5	Water barrier effect	Mohan and Kanny (2011)
Polypropylene	Hemp	1–3–5	Tensile properties up to 3 % nanoclay	Kord (2012)
Polypropylene	Bleached red algae fiber (BRAF), kenaf fiber (KF), and cotton pulp fiber (CPF) all 40 %	5–10	Dimensional stability and dynamical mechanical properties (highest with BRAF)	Sim and Han (2013)
Polypropylene (with 5 % maleic anhydride)	Pineapple leaf 30 %	1–3	Tensile and flexural strength	Biswal et al. (2009)
HDPE (with 2, 4, 6 or 12 % maleic anhydride) (MAPE)	Bamboo up to ca. 30 %	1–3–5	Dynamical and static flexural modulus. Exfoliation with MAPE	Han et al. (2008)
Poly(lactic acid)	Short kenaf fibres 20 %	3	Impact and tensile properties	Kaiser et al. (2013)

6 Thermal Properties of Nanoclay-Filled PFCs

In general terms, as it is usually the case with ceramics, the addition of nanoclay to polymer matrices does improve their thermal stability, which can be generally indicated by the amount of residual char obtained at high temperature (such as

600–800 °C) (Alamri et al. 2012). With the introduction of vegetable fibres the problem of thermal degradation becomes even more crucial: the initial perspective was to investigate whether processing of the polymers at the usual temperatures for example for polyolefines would substantially damage the fibre structure, hence making them unsuitable for matrix reinforcement. In practice, for example on jute and flax fibres it was found that heating them at a temperature exceeding 170 °C for a period of time of 120 min resulted in a significant decrease of tenacity and degree of polymerization for the fibres (Gassan and Bledzki 2001). This was the case also on biomatrices, such as polylactic acid (PLA) in a hybrid composite with montmorillonite and kenaf fibres. Dynamic mechanical analysis at temperature ranging from −80 to 100 °C with loading frequency of 1 Hz demonstrated the increase of all parameters, including storage modulus, loss modulus and damping by an addition of up to 7 % MMT (Kaiser et al. 2011). Another study proposed that nanoclay could even act as a compatibiliser for improving the adhesion between poly(lactic acid) matrix and flax fibres (Kumar et al. 2010b). In this case the achievement of a higher compatibility by the addition of MMT was indicated by dynamical-mechanical analysis and in particular by the decrease of damping factor. The flax-polylactide composites had been previously prepared with amphiphilic additives that served to accelerate their biodegradation to be used as compostable materials at end-of-life (Kumar et al. 2010a). Values of dynamic loss (tan δ) as a function of the temperature for these composites are shown in Fig. 7: tan δ reduction demonstrated effective in revealing the quality of interface of the nanocomposites obtained, also for example in the case of interaction between different types of Malaysian wood fibres with nanoclay in a phenol-formaldehyde resin, as from Rahman et al. 2012. Coming back to Kumar et al. (2010a, b), the efficacy of the different additives, which are listed in the caption of Fig. 7, has been compared: this allowed demonstrating that apart from a general increase in stiffness, mandelic acid proved particularly effective owing to low water uptake, high mechanical properties and high storage modulus. This statement also holds true for

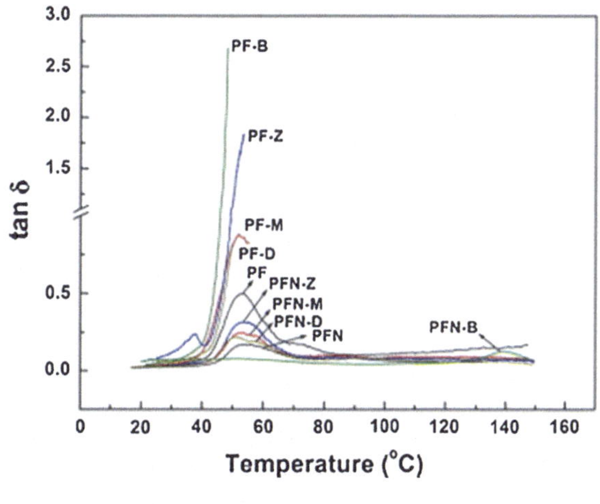

Fig. 7 Dynamic loss versus tempeature of composites with different additives: PF means poly(lactic acid)-flax with no nanoclay, PFN poly (lactic acid)-flax with added 2.5 wt% of nanoclay, while B, M, Z and D represent benzilic acid, mandelic acid, zein and dicumyl peroxide (DCP), respectively

all the additives in presence of nanoclay. To confirm the interfacial adhesion between PLA, fibres and additives, morphology studies of the bio-composites were also conducted (Kumar et al. 2010a, b).

With progression of use and development of PFCs, other more complex requirements were suggested as far as thermal stability is concerned: in practice, it is considered important to get sounder information about the physical interaction between the matrix, the vegetable fibres present and the nanoclay introduced. Particularly important are the observations obtained on kenaf/coir hybrid polypropylene composites with the addition of organic montmorillonite, in that these really clarify the extent and importance of respective interactions within the material: compared Fourier transform infrared spectroscopy (FTIR) are reported in Fig. 8 (Saiful et al. 2015). In this case, the spectrum of pure polypropylene (Fig. 8a) displays the characteristic absorption peaks in the region of 2936 cm^{-1} for alkyl C–H. Other peaks in the further spectra (at 3400, 1700, 1150–950 and 470 are most likely to be related to the stretching vibrations for hydroxide (–OH), carbonyl (C=O), C–O in alcohol and Si–O–Si, respectively. The absorption peaks of polypropylene were slightly shifted after fibre loading as clearly observed from Fig. 8c, e, g. Spectra in Fig. 8b, d, f, h indicated that polypropylene peaks changed after blending with nanoclay, so that another peak was observed at around 470 cm^{-1} due to Si–O–Si stretching vibration for all nanocomposites. This peak is expected, due to the

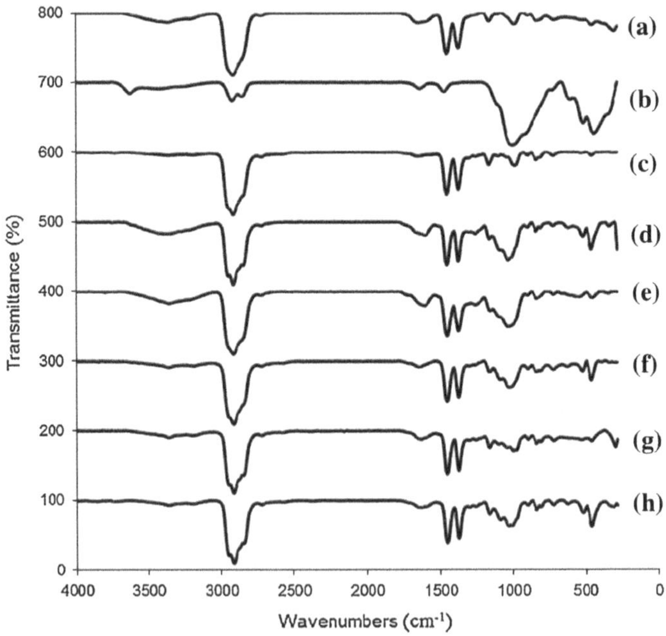

Fig. 8 FTIR spectra of **a** polypropylene, **b** nanoclay, **c** coir/PP composite, **d** coir/PP/MMT nanocomposite, **e** kenaf/PP composite and **f** kenaf/PP/MMT nanocomposite, **g** kenaf/coir/PP hybrid composite and **h** kenaf/coir/PP/MMT hybrid nanocomposites

incorporation of MMT into the composites, and was reported by Ray and Okamoto (2003) and is a sign of good incorporation. Another sign of good incorporation of MMT improving fibre–matrix interfacial adhesion in the composites was the slight shift of C=O C–O peaks and –OH peaks toward higher wave numbers after addition of MMT (Islam et al. 2012a, b). 2

7 Conclusions and Future Perspective

Plant fibre composites, though their production may be desirable for a number of reasons, including the possible obtainment of fully biodegradable composites with the use of biomatrices, present a number of issues that are not completely resolved, so that their use in semi-structural composites and possible replacement, even if partial, for glass fibres is not always practicable. In terms of mechanical properties, what is particularly required is the achievement of sufficient impact properties and effective dynamical-mechanical behaviour, so that damping is not excessive, when inserted in visco-elastic matrices, which would lead to limited dimensional stability.

Factors for concern include in particular the significant sensitivity to moisture content, which is also linked to the presence of internal voids in the fibres, known as lumens or lacunae, and the dimensional variability of the fibres. These can be arranged in different architectures, among which the incorporation of short fibres in pellets is desirable for the purpose of injection moulding, the preferred manufacturing process for commodity items. Dimensional variability, which has the effect of limiting the amount of vegetable fibres that can be introduced in the composite, can be reduced by the application of a number of different surface treatments on the fibres, among which are alkali treatment, typically with sodium hydroxide, bleaching or less aggressive ones, such as silane treatment. These are adapted for the purpose, although they introduce yet another factor of variability in the fibre properties and on the composites produced by their use. With the idea of end-of-life recycling and reprocessing, the thermal degradation of plant fibres has also been widely studied and represent another factor for concern. In practice, the most used fibres are flax, hemp, jute, kenaf, sisal and coir, for their inherent properties and the availability of different fibre arrangements: however, in principle the number of possible crops from which fibres for composites can be obtained is very large, being based on specific local productive systems.

All of the above reasons contribute to suggesting that the introduction of nanoclay in small amounts, normally not exceeding 5 wt%, but in some cases considerably less, may be beneficial for plant fibre composites. The main improvements were obtained as far as dimensional stability and ease of processing is concerned. This had also normally a non negligible effect on dynamical-mechanical properties and on tensile and flexural strength, although the balance between improved rigidity and avoiding the possible induction of a brittle behaviour was not always obvious. This difficulty was observed often in impact loading, usually performed by Izod or

Charpy tests, where the variability of the absorbed energy could even conceal the advantage obtained in terms of properties. It is noteworthy that the improvement is notable only in the case a sufficient interaction between the matrix and nanoclay exists. This is not always easy due to the fact that most frequently used matrices are not polar, such as polyolefins, whilst nanoclay is quite strongly polar. On the other side, the use of biomatrices, both based on polysaccharides and on vegetable oils, would pose other problems, such as the possible penetration of water in the material, which would have an effect also on agglomeration of nanoclay. This is a very sensitive issue at amounts of these exceeding a few percents.

Further developments of plant fibre composites including nanoclay would possibly go along the following lines:

- Combined use of different vegetable fibres possibly arranged in more complex and 3-D geometries so to have more effective interaction even in presence of a limited amount of nanoclay
- More widespread studies of loading simulating service (e.g., fatigue testing, falling weight impact) with the idea of extending their useful life even in harsh conditions (measurement of residual properties after being subjected to water absorption would e.g., be useful)
- The development of new biomatrices (particularly, real thermosetting matrices from vegetable oils) would improve the profile of use of plant fibre composites and this could include more diffuse application of nanoclay

References

Alamri, H., Low, I.M., Alothman, Z.: Mechanical, thermal and microstructural characteristics of cellulose fibre reinforced epoxy/organoclay nanocomposites. Compos. B **43**(7), 2762–2771 (2012)

Ali, E.S., Ahmad, S.: Bionanocomposite hybrid polyurethane foam reinforced with empty fruit bunch and nanoclay. Compos. B **43**(7), 2813–2816 (2012)

Aly, M., Hashmi, M.S.J., Olabi, A.G., Messeiry, M., Hussain, A.I., Abadir, E.F.: Effect of nano-clay and waste glass powder on the properties of flax fibre reinforced mortar. ARPN J. Eng. Appl. Sci. **6**(10), 19–28 (2011)

Arrieta, M.P., López, J., Ferrándiz, S., Peltzer, M.A.: Characterization of PLA-limonene blends for food packaging applications. Polym. Testing **32**(4), 760–768 (2013)

Assaedi, H., Shaikh, F.U.A., Low, I.M.: Utilization of nanoclay to reinforce flax fabric-geopolymer composites. Int. J. Chem. Mol. Nucl. Mater. Metall. Eng. **9**(12), 1297–1305 (2015)

Barmar, M., Barikani, M., Fereidounnia, M.: Study of polyurethane/clay nanocomposites produced via melt intercalation method. Iran. Polym. J. **15**(9), 709–714 (2006)

Bensadoun, F., Kchit, N., Billotte, C., Bickerton, S., Trochu, F., Ruiz, E.: A study of nanoclay reinforcement of biocomposites made by liquid composite molding. Int. J. Polym. Sci. 964193 (2011)

Biswal, M., Mohanty, S., Nayak, S.K.: Influence of organically modified nanoclay on the performance of pineapple leaf fiber-reinforced polypropylene nanocomposites. J. Appl. Polym. Sci. **114**(6), 4091–4103 (2009)

Bordes, P., Pollet, E., Avérous, L.: Nano-biocomposites: biodegradable polyester/nanoclay systems. Prog. Polym. Sci. **34**(2), 125–155 (2009)

Chan, M., Lau, K., Wong, T., Ho, M., Hui, D.: Mechanism of reinforcement in a nanoclay/polymer composite. Compos. B **42**(6), 1708–1712 (2011)

Chen, L., Wong, S.C., Pisharath, S.: Fracture properties of nanoclay-filled polypropylene. J. Appl. Polym. Sci. **88**(14), 3298–3305 (2003)

Deepak, K., Reddy, N.S., Naidu, T.V.S.: Thermosetting polymer and nano clay based natural fiber bio- composites. Procedia Mater. Sci. **10**, 626–631 (2015a)

Deepak, K., Prabhakar Vattikuti, S.V., Venkatesh, B.: Experimental investigation of jute fiber reinforced nanoclay composite. Procedia Mater. Sci. **10**, 238–242 (2015b)

Deka, H., Karak, N.: Vegetable oil-based hyperbranched thermosetting polyurethane/clay nanocomposites. Nanoscale Res. Lett. **4**, 758–765 (2009)

Dewan, M.W., Hossain, M.K., Hosur, M., Jeelani, S.: Thermomechanical properties of alkali treated jute-polyester/nanoclay biocomposites fabricated by VARTM process. J. Appl. Polym. Sci. **128**(6), 4110–4123 (2013)

Divya, G.S., Kakhandaki, A., Suresha, B.: Wear behavior of coir reinforced treated and untreated hybrid composites. Int. J. Innovative Res. Dev. **3**(5), paper no. 50754 (2014)

Faruk, O., Matuana, L.M.: Nanoclay reinforced HDPE as a matrix for wood-plastic composites. Compos. Sci. Technol. **68**, 2073–2077 (2008)

Faruk, O., Bledzki, A.K., Fink, H.-P., Sain, M.: Biocomposites reinforced with natural fibers: 2000–2010. Prog. Polym. Sci. **37**(11), 1552–1596 (2012)

Gassan, J., Bledzki, A.K.: Effect of moisture content on the properties of silanized jute-epoxy composites. Polym. Compos. **18**(2), 179–184 (1997)

Gassan, J., Bledzki, A.K.: Thermal degradation of flax and jute fibers. J. Appl. Polym. Sci. **82**(6), 1417–1422 (2001)

George, J., Sreekala, M.S., Thomas, S.: A review on interface modification and characterization of natural fiber reinforced plastic composites. Polym. Eng. Sci. **41**(9), 1471–1485 (2001)

Hakamy, A., Shaikh, F.U.A., Low, I.M.: Microstructures and mechanical properties of hemp fabric reinforced organoclay–cement nanocomposites. Constr. Build. Mater. **49**, 298–307 (2013)

Hakamy, A., Shaikh, F.U.A., Low, I.M.: Characteristics of hemp fabric reinforced nanoclay–cement nanocomposites. Cement Concr. Compos. **50**, 27–35 (2014a)

Hakamy, A., Shaikh, F.U.A., Low, I.M.: Thermal and mechanical properties of hemp fabric-reinforced nanoclay-cement nano-composites. J. Mater. Sci. **49**(4), 1684–1694 (2014b)

Hakamy, A., Shaikh, F.U.A., Low, I.M.: Thermal and mechanical properties of NaOH treated hemp fabric and calcined nanoclay-reinforced cement nanocomposites. Mater. Des. **80**, 70–81 (2015)

Han, G., Lei, Y., Wu, Q., Kojima, Y., Suzuki, S.: Bamboo–fiber filled high density polyethylene composites: effect of coupling treatment and nanoclay. J. Polym. Environ. **16**(2), 123–130 (2008)

Hapuarachchi, T.D., Peijs, T.: Multiwalled carbon nanotubes and sepiolite nanoclays as flame retardants for polylactide and its natural fibre reinforced composites. Compos. A **41**(8), 954–963 (2010)

Haq, M., Burgueño, R., Mohanty, A.K., Misra, M.: Hybrid bio-based composites from blends of unsaturated polyester and soybean oil reinforced with nanoclay and natural fibers. Compos. Sci. Technol. **68**(15–16), 3344–3351 (2008)

Herrera-Franco, P.J., Valadez-González, A.: A study of the mechanical properties of short natural-fiber reinforced composites. Compos. B **36**(8), 597–608 (2005)

Holbery, J., Houston, D.: Natural-fiber-reinforced polymer composites in automotive applications. JOM **58**(11), 80–86 (2006)

Hossain, M.K., Dewan, M.W., Hosur, M., Jeelani, S.: Mechanical performances of surface modified jute fiber reinforced biopol nanophased green composites. Compos. B **42**(6), 1701–1707 (2011)

Huang, X., Netravali, A.: Characterization of flax fiber reinforced soy protein resin based green composites modified with nano-clay particles. Compos. Sci. Technol. **67**(10), 2005–2014 (2007)

Iman, M., Maji, T.K.: Effect of crosslinker and nanoclay on starch and jute fabric based green nanocomposites. Carbohydr. Polym. **89**(1), 290–297 (2012)

Iman, M., Bania, K.K., Maji, T.K.: Green jute-based cross-linked soy flour nanocomposites reinforced with cellulose whiskers and nanoclay. Ind. Eng. Chem. Res. **52**(21), 6969–6983 (2013)

Iman, M., Mandal, M., Maji, T.K.: Effect of zinc oxide and nanoclay on crosslinked jute-reinforced soy flour green nanocomposites. J. Compos. Mater. Published online before print, April 15, 2015. doi: 10.1177/0021998315580831

Islam, M.S., Hamdan, S., Talib, Z.A., Ahmed, A.S., Rahman Md, R.: Tropical wood polymer nanocomposite (WPNC): the impact of nanoclay on dynamic mechanical thermal properties. Compos. Sci. Technol. **72**(16), 1995–2001 (2012a)

Islam, M.S., Khan, T., Park, J.K.: Nanoreinforced bacterial cellulose–montmorillonite composites for biomedical applications. Carbohyd. Polym. **89**(4), 1189–1197 (2012b)

Islam, M.S., Ahmad, M.B., Hasan, M., Aziz, S.A., Jawaid, M., Haafiz, M.K.M., Zakaria, S.A.H.: Natural fiber-reinforced hybrid polymer nanocomposites: effect of fiber mixing and nanoclay on physical, mechanical, and biodegradable properties. Bioresources **10**(1), 1394–1407 (2015)

Islam, M.S., Talib, Z.A., Hasan, M., Ramli, I., Haafiz, M.K.M., Jawaid, M., Islam, A., Inuwa, I. M.: Evaluation of mechanical, morphological, and biodegradable properties of hybrid natural fiber polymer nanocomposites. Polym. Compos. Available online, doi: 10.1002/pc.23616

Kaiser, M.R., Anuar, H.: Effect of nanoclay on thermal properties of polylactic acid-kenaf hybrid bio-composite. In: International Conference on Chemical Engineering (ICChE 2011), 29–30 December 2011, Dhaka, Bangladesh, from http://irep.iium.edu.my/id/eprint/17072. Accessed 24 Jan 2016

Kaiser, M.R., Anuar, H., Samat, N.B., Razak, S.B.A.: Effect of processing routes on the mechanical, thermal and morphological properties of PLA-based hybrid biocomposite. Iran. Polym. J. **22**(2), 123–131 (2013)

Kalia, S., Kaith, B.S., Kaur, I.: Pretreatments of natural fibers and their application as reinforcing material in polymer composites—a review. Polym. Eng. Sci. **49**(7), 1253–1272 (2009)

Kamardin, N.K., Taib, Y.M., Kalam, A.: The effects of kenaf fiber loading reinforced polypropylene and nanoclay. Jurnal Teknologi **76**(3), 53–56 (2015)

Katschnig, M., Battisti, M.: Processing of polymer-nanoclay composites, Chapter 3 in "Polymer nanoclay composites". In: Laske, S. (ed.) William Andrew Publishing, ISBN 978-0323299626 (2015)

Kiliaris, P., Papaspyrides, C.D.: Polymer/layered silicate (clay) nanocomposites: an overview of flame retardancy. Prog. Polym. Sci. **35**(7), 902–958 (2010)

Klemm, D., Heublein, B., Fink, H.-P., Bohn, A.: Cellulose: fascinating biopolymer and sustainable raw material. Angew. Chem. **44**(22), 3358–3393 (2005)

Kord, B.: Effect of nanoparticles loading on properties of polymeric composite based on hemp fiber/polypropylene. J. Thermoplast. Compos. Mater. **25**(7), 793–806 (2012)

Ku, H., Wang, H., Pattarachaiyakoop, N., Trada, M.: A review on the tensile properties of natural fiber reinforced polymer composites. Compos. B **42**(4), 856–873 (2011)

Kumar, R., Yakubu, M.K., Anandjiwala, R.D.: Biodegradation of flax fiber reinforced poly lactic acid. Express Polym. Lett. **4**(7), 423–430 (2010a)

Kumar, R., Yakabu, M.K., Anandjiwala, R.D.: Effect of montmorillonite clay on flax fabric reinforced poly lactic acid composites with amphiphilic additives. Compos. A **41**(11), 1620–1627 (2010b)

La Mantia, F.P., Morreale, M.: Green composites: a brief review. Compos. A **42**(6), 579–588 (2011)

Lashgari, A., Eshghi, A., Farsi, M.A.: Study on some properties of polypropylene based nanocomposites made using almond shell flour and organoclay. Asian J. Chem. **25**(2), 1043–1049 (2013)

Li, H., Huneault, M.A.: Comparison of sorbitol and glycerol as plasticizers for thermoplastic starch in TPS/PLA blends. J. Appl. Polym. Sci. **119**(4), 2439–2448 (2011)

Li, X., Tabil, L.G., Panigrahi, S.: Chemical treatments of natural fiber for use in natural fiber-reinforced composites: a review. J. Polym. Environ. **15**(1), 25–33 (2007)

Lim, L.T., Auras, R., Rubino, M.: Processing technologies for poly(lactic acid). Prog. Polym. Sci. **33**(8), 820–852 (2008)

Majeed, K., Jawaid, M., Hassan, A., Bakar, A.A., Khalil, H.P.S.A., Salem, A.A., Inuwa, I.: Potential materials for food packaging from nanoclay/natural fibres filled hybrid composites. Mater. Des. **46**, 391–410 (2013)

Mohan, T.P., Kanny, K.: Water barrier properties of nanoclay filled sisal fibre reinforced epoxy composites. Compos. A **42**(4), 385–393 (2011)

Mohanty, S., Verma, S.K., Nayak, S.K.: Dynamic mechanical and thermal properties of MAPE treated jute/HDPE composites. Compos. Sci. Technol. **66**(3–4), 538–547 (2006)

Mosiewicki, M.A., Aranguren, M.I.: A short review on novel biocomposites based on plant oil precursors. Eur. Polymer J. **49**(6), 1243–1256 (2013)

Pillin, I., Montrelay, N., Grohens, Y.: Thermo-mechanical characterization of plasticized PLA: is the miscibility the only significant factor? Polymer **47**(13), 4676–4682 (2006)

Rahman Md, R., Hamdan, S., Islam Md, S., Ahmed, A.S.: Main content areaInfluence of Nanoclay/Phenol Formaldehyde Resin on Wood Polymer Nanocomposites. J. Appl. Sci. **12** (14), 1481–1487 (2012)

Rajini, N., Jappes, J.T.W., Jeyaraj, P., Rajakarunakaran, S., Bennet, C.: Effect of montmorillonite nanoclay on temperature dependence mechanical properties of naturally woven coconut sheath/polyester composite. J. Reinf. Plast. Compos. **32**(11), 811–822 (2013)

Ray, S.S., Bousmina, M.: Biodegradable polymers and their layered silicate nanocomposites: in greening the 21st century materials world. Prog. Mater. Sci. **50**(8), 962–1079 (2005)

Ray, S.S., Okamoto, M.: Polymer/layered silicate nano-composites: a review from preparation to processing. Prog. Polym. Sci. **28**, 1539–1641 (2003)

Rozman, H.D., Rozyanty, A.R., Musa, L., Tay, G.S.: Ultra-violet radiation-cured biofiber composites from kenaf: the effect of montmorillonite on the flexural and impact properties. J. Wood Chem. Technol. **30**(2), 152–163 (2010)

Saheb, D.N., Jog, J.P.: Natural fiber polymer composites: a review. Adv. Polym. Technol. **18**(4), 351–363 (1999)

Saiful Md, I., Hasbullah, N.A.B., Hasan, M., Talib, Z.A., Jawaid, M., Haafiz, M.K.: Physical, mechanical and biodegradable properties of kenaf/coir hybrid fiber reinforced polymer nanocomposites. Mater Today Commun. **4**, 69–76 (2015)

Santulli, C.: Review. Impact properties of glass/plant fibre hybrid laminates. J. Mater. Sci. **42**, 3699–3707 (2007)

Satyanarayana, K.G., Kulkarni, A.G., Rohatgi, P.K.: Structure and properties of coir fibres. In: Proceedings of the Indian Academy of Sciences C **4**(4), 419–436 (1981)

Saw, S.K.: Static and dynamic mechanical analysis of coir fiber/montmorillonite nanoclay-filled novolac/epoxy hybrid nanocomposites, Eco-friendly Polymer Nanocomposites. In: Volume 75 of the series Advanced Structured Materials, 2015, 137–154, ISBN 978-81-322-2469-3 (2015)

Sim, I.N., Han, S.O.: Effect on modified nanoclay on dynamic mechanical and thermomechanical properties of natural fiber/polypropylene biocomposites. J. Adhes. Sci. Technol. **27**(12), 1313–1323 (2013)

Singh, R.P., Ngo, T.D., Hu, W., Ton-That, T.M., Denault, J.: Performance improvement of green composites using nanoclays, Report of National Research Council, Canada, from nrc-cnrc.gc.ca. Accessed 10 Feb 2016 (2009)

Tateyama, H., Tsunematsu, K., Kimura, K., Hirosue, H., Jinnai, K., Furusawa, T.: Method for producing fluorine mica, Patent n. US 5204078 A, 20 Apr 1993

Effects of Micro- and Nano-cellulose on Tensile and Morphological Properties of Montmorillonite Nanoclay Reinforced Polylactic Acid Nanocomposites

Reza Arjmandi, Azman Hassan, M.K. Mohamad Haafiz and Zainoha Zakaria

Abstract In this chapter, the effects of micro- and nano-cellulose; microcrystalline cellulose (MCC) and cellulose nanowhiskers (CNW), on the properties of montmorillonite (MMT)/polylactic acid (PLA) nanocomposites fabricated using solution casting method were investigated. CNW were isolated from MCC using a chemical swelling method. An initial study showed that the optimum MMT content, based on tensile properties, in PLA/MMT nanocomposite is 5 phr (parts per hundred parts of resin). PLA/MMT/MCC hybrid composites and PLA/MMT/CNW hybrid nanocomposites were produced by incorporating of various contents of MCC and CNW into the optimum formulation of PLA/MMT nanocomposites (P/MT5), respectively. The Young's modulus of PLA/MMT/MCC hybrid composites and PLA/MMT/CNW hybrid nanocomposites increased gradually with increasing MCC and CNW content, respectively. The tensile strength of hybrid composites decreased compared to P/MT5 nanocomposite. However, the tensile strength for the hybrid nanocomposites increased with incorporation of CNW and reached the highest value at 1 phr CNW content. The ductility of hybrid composites and hybrid nanocomposites was also increased significantly from ~10 to ~60 and ~90 % with the incorporation of 1 phr MCC and CNW fillers, respectively. This increase in ductility was proposed to be due to the nucleation of crazes and the formation of shear bands in the PLA. Transmission electron microscopy and X-ray diffraction analyses showed that MCC slightly decreased the MMT interlayer spacing in hybrid composites, while the MMT interlayer spacing in hybrid nanocomposites increased by incorporation of CNW into P/MT5 nanocomposite.

R. Arjmandi · A. Hassan (✉)
Department of Bioprocess and Polymer Engineering, Faculty of Chemical and Energy Engineering, Universiti Teknologi Malaysia, 81310 Skudai, Johor Bahru, Malaysia
e-mail: azmanh@cheme.utm.my

M.K. Mohamad Haafiz
School of Industrial Technology, Universiti Sains Malaysia, 11800 Gelugor, Penang, Malaysia

Z. Zakaria
Faculty of Science, Universiti Teknologi Malaysia, 81310 Skudai, Johor Bahru, Malaysia

M. Jawaid et al. (eds.), *Nanoclay Reinforced Polymer Composites*,
Engineering Materials, DOI 10.1007/978-981-10-0950-1_5

Keywords Polylactic acid · Montmorillonite · Microcrystalline cellulose · Cellulose nanowhiskers · Hybrid nanocomposites · Tensile properties

1 Introduction

In recent years, biodegradable polymers produced from renewable resources have become attractive for practical applications such as medical devices and food packaging (Vroman and Tighzert 2009). Polylactic acid (PLA) is a commercially available biopolymer. It is biodegradable thermoplastic polyester produced from L- and D-lactic acid, which is derived from the fermentation of corn starch (Petersson and Oksman 2006). PLA is a "green polymer" because it can be finally degraded into CO_2 and H_2O (Liu et al. 2013a). Due to its high strength/stiffness and ease of process, PLA offers a potential alternative to petrochemical plastics in many applications (Huda et al. 2008). However, there are several obstacles for the use of PLA, including low toughness, low impact resistance and moderate gas barrier properties, its inherent brittleness, high production cost and undesirable thermal stability, which are all important properties for packaging applications (Oksman et al. 2006; Sanchez-Garcia and Lagaron 2010). Hence, many approaches have been adopted to overcome these inherent limitations, including the use of plasticizers, polymers blending, layered silicates and other inorganic particles (Sanchez-Garcia and Lagaron 2010; Petersson et al. 2007).

Clay reinforced polymer nanocomposites have received significant attention compared to conventional composites because polymer/clay nanocomposites often exhibit considerable improvement in physical, thermal, mechanical and optical properties and fire retardancy at low filler contents (Hong and Kim 2013). PLA/layered silicate nanocomposites have been shown to exhibit improved tensile strength and Young's modulus compared to neat PLA (Sinha Ray and Okamoto 2003). Montmorillonite (MMT) which consists of two fused silica tetrahedral sheets sandwiching an edge shared octahedral sheet is the most commonly used layered silicate in this respect. However, raw MMT often produces immiscible phases with hydrophobic polymers such as PLA due to its hydrophilic character. For this reason, the ammonium ions are exchanged by more voluminous organic cations, which can strengthen the chemical interaction between PLA and MMT and also induces the formation of layered or exfoliated organic/inorganic nanocomposites (Sinha Ray and Okamoto 2003). MMT reinforced PLA nanocomposites can be prepared by two different methods; namely melt processing and solution casting techniques (Sinha Ray and Okamoto 2003). Based on previous studies, there is no clear optimum content for MMT-based nanocomposites. Thellen et al. (2005) reported that an optimum tensile strength was achieved at a 5 wt% MMT content, while He et al. (2011) and Chang et al. (2003a) showed this occurred at 3 and 4 wt% contents respectively. The Young's modulus has also been shown to increase gradually with an increase in the MMT content. However, these property enhancements are usually accompanied by a severe decrease in ductility. Therefore, there is a need to investigate ways to produce PLA nanocomposites with both high strength/stiffness and ductility. Besides that, to keep PLA nanocomposites as green as possible and minimize production costs, the addition of renewable, biodegradable and low price reinforcements such as cellulose could be a promising technique (Shayan et al. 2015).

Cellulose is a naturally abundant material which can enhance the biodegradability, mechanical properties, thermal stability and also reduce the production costs of PLA (Herrera et al. 2015; Wang et al. 2014; Karande et al. 2014). Cellulose can be obtained from several sources including bacteria, algae, marine creatures and plants. However, cotton served the highest percentage of cellulose as compared to other sources (>95 %). According to Satyamurthy et al. (2011), a single cotton fiber is consisting of fibrils with the thickness of 20–30 μm and nanometer scale diameters. Microcrystalline cellulose (MCC) is one of the cellulosic derivatives that could be used as fillers in the polymer matrices which can be produced through a combination of mechanical and chemical processing (Chen et al. 2011a; Iwatake et al. 2008; Lu et al. 2008; Mathew et al. 2005). One of the most important properties that could influence the performance of MCC in various applications is the degree of crystallinity. The crystallinity degree of MCC is depending on several factors including displayed the source of cellulose, and processing variables such as the reaction duration and temperature, the extent mechanical agitation applied to the slurry and the drying condition, where usually ranges 55–80 % (Haafiz et al. 2013; Chuayjuljit et al. 2010).

Cellulose nanofibers have been used as a reinforcing filler to enhance the performance of the polymer matrices (Azizi Samir et al. 2005). In terms of performance they possess high specific strength and modulus, easy processability and a relatively reactive surface. Cellulose nanowhiskers (CNW) have been gaining considerable interest among scientists in the last 10 years as potential nano-reinforcement in a number of various polymers. The tensile properties of PLA nanocomposites have been shown to increase when CNW from grass were used as a reinforcement (Pandey et al. 2009). Additionally, CNW obtained from commercial MCC has been shown to improve the storage modulus of neat PLA (Cho et al. 2013). Petersson et al. (2007) reported that CNW improved the storage modulus of PLA at higher temperatures. A study to compare the effect of MMT and CNW on PLA was reported by Petersson and Oksman (2006). They found that MMT increased both yield strength and tensile modulus, whereas CNW only improved the yield strength. Petersson and Oksman (2006) also reported that PLA/CNW nanocomposites have a higher ductility than PLA/MMT nanocomposites. This formed the basis to enhanced the ductility of PLA nanocomposites by incorporation of hybridized fillers. Hybrid fillers can be promising for a number of applications where high strength/stiffness and ductility are required. limited studies have been conducted on the use of hybrid fillers in PLA polymer. Hong and Kim (2013) showed that the simultaneous use of CNW and nanoclay as nanofillers improved the tensile strength and glass transition temperature of PLA using melt mixing.

In this chapter, the effects of micro size MCC and nano size CNW on the tensile and morphological properties of PLA/MMT/MCC hybrid composites and PLA/MMT/CNW hybrid nanocomposites produced by solution casting have been investigated. The main reason for incorporation MCC and CNW into PLA/MMT nanocomposites is to enhance the Young's modulus and elongation at break of the nanocomposites which are the main properties for packaging materials.

2 Materials and Methods

2.1 *Materials*

PLA pellets (3001D) was supplied from NatureWork® LLC, Minnetonka, MN USA with a density and melt flow index of 1.24 g/cm and *ca.* 15 g/10 min (190 °C/2.16 kg), respectively. MMT clay was supplied from Nanocor, Inc., Arlington Heights IL, USA (Nanomer 1.30TC), where it was organically modified with octadecylamine (~30 wt%) and has an average particle size of 16–22 μm. MCC filler (Avicel; PH-101) was purchased from Sigma-Aldrich, Malaysia, which was produced from a cotton linter with average particle size of 50 μm. Other reagents such as chloroform, *N*, *N*-dimethylacetamide (DMAc) (99 % purity) and lithium chloride (LiCl) (99 % Purity) were supplied from Merck, Malaysia.

2.2 *Production of CNW Using Chemical Swelling*

As described by Pereda et al. (2011) and Arjmandi et al. (2015a), CNW were prepared by swelling MCC and separating to whiskers by a sequence of chemical treatment using DMAc and 0.5 % LiCl solution as swelling agents followed by mechanical treatment using a sonication bath (Branson 2510). 10 wt% was the initial concentration of MCC in DMAc/LiCl. Then, MCC was vigorously stirred by a mechanical stirrer inside a water bath at 70 °C for 12 h in order to swell the particles. Subsequently, the swelled MCC particles were sonicated using the same sonication bath for 3 h over a period of 5 days, where long intervals between each sonication period was performed in order to separate the CNW nanofillers. Finally, the gel-like suspension of CNW were repeatedly washed with distilled water, then refrigerated and designated as CNW.

2.3 *Production of PLA Film*

As described in our recent publications (Arjmandi et al. 2014, 2015a, b), PLA pellets (10 g) were dissolved in chloroform (64 ml) at 60 °C for 2 h with constant stirring using a mechanical stirrer. Subsequently, the dissolved PLA was casted on the glass plates and was left at ambient temperature (48 h) to evaporate the solvent and produce the PLA film with the thickness of approximately 100 μm.

2.4 Production of PLA/MMT Nanocomposites Film

10 g of PLA pellets with various content of MMT [1–9 phr (parts per hundred parts of resin)] were mixed in order to produce PLA/MMT nanocomposites, as reported in our recent study (Arjmandi et al. 2015a). 64 ml of chloroform were added to the various mixtures of PLA/MMT and stirred using a mechanical stirrer at 60 °C for 2 h. The PLA/MMT suspensions were then sonicated using the same sonication bath (as described earlier) for 5 min and casted on the glass plates to produce PLA/MMT nanocomposite films with ~100 μm thickness. The nanocomposites with 1–9 phr MMT content were produced and designated as P/MT1, P/MT3, P/MT5, P/MT7 and P/MT9. Table 1 presents the formulation of PLA nanocomposites with various MMT contents.

2.5 Production of PLA/MMT/MCC Hybrid Composites Film

10 g of PLA pellets, 5 phr MMT and various contents of MCC (1, 3 5 and 7 phr) were mixed in order to fabricate the PLA/MMT/MCC hybrid composites, as described by Arjmandi et al. (2015b). 64 ml of chloroform were added into various mixtures of PLA/MMT/MCC and then stirred using a mechanical stirrer at 60 °C for 2 h. Subsequently, the PLA/MMT/MCC suspensions were sonicated using the same sonication bath (as described earlier) for 5 min and then casted on the glass plates (as described above) to obtain PLA/MMT/MCC hybrid composite films with ~100 μm thickness. The hybrid nanocomposites with 1–7 phr MCC were

Table 1 Polylactic acid nanocomposites formulation

Designation	PLA (wt%)	MMT (phr)	MCC (phr)	CNW (phr)
PLA	100	0	0	0
P/MT1	100	1	0	0
P/MT3	100	3	0	0
P/MT5	100	5	0	0
P/MT7	100	7	0	0
P/MT9	100	9	0	0
P/MT5/MC1	100	5	1	0
P/MT5/MC3	100	5	3	0
P/MT5/MC5	100	5	5	0
P/MT5/MC7	100	5	7	0
P/MT5/CW1	100	5	0	1
P/MT5/CW3	100	5	0	3
P/MT5/CW5	100	5	0	5
P/MT5/CW7	100	5	0	7
P/MT5/CW9	100	5	0	9

produced and designated as P/MT5/MC1, P/MT5/MC3, P/MT5/MC5 and P/MT5/MC7, as summarized in Table 1.

2.6 Production of PLA/MMT/CNW Hybrid Nanocomposites Film

10 g of PLA pellets, 5 phr of MMT and various content of CNW (1, 3, 5, 7 and 9 phr) were mixed in order to produce the PLA/MMT/CNW hybrid nanocomposites, as explained in our recent publication (Arjmandi et al. 2015a). The CNW nanofiller used was in water suspension form. Solvent exchange was then carried out via centrifugation using a Universal 32 Hettich centrifuge (Newport Pagnell, England). Water was then exchanged with acetone and acetone exchanged with chloroform. The CNW filler was then sonicated using an ultrasonic bath (Branson 2510) for 5 min to ensure a homogenous dispersion inside the chloroform. The dispersed CNW was then transferred into a reaction flask containing PLA pellets and 5 phr of MMT. Subsequently, 64 ml of chloroform added into the various mixtures of PLA/MMT/CNW and the stirred using mechanical stirrer at 60 °C for 2 h. The dissolved PLA containing MMT/CNW was then sonicated for another 5 min. Finally, the PLA/MMT/CNW suspension was casted on the glass plates to obtain nanocomposite films with ~ 100 μm thickness. The hybrid nanocomposites with 1–9 phr CNW were produced and designated as P/MT5/CW1, P/MT5/CW3, P/MT5/CW5, P/MT5/CW7 and P/MT5/CW9, as summarized in Table 1.

3 Characterization

3.1 Tensile Properties

Tensile testing was performed using an Instron 4400 Universal Tester to measure the tensile strength at the break point and Young's modulus for each formulation. Rectangular specimens were cut from the obtained film with dimension of $60 \times 13 \times 0.1\,\text{mm}^3$. The tensile tests were carried out at ambient temperature (ASTM D882-12). A fixed crosshead rate and gauge length of $12.5\,\text{mm min}^{-1}$ and 30 mm were utilized for all formulations, respectively, and the reported results are the average of 10 samples.

3.2 Fourier Transform Infrared Spectroscopy

A Perkin Elmer 1600 Infrared spectrometer manufactured by USA was used in order to perform the Fourier transform infrared spectroscopy (FTIR). 1 wt% of KBr was mixed with the specimens which were made in powder form. Also, neat KBr was used as FTIR spectral reference. FTIR of the specimens were recorded at constant scan and resolution of 32 and 4 cm^{-1}, respectively, using Nicolet's AVATAR 360 and the wave range of 4000–370 cm^{-1}. The position of significant transmittance peaks was determined using the "find peak tool" provided by Nicolet OMNIC 5.01 software.

3.3 Morphological Properties

Field emission scanning electron microscopy (FESEM) and transmission electron microscopy (TEM) were used in order to characterize the morphologies of specimens. FESEM was carried out using a Carl Zeiss manufactured by Germany (Supra 35 VP microscope) with an accelerating voltage in the range of 8–10 kV. The samples were placed onto specimen stubs and sputter-coated with gold prior observation to prevent from the accumulation of electrostatic charges on the surface of the samples during electron irradiation. Then, the coat samples were then examined in a FESEM. In addition to FESEM, the microstructure of the neat PLA and its hybrid composites and hybrid nanocomposite in order to investigate the fillers dispersion in the PLA matrix were carried out using a TEM, JEOL JEM-2010, USA. The TEM samples were sectioned using an ultra-microtome, RMC (model MTXL) to obtain thick slices ($\sim$70 nm). Then the samples were negatively stained by allowing the grids to float in a 1 % uranyl diacetate solution for 3 min. The samples were examined at an accelerating voltage of 120 k. The TEM images were obtained using soft imagine system software.

3.4 X-Ray Diffraction

X-ray diffraction (X-RD) analysis was carried out to study the structures of hybrid composites and hybrid nanocomposites using a Siemens D5000 X-ray diffractometer manufactured by Berlin, Germany. The diffraction patterns (Ni-filtered Cu Kα radiation) were recorded using a step size of 0.02°, $2\theta = 2.0 - 10°$ (2θ angle range). The operating voltage and current were 40 kV and 50 mA, respectively. According to Bragg's Law, the interlayer distances (*d-spacing*) of the MMT layers in the hybrid composites and hybrid nanocomposites were obtained from the peak positions in the X-RD patterns and calculated using Eq. (1).

$$d\text{-}spacing = \frac{n\lambda}{2\sin\theta} \tag{1}$$

where n is an integer ($n = 1$), θ is the diffraction angle and λ is the X-ray wavelength, $\lambda = 0.15406\,\text{nm}$ (Cu Kα).

4 Results and Discussion

4.1 Tensile Properties

Figure 1a–c reports the tensile properties of PLA/MMT nanocomposites, PLA/MMT/MCC hybrid composites and PLA/MMT/CNW hybrid nanocomposites, respectively. Figure 1a presents the tensile strengths and Young's modulus of the PLA and PLA/MMT nanocomposites at various MMT contents (Arjmandi et al. 2015a). The tensile strength of PLA/MMT nanocomposites is observed to increase with the addition of MMT, reaching a plateau at a filler content of 5 phr, then declining at higher contents. The tensile strength of PLA/MMT nanocomposites at a 5 phr filler content is ~63 % higher than that of the neat PLA, suggesting a significant reinforcing effect of the inorganic phase. It is thought that at this filler content the MMT is distributed uniformly within the PLA matrix. As a result, at this filler content, the tensile strength reaches a maximum value. However, when the filler content exceeds 5 phr, the tensile strength dramatically decreases probably due to the aggregation of the filler resulting in stress concentration points for fracture. This result is in agreement with Lee et al. (2008), where the agglomeration

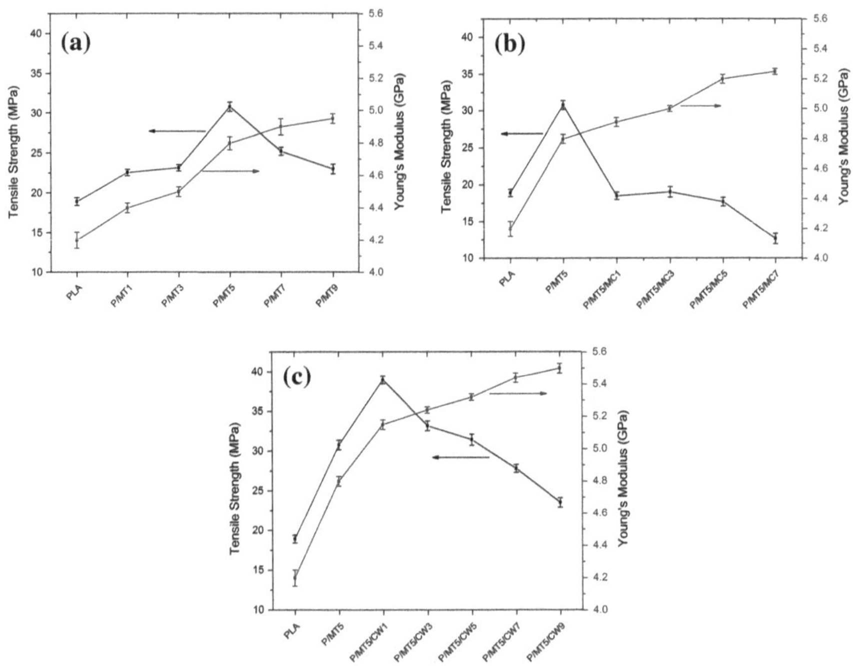

Fig. 1 Tensile strength and Young's modulus of **a** PLA/MMT nanocomposites, **b** PLA/MMT/MCC hybrid composites and **c** PLA/MMT/CNW hybrid nanocomposites. Errors presented represent standard deviations from the mean (n = 10) (Arjmandi et al. 2014, 2015a)

of MMT at high filler contents (>5 wt%) resulted in a decrease in tensile strength. However, the Young's modulus of MMT reinforced PLA nanocomposites increased gradually with increasing MMT content. It is notable that the Young's modulus of nanocomposites at 9 phr MMT content increased by approximately 18 % compared to neat PLA film. This increase is attributed to the rigid MMT fillers which constrict segmental chain motion in the polymer matrix. This result is consistent with a previous reported study by He et al. 2011.

Tensile properties (tensile strength and Young's modulus) of MCC filled PLA/MMT nanocomposites have shown in Fig. 1b, as reported in our recent publication (Arjmandi et al. 2014). As can be observed from Fig. 1b, the Young's modulus of PLA/MMT/MCC hybrid composites increased gradually with increasing MCC content (from 4.8 to 5.31 GPa), where it was higher than PLA/MMT nanocomposites. This improvement in modulus of hybrid composites was likely due to the stiffening effect of the high modulus MCC fillers. Haafiz et al. (2013) were reported similar trend, where MCC filled PLA composites. In contrast, the tensile strength of PLA/MMT/MCC hybrid composites decreased by incorporating of MCC into PLA/MMT nanocomposites, where the tensile strength decreased by ~38 % with incorporation of 1 phr MCC compared to optimum formulation of PLA/MMT nanocomposites (P/MT5). The decrease in tensile strength could be attributed to the aggregation of MCC in PLA hybrid nanocomposites which induced by Van der Waal's forces. From Fig. 1a, b, it is interesting to compare the relative increase in modulus for nanocomposites and hybrid composites. The addition of 7 phr MMT to PLA resulted in a modulus increase to 4.90 GPa, while by addition of MCC (1 phr) to the optimized nanocomposite sample (P/MT5 nanocomposite) produced a slightly higher increase in modulus (4.91 GPa). In addition, it is interesting to note that the Young's modulus of P/MT5/MC3 hybrid composite with the total of 8 phr filler content (5 phr MMT + 3 phr MCC) is higher than the Young's modulus of the P/MT9 nanocomposite with 9 phr MMT content.

On the other hand, Fig. 1c presents the tensile properties (tensile strength and Young's modulus) of CNW reinforced PLA/MMT nanocomposites, as reported in our recent study (Arjmandi et al. 2015a). It was observed that these hybrid nanocomposites (at 1 phr CNW content) had the highest tensile strengths; increases of ~27 and ~106 % were observed compared to those of the P/MT5 nanocomposite and neat PLA, respectively. The highest tensile strength at 1 phr CNW content might be attributed to the homogeneous dispersion and specific interaction that took place between the PLA and both MMT and CNW. This also indicates a positive synergistic effect of MMT and CNW in the enhancement of the tensile strength of the hybrid nanocomposites through the formation of good interactions with the PLA. Similar conclusion has been reported by Wang et al. (2006), where CNW reinforced PLA nanocomposites. When the CNW content was increased beyond 1 phr the tensile strength however decreased. The decrease in tensile strength is thought to be due to the aggregation of MMT platelets induced by Van der Waal's forces and thereby a reduced surface area of interaction between the PLA and both MMT and CNW. This result is in agreement with Liu et al. (2013b) for a halloysite filled PLA polymer. In addition, the Young's modulus of

PLA/MMT/CNW hybrid nanocomposites (Fig. 1c) gradually increased with increasing CNW content (from 4.8 to 5.5 GPa). The Young's modulus of the hybrid nanocomposites increased by ~23, ~27 and ~31 % above that of the neat PLA by incorporation of 1, 5 and 9 phr CNW content, respectively. This improvement in Young's modulus could be attributed to the high modulus of both MMT and CNW fillers. Cheng et al. (2009) and Haafiz et al. (2013) were reported similar conclusion, where cellulose fibers and microcrystalline cellulose reinforced PLA composites, respectively.

The elongation at break of PLA/MMT nanocomposites, PLA/MMT/MCC hybrid composites and PLA/MMT/CNW hybrid nanocomposites were summarized in Table 2, as reported in our previous publications (Arjmandi et al. 2015a, b). The hybrid composites and hybrid nanocomposites were observed to have higher ductility compared to MMT reinforced PLA nanocomposites. The elongation at break of PLA/MMT/MCC hybrid composites and PLA/MMT/CNW hybrid nanocomposites significantly increased from ~10 to ~60 and ~90 % by incorporation of 1 phr MCC and CNW into the optimized PLA/MMT nanocomposites (P/MT5), respectively. Petersson and Oksman (2006) reported similar observation, where they found that PLA/CNW nanocomposites have much better ductility compared to PLA/MMT nanocomposites. However, incorporation of higher content of MCC and CNW (>1 phr) decreased the elongation at break of hybrid composites and hybrid nanocomposites significantly. The improvement in ductility of hybrid composites and hybrid nanocomposites is due to the ability of MCC and CNW to restrict the polymer chains' mobility. Due to the aggregation and poor adhesion of MCC in PLA matrix, incorporation of 1 phr MCC filler into P/MT5 nanocomposite decreased the tensile strength of hybrid composites significantly, as discussed earlier. Therefore, it can be concluded that at this filler content (1 phr), aggregation

Table 2 Elongation at break of (a) PLA/MMT nanocomposites, (b) PLA/MMT/MCC hybrid composites and (c) PLA/MMT/CNW hybrid nanocomposites (Arjmandi et al. 2014, 2015a)

	Designation	Elongation at break (%)
(a)	PLA	124.8 ± 0.5
	P/MT1	55.0 ± 0.3
	P/MT3	29.5 ± 0.5
	P/MT5	10.6 ± 0.4
	P/MT7	6.4 ± 0.5
	P/MT9	4.6 ± 0.3
(b)	P/MT5/MC1	60.1 ± 0.5
	P/MT5/MC3	35.6 ± 0.3
	P/MT5/MC5	11.5 ± 0.4
	P/MT5/MC7	8.6 ± 0.5
(c)	P/MT5/CW1	89.6 ± 0.4
	P/MT5/CW3	56.8 ± 0.3
	P/MT5/CW5	37.7 ± 0.5
	P/MT5/CW7	33.5 ± 0.4
	P/MT5/CW9	29.4 ± 0.3

and poor adhesion of MCC is detrimental to tensile strength but not to elongation at break. In contrast, incorporation of 1 phr CNW filler into P/MT5 nanocomposite increased the tensile strength as well because of good dispersion/distribution and adhesion of CNW in PLA matrix, which shows the advantages of nano-size CNW compared to micro-size MCC. Interestingly, although the incorporation of higher content of MCC and CNW (>1 phr) into optimized formulation of PLA/MMT nanocomposites (P/MT5) decreased the elongation at break of hybrid composites and nanocomposites compared to lower content (1 phr), the elongation at break of hybrid composites and hybrid nanocomposites at 3 phr MCC and CNW content was still higher than optimized PLA/MMT nanocomposites. From the current results it can be concluded that the MCC and CNW could play an important role in enhancing the elongation at break of MMT reinforced PLA nanocomposites.

Additionally, increasing ductility seemed likely to be due to their enhanced hydrophilic character as will be confirmed by the FT-IR analysis. Thus, the existence of a proportional content of hydrophilic groups may contribute to plasticization leading to enhanced deformability. It may also be the case that the addition of MCC and CNW leads to nucleation of crazes in the amorphous fraction of the PLA resin, as has been previously observed for a different system (Bulota and Hughes 2012). Typical images of the test specimens of the P/MT5 nanocomposite and P/MT5/CW1 hybrid nanocomposite films before and after deformation (Fig. 2a–d) was reported in our recent study (Arjmandi et al. 2015a). According to Bulota and Hughes (2012), the high strain-to-failure arising from craze nucleation is indicated by a whitening of the polymer matrix due to air scattering. As can be clearly observed from Fig. 2d, P/MT5/CW1 hybrid nanocomposites exhibit much greater ductility compared to P/MT5 nanocomposite, accompanied by pronounced stress whitening. The stress-whitened zone observed in the test specimens suggests the presence of shear bands in the PLA.

Besides that, it is worth noting that the PLA/MMT/CNW hybrid nanocomposites provided higher tensile properties compared to PLA/MMT nanocomposites and PLA/MMT/MCC hybrid composites, as confirmed by typical stress–strain curves in our previous publication (Arjmandi et al. 2015a). Figure 3 presents the typical stress–strain curves of P/MT5 nanocomposite, P/MT5/MC1 hybrid composite and P/MT5/CW1 hybrid nanocomposite. As can be observed from this Fig. 3, the tensile increased compared to P/MT5 nanocomposite and P/MT5/MC1 hybrid composite. An interesting observation from Fig. 3 is that the elongation at break of P/MT5/CW1 hybrid nanocomposite also increased compared to P/MT5/MC1 hybrid composite, which shows the advantages of nano-size CNW compared to micro-size MCC.

4.2 *FT-IR Spectroscopy Analysis*

Figure 4a–d illustrate typical IR spectra of the neat PLA, P/MT5 nanocomposite, P/MT5/MC1 hybrid composite and P/MT5/CW1 hybrid nanocomposites,

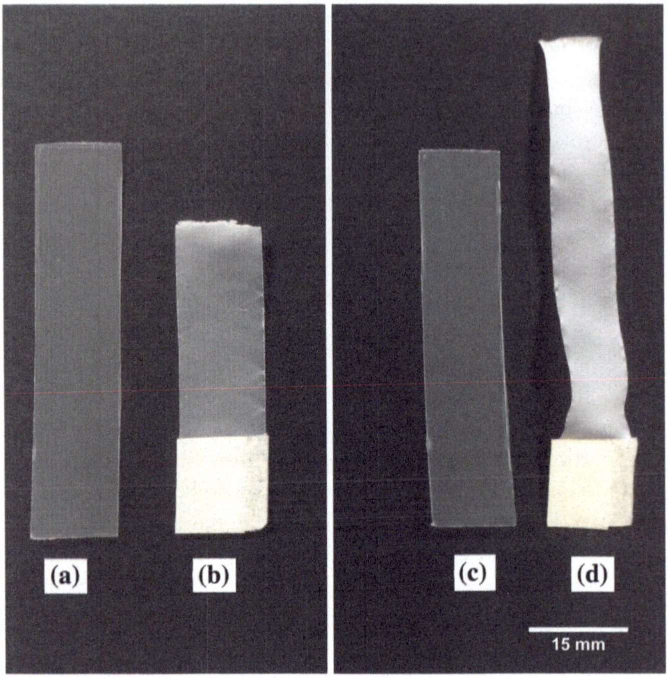

Fig. 2 Photograph images of the test specimens. **a** P/MT5 nanocomposite film prior deformation, **b** P/MT5 nanocomposite film at ~10 % strain, **c** P/MT5/CW1 hybrid nanocomposite film prior deformation and **d** P/MT5/CW1 hybrid nanocomposite film at ~90 % strain (Arjmandi et al. 2015a)

Fig. 3 Stress–strain curves of P/MT5 nanocomposite, P/MT5/MC1 hybrid composite and P/MT5/CW1 hybrid nanocomposite (Arjmandi et al. 2015a)

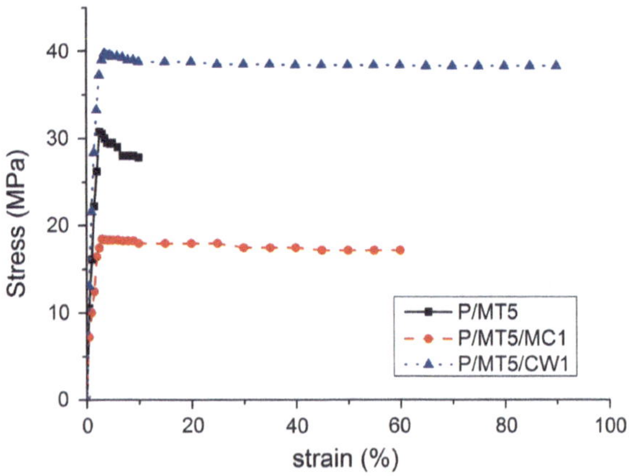

respectively (Arjmandi et al. 2015a, b). As shown in Fig. 4a, the absorption peaks at 3508 and 3644 cm^{-1} attributed to a bending vibration of two terminal hydroxyl groups and the absorption peaks at 2948 and 2997 cm^{-1} are corresponded to an

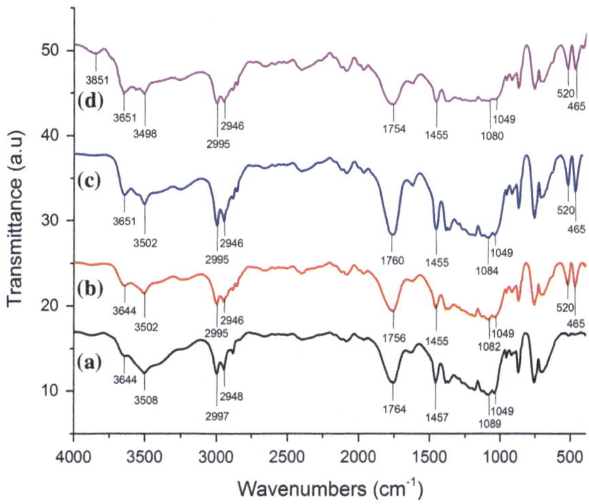

Fig. 4 FT-IR spectra of **a** neat PLA, **b** P/MT5 nanocomposite, **c** P/MT5/MC1 hybrid composite and **d** P/MT5/CW1 hybrid nanocomposite (Arjmandi et al. 2015a, b)

asymmetrical stretching vibration of a –CH moiety. The absorption peaks at 1764 and 1089 cm^{-1} are assigned to the stretching vibration of C=O groups and stretching vibration of C–O groups, respectively. In addition, the peaks at 1049 and 1457 cm^{-1} attributed to –OH and $–CH_3$ bending vibrations (Field et al. 2012).

Upon the addition of the MMT into the PLA (Fig. 4b), the absorption peaks at 1089, 1764 and 3508 cm^{-1} of the PLA shifted to lower wave number positions of (1082, 1756 and 3502 cm^{-1}, respectively). Moreover, two new peaks appeared at 465 and 520 cm^{-1} that correspond to stretching vibrations of the Si–O groups of the MMT filler. These changes in the spectra are thought to be due to the good interactions an adhesion between the PLA matrix and the MMT filler which caused by formation of polar interactions between the functional groups of both components (Chen et al. 2011b). It is noticeable that the MMT layers are contains huge numbers of polar sites along the structure which distributed uniformly, indicating an electron density content around the surfaces and interlayer spaces. As a result, polar interactions between the MMT filler and PLA matrix thought to readily occur. Liu et al. (2013b) indicated that the shifting of absorption peaks in the IR spectrum of PLA nanocomposites could be due to the formation of hydrogen bonding between the Si–O groups of the MMT and hydroxyl groups of the PLA matrix.

As can be observed from Fig. 4c, the addition of MCC into optimized formulation of PLA/MMT nanocomposites (P/MT5) did not influence the chemical structure of P/MT5 nanocomposite since there is no any new functional group was observed in the IR spectrum of P/MT5/MC1 hybrid composite compared to P/MT5 nanocomposite. This indicate that the MCC fillers did not chemically interact with PLA and only physical interactions occurred between MCC and PLA matrix. Similar observation was reported by Qu et al. (2010). In addition, due to the interactions between the hydroxyl groups of MCC and the carbonyl groups of PLA matrix, the absorption peak at 3644 cm^{-1} (Fig. 4b) was shifted to higher wavenumber (3651 cm^{-1}) by incorporation of MCC into P/MT5 nanocomposite

(Fig. 4c). Besides that, the interactions between the –OH groups of MCC and PLA could have also contributed to this shift.

In contrast, the effect of the addition of CNW on the PLA/MMT nanocomposites is shown in Fig. 4d. A new absorption peak at 3851 cm^{-1} was observed in the P/MT5/CW1 IR spectrum as compared to the optimized formulation of PLA/MMT nanocomposites (P/MT5). The shift of the peak from 3644 to 3651 cm^{-1} was also observed, which indicating a good interaction between the –OH groups of the CNW and the –OH/C=O groups of the PLA. Meanwhile, the presence of a new peak at 3851 cm^{-1} assigned to a bending vibration of the hydroxyl groups indicating the presence of CNW in the PLA matrix, thereby contributing to the increased hydrophilic character of the hybrid nanocomposites. These findings clearly imply that either the addition of a single filler or the addition of hybrid fillers has changed the molecular conformation of the PLA, thus altering the properties of the nanocomposites.

4.3 Optical Microscopy

in our recent study (Arjmandi et al. 2015a), further evidence of crazing and shear banding in the samples was obtained using optical microscopy. The typical light microscope images of the P/MT5/MC1 hybrid composite and P/MT5/CW1 hybrid nanocomposite before and after deformation are presents in Fig. 5a, b, respectively. To better observe the crazes and shear bands, light microscopy images were conducted using back-illumination. Figure 5a, b, clearly shows crazes forming in the sample after deformation (right side images), which look like cracks but characteristically form perpendicular to the draw direction. In addition, the formation of shear bands (at 45° to the direction orthogonal to the tensile axis) were also clearly observed. These shear bands may also increase the toughness of the nanocomposites through localized yielding of the polymer. High strain within a craze zone and the formation of shear bands could result in an overall increase in the strain to failure of the sample.

4.4 Field Emission Scanning Electron Microscopy

FESEM micrographs of the fractured cross-sectional surfaces of neat PLA, P/MT5 nanocomposite and P/MT7 nanocomposite are presents in Fig. 6a–c, respectively, as reports in our previous publications (Arjmandi et al. 2015a, b). As can be observed from Fig. 6a, neat PLA has a typical smooth surface (Jiang et al. 2007). Figure 6b reports the fractured cross-sectional surfaces of P/MT5 nanocomposite which shows that the MMT layers are uniformly dispersed in the PLA matrix because of its finer particle size, as indicated by the arrows. Figure 6c reports the fractured cross-sectional surfaces of P/MT7 nanocomposite which clearly shows

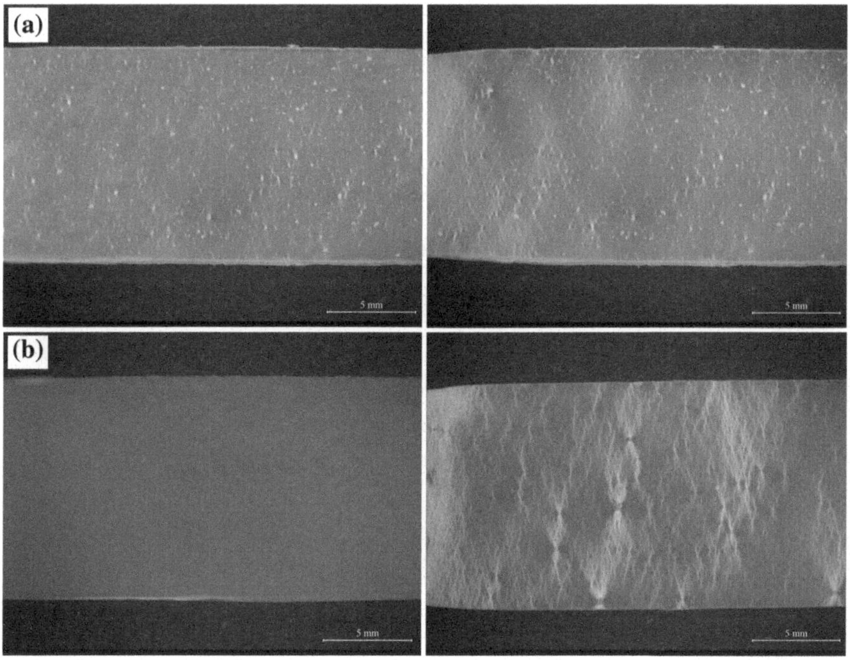

Fig. 5 Optical microscopy images prior (*left*) and after (*right*) deformation. **a** P/MT5/MC1 hybrid composite film and **b** P/MT5/CW1 hybrid nanocomposite film (Arjmandi et al. 2015a)

that further addition of MMT fillers (>5 phr) can resulted in MMT aggregation, as indicated by the circles. Chang et al. (2003a) also reported similar observation, where the aggregation of MMT was found at high MMT content (>5 wt%).

Figure 7 reports the FESEM micrographs of fractured cross-sectional surfaces of PLA/MMT/MCC hybrid composites and PLA/MMT/CNW hybrid nanocomposites, as reported in our previous studies (Arjmandi et al. 2015a, b). The fractured cross-sectional surfaces micrographs of P/MT5/MC1 hybrid composite, P/MT5/MC3 hybrid composite, P/MT5/CW1 hybrid nanocomposite and P/MT5/CW3 hybrid nanocomposite are presents in Fig. 7a–d, respectively. As can be observed from Fig. 7a, incorporation of MCC fillers (1 phr) into optimized PLA/MMT nanocomposites (P/MT5) led to aggregation of MCC in the PLA matrix, as indicated by the circle. This MCC aggregation likely resulted in the reduction of tensile strength compared to P/MT5 nanocomposite, as discussed earlier. As the concentration of MCC fillers increased in the hybrid composites, the aggregation of MCC was also increased (Fig. 7b), as indicated by the circle. Haafiz et al. (2013) and Mathew et al. (2005) were reported similar results, where MCC filled PLA composites using solution casting and melt mixing process, respectively.

In contrast, CNW disperses uniformly throughout the PLA/MMT nanocomposites; MMT fillers are indicated by the white arrows and the CNW particles by the red arrows (Fig. 7c). Accordingly, the interaction between the matrix and the

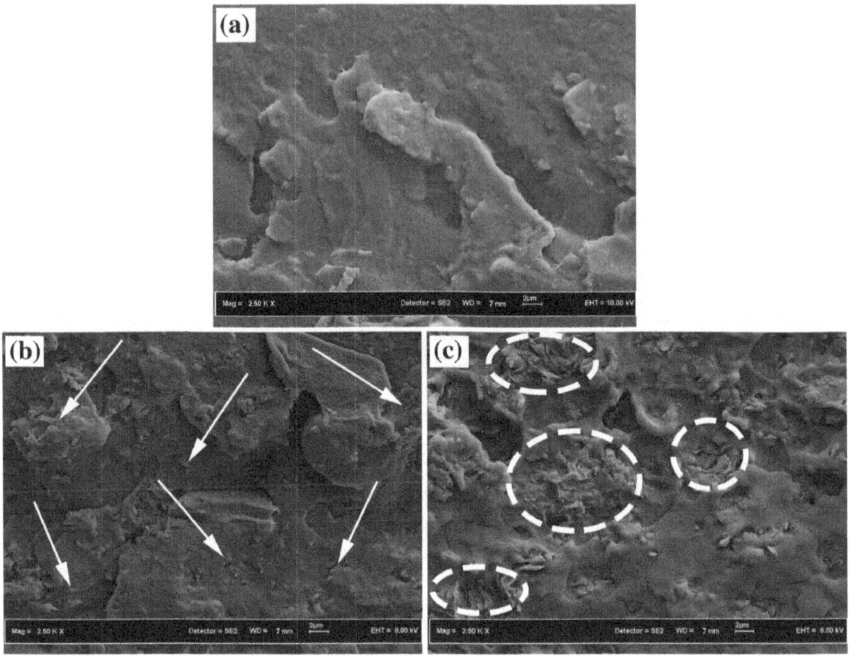

Fig. 6 FESEM micrographs of fractured cross-sections of **a** neat PLA, **b** P/MT5 and **c** P/MT7 nanocomposites (Arjmandi et al. 2015a, b)

fillers has enhanced considerably, as reflected by an increased tensile strength. Meanwhile, at higher filler content filler-filler interactions are more favorable than filler-matrix interactions, leading to filler agglomeration. From Fig. 7d, agglomeration of the MMT filler (circled area) and the CNW particles (red arrows) in P/MT5/CW3 hybrid nanocomposite can be seen. This agglomeration plays a major role in decreasing the tensile properties of the PLA/MMT/CNW hybrid nanocomposites at high concentrations of CNW (>1 phr).

4.5 Transmission Electron Microscopy

TEM micrographs of the neat PLA, P/MT5 nanocomposite, P/MT5/MC1 hybrid composite and P/MT5/CW1 hybrid nanocomposite have shown in Fig. 8a–d, respectively, as reported in our recent works (Arjmandi et al. 2015a, b). As can be observed from Fig. 8a, neat PLA clearly shows a smooth and clean surface with no obvious features. TEM micrograph of P/MT5 nanocomposite (Fig. 8b) shows that the MMT layers are intercalated in the PLA matrix, where the spaces between the dark lines are the interlayer spaces, as indicated by the white arrows. Similar observations were reported for PLA reinforced with MMT filler (Chang et al.

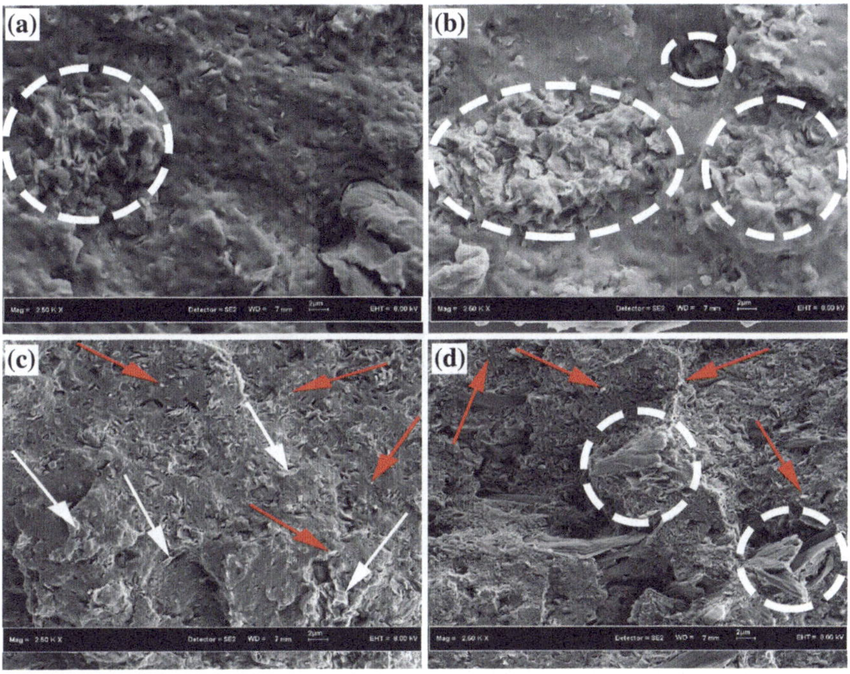

Fig. 7 FESEM micrographs of fractured cross-sections of **a** P/MT5/MC1 hybrid composite **b** P/MT5/MC3 hybrid composite, **c** P/MT5/CW1 hybrid nanocomposite and **d** P/MT5/CW3 hybrid nanocomposite. MMT fillers are indicated by the *white arrow*, the CNW particles are indicated by a *red arrow* and the agglomerated MCC and MMT are indicated by *circles* (Arjmandi et al. 2015a, b)

2003a, b). The TEM micrograph of P/MT5 nanocomposie along with X-RD analysis (will be disscused in the next section) confirmed the formation of an intercalated structure system at 5 phr MMT content. TEM micrograph of P/MT5/MC1 hybrid composite (Fig. 8c) shows the aggregation of MCC particles (indicated by the circle) due to the unevenly dispersion and distribution of MCC fillers in the PLA matrix, where the MMT layers dispersed uniformly in the hybrid composite (indicated by the arrows). Cheng et al. (2012) reported similar observation for PLA/MMT/aluminum trihydrate hybrid composites. Besides that, the intercalation of MMT layers affected by aggregation of MCC particles, where the intercalation level reduced compared to P/MT5 nanocomposite.

However, as can be seen in Fig. 8d, the silicate-layered MMT evenly disperses throughout the PLA (white arrow), and most probably nanorod-like lines correspond to the CNW filler that uniformly disperses in the PLA matrix (red arrow). As can be observed from Fig. 8d, the CNWs stuck to the MMT and were also then covered by MMT platelets. Due to this, it is quite difficult to uniquely identify CNWs in the PLA/MMT/CNW hybrid nanocomposites. The good dispersion of MMT and CNW could be the reason for tensile strength improvements as

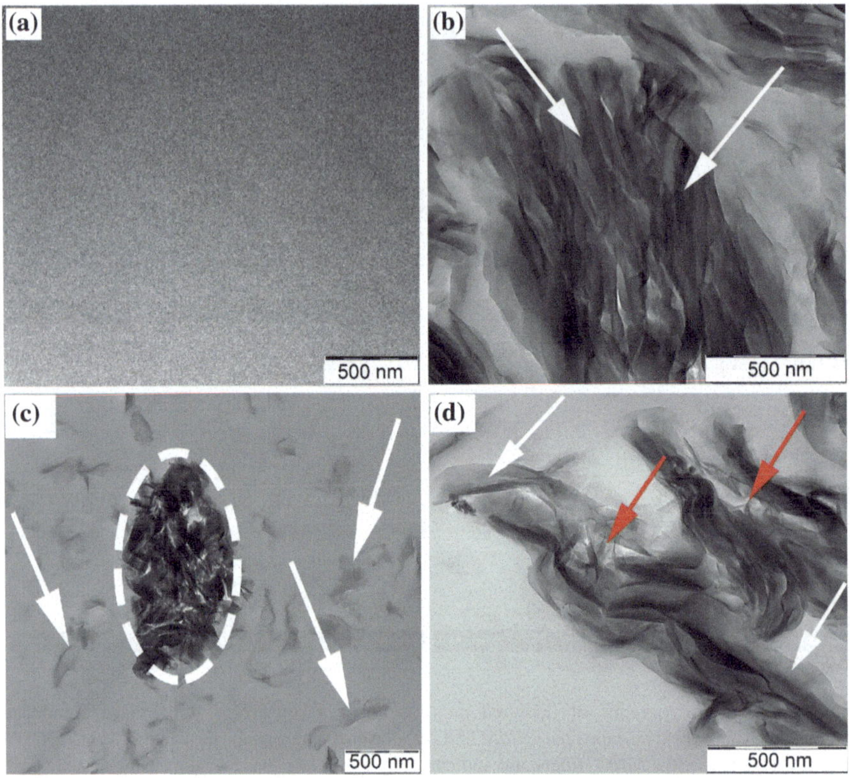

Fig. 8 TEM micrographs of **a** neat PLA, **b** P/MT5 nanocomposite, **c** P/MT5/MC1 hybrid composite and **d** P/MT5/CW1 hybrid nanocomposite. MMT fillers are indicated by the *white arrows* and the CNW particles by the *red arrows* and the agglomerated MCC is indicated by a *circle* (Arjmandi et al. 2015a, b)

previously discussed. Additionally, it is speculated that given the evidence presented in Fig. 8d that the CNWs bind closely to the clay particles, increasing the available surface area for interaction with the PLA matrix. This effect could further enhance the stiffness of the nanocomposites. This finding indicates that the presence of a high content of CNW may induce the agglomeration of MMT filler and adversely affect the intercalation process of the MMT, leading to a decrease in the tensile strength.

4.6 X-ray Diffraction Analysis

In order to further confirm the existence of the interaction between the fillers and the matrix, X-RD analysis of the nanocomposites was carried out. Figure 9 shows the X-RD patterns of the neat PLA and MMT, P/MT5 nanocomposite, P/MT5/MC1

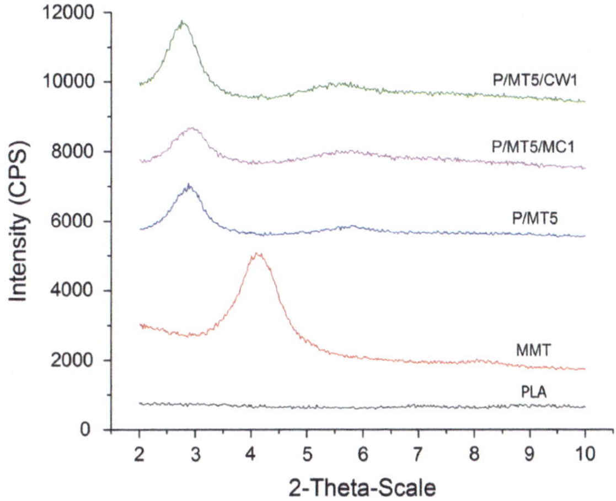

Fig. 9 X-RD pattern of neat PLA, neat MMT, P/MT5 nanocomposite, P/MT5/MC1 hybrid composite and P/MT5/CW1 hybrid nanocomposite (Arjmandi et al. 2015a, b)

hybrid composite and P/MT5/CW1 hybrid nanocomposite, as reported in our previous publications (Arjmandi et al. 2015a, b). Table 3 tabulates the relative *d*-spacing of the nanocomposites. The MMT filler exhibits a diffraction peak at $2\theta = 4.15°$, corresponding to the 001 lattice plane which the *d*-spacing is dedicated to 2.14 nm. This peak shifts to a lower 2θ value as the MMT filler is incorporated into the PLA; the peak shifted to $2\theta = 2.82°$ which the *d*-spacing is dedicated to 3.15 nm. This result indicates that the scattering is due to an intercalation of the mineral phase by the PLA resin (Petersson and Oksman 2006; Shyang and Kuen 2008). Moreover, this intercalation in the interlayer spacing of the MMT may be due to the formation of interactions with the PLA, thus providing the possibility for PLA chains to diffuse between the gallery layers of the MMT during processing (Balakrishnan et al. 2010). According to Balakrishnan et al. (2010), diffusion of PLA chains in the layers of MMT could increase the inter-gallery distance and also reduce the electrostatic attractions between abutting platelets. Besides that, the distribution of MMT layers in the PLA could be due to the intractions between ammonioum groups of MMT and C=O groups of PLA chain segments which resulted in MMT intercatlation. Other possible interactions between the PLA and the MMT filler may also occur via the –OH groups of the PLA and MMT (Jiang et al. 2007).

The *d*-spacing for the 001 peak of P/MT5/MC1 hybrid composite (Fig. 9) slightly decreased to 3.08 nm compared to optimized formulation of PLA/MMT

Table 3 The 2θ angle and *d*-spacing of MMT, a P/M5 nanocomposite, a P/M5/C1 hybrid nanocomposite and a P/M5/C3 hybrid nanocomposite (Arjmandi et al. 2015a, b)

Sample	MMT	P/MT5	P/MT5/MC1	P/MT5/CW1
	001	001	001	001
2θ°	4.15	2.82	2.90	2.76
d-spacing (nm)	2.14	3.15	3.08	3.21

nanocomposites (P/MT5) which dedicated to 3.15 nm, as summarized in Table 3. The decrease in *d*-spacing is confirmed the morphological properties which discussed earlier, where the incorporation of MCC into P/MT5 nanocomposite caused to aggregation of MCC, leading to fewer the intercalation of MMT in the hybrid nanocomposites. The effect of MCC agglomeration resulted in significant decrease in tensile strength compared to P/MT5 nanocomposite, although the decrease in *d*-spacing is small. However, incorporation of nano-size CNW into P/MT5 nanocomposite showed different behavior compared to micro-size MCC. The *d*-spacing for the 001 peak of P/MT5/CW1 hybrid nanocomposite (Fig. 9) slightly increased compared to P/MT5 nanocomposite (from 3.15 to 3.21 nm), as summarized in Table 3. This increase is an indication of a further enlarged interlayer space in the MMT nanostructure that could be due to the chain conformation of the matrix caused by a synergistic interaction with the CNW. In general, X-RD analysis of the PLA/MMT nanocomposites, PLA/MMT/MCC hybrid composites and PLA/MMT/CNW hybrid nanocomposites is well correlated with the tensile and morphological properties.

5 Conclusion and Future Perspective

Neat PLA, PLA/MMT nanocomposites, PLA/MMT/MCC hybrid composites and PLA/MMT/CNW hybrid nanocomposites were fabricated using solution casting method. Based on tensile strength, the optimum formulation of PLA/MMT nanocomposites was found at 5 phr MMT content. The Young's modulus of PLA/MMT nanocomposites, PLA/MMT/MCC hybrid composites and PLA/MMT/CNW hybrid nanocomposites increased gradually with increasing MMT, MCC and CNW content, respectively. The good dispersion and distribution of MMT fillers contributed to the enhanced tensile properties of PLA/MMT nanocomposites. Incorporation of MCC into P/MT5 nanocomposite led to aggregation of MCC in the hybrid composites. Due to the MCC aggregation in the hybrid composites, the *d*-spacing of MMT layers slightly decreased compared to P/MT5 nanocomposite, leading to less intercalation of MMT layers in the hybrid composites. It can be said that the incorporation of MCC into P/MT5 nanocomposite led to significant decrease in tensile strength, whereas the Young's modulus and ductility of hybrid composites increased significantly compared to P/MT5 nanocomposite. Nevertheless, the tensile strength of PLA/MMT/CNW hybrid nanocomposites increased significantly by incorporation of 1 phr CNW. Due to the possible nucleation of crazes in the amorphous fraction of the PLA and the formation of shear bands, the ductility of hybrid composites and hybrid nanocomposites increased significantly compared to PLA/MMT nanocomposites. The presence of polar interactions between the both fillers and PLA matrix were confirmed by FT-IR analysis. The formation of an intercalated nanostructure in PLA/MMT nanocomposites and PLA/MMT/CNW hybrid nanocomposites was confirmed by TEM and X-RD analysis. A uniform dispersion of CNW was readily apparent in the hybrid nanocomposites by incorporation of 1 phr CNW. Due to the presence of dispersed CNW in the

PLA/MMT/CNW hybrid nanocomposites, the interlayer spacing of the MMT slightly increased, leading to more intercalation, which in turn increased the tensile properties of the hybrid nanocomposites. The remaining challenge is to find an effective method to further enhance adhesion between PLA and both MMT and CNW. Overcoming this challenge is expected to further enhance the ductility, Young's modulus and tensile strength of these interesting materials.

Acknowledgments Authors wish to acknowledge the Universiti Teknologi Malaysia (UTM) and Research University Grant 05H22, sub-code: Q.J130000.2509.05H22 for financial support.

References

Arjmandi, R., Hassan, A., Haafiz, M.K.M., Zakaria, Z., Inuwa, I.M.: Characterization of polylactic acid/microcrystalline cellulose/montmorillonite hybrid composites. Malays. J. Anal. Sci. **18**, 642–650 (2014)

Arjmandi, R., Hassan, A., Eichhorn, S.J., Haafiz, M.K.M., Zakaria, Z., Tanjung, F.A.: Enhanced ductility and tensile properties of hybrid montmorillonite/cellulose nanowhiskers reinforced polylactic acid nanocomposites. J. Mater. Sci. **50**, 3118–3130 (2015a)

Arjmandi, R., Hassan, A., Haafiz, M.K.M., Zakaria, Z.: Effect of microcrystalline cellulose on biodegradability, tensile and morphological properties of montmorillonite reinforced polylactic acid nanocomposites. Fiber Polym. **16**, 2284–2293 (2015b)

Azizi Samir, M.A.S., Alloin, F., Dufresne, A.: Review of recent research into cellulosic whiskers, their properties and their application in nanocomposite field. Biomacromolecules **6**, 612–626 (2005)

Balakrishnan, H., Hassan, A., Wahit, M.U., Yussuf, A.A., Razak, S.B.A.: Novel toughened polylactic acid nanocomposite: mechanical, thermal and morphological properties. Mater. Design **31**, 3289–3298 (2010)

Bulota, M., Hughes, M.: Toughening mechanisms in poly(lactic) acid reinforced with TEMPO-oxidized cellulose. J. Mater. Sci. **47**, 5517–5523 (2012)

Chang, J.H., An, Y.U., Cho, D., Giannelis, E.P.: Poly(lactic acid) nanocomposites: comparison of their properties with montmorillonite and synthetic mica (II). Polymer **44**, 3715–3720 (2003a)

Chang, J.H., An, Y.U., Sur, G.S.: Poly(lactic acid) Nanocomposites with Various Organoclays. I. Thermomechanical Properties, Morphology, and Gas Permeability. J Polym Sci B **41**, 94–103 (2003b)

Chen, N.L., Feng, H.X., Guo, J.W., Luo, H.M., Qiu, J.H.: Biodegradable Poly(lactic acid)/TDI-montmorillonite nanocomposites: preparation and characterization. Adv. Mat. Res. **221**, 211–215 (2011a)

Chen, W., Yu, H., Liu, Y., Chen, P., Zhang, M., Hai, Y.: Individualization of cellulose nanofibers from wood using high-intensity ultrasonication combined with chemical pretreatments. Carbohydr. Polym. **83**, 1804–1811 (2011b)

Cheng, K.C., Yu, C.B., Guo, W., Wang, S.F., Chuang, T.H., Lin, Y.H.: Thermal properties and flammability of polylactide nanocomposites with aluminum trihydrate and organoclay. Carbohydr. Polym. **87**, 1119–1123 (2012)

Cheng, Q., Wang, S., Rials, T.G.: Poly (vinyl alcohol) nanocomposites reinforced with cellulose fibrils isolated by high intensity ultrasonication. Compos. Part A **40**, 218–224 (2009)

Cho, S.Y., Park, H.H., Yun, Y.S., Jin, H.J.: Cellulose nanowhisker-incorporated poly (lactic acid) composites for high thermal stability. Fiber Polym. **14**, 1001–1005 (2013)

Chuayjuljit, S., Su-uthai, S., Charuchinda, S.: Poly(vinyl chloride) film filled with microcrystalline cellulose prepared from cotton fabric waste: properties and biodegradability study. Waste Manag. Res. **28**, 109–117 (2010)

Field, L.D., Sternhell, S., Kalman, J.R.: Organic structures from spectra, 5th edn. Wiley, Chichester (2012)

Haafiz, M.K.M., Hassan, A., Zakaria, Z., Inuwaa, I.M., Islam, M.S., Jawaid, M.: Properties of polylactic acid composites reinforced with oil palm biomass microcrystalline cellulose. Carbohydr. Polym. **98**, 139–145 (2013)

He, Y.N., Guo, K., Chen, J.Z., Niu, M.J., Wang, W.J., Li, X.F.: Preparation and characterization of poly(lactic acid)/montmorillonite nanocomposites via a masterbatching method. Adv. Mat. Res. **335–336**, 1493–1498 (2011)

Herrera, N., Mathew, A.P., Oksman, K.: Plasticized polylactic acid/cellulose nanocomposites prepared using melt-extrusion and liquid feeding: mechanical, thermal and optical properties. Compos. Sci. Technol. **106**, 149–155 (2015)

Hong, J., Kim, D.S.: Preparation and physical properties of polylactide/cellulose nanowhisker/nanoclay composites. Polym. Compos. **34**, 293–298 (2013)

Huda, M.S., Drzal, L.T., Mohanty, A.K., Misra, M.: Effect of fiber surface-treatments on the properties of laminated biocomposites from poly (lactic acid) (PLA) and kenaf fibers. Compos. Sci. Technol. **68**, 424–432 (2008)

Iwatake, A., Nogi, M., Yano, H.: Cellulose nanofiber-reinforced polylactic acid. Compos. Sci. Technol. **68**, 2103–2106 (2008)

Jiang, L., Zhang, J., Wolcott, M.P.: Comparison of polylactide/nanosized calcium carbonate and polylactide/montmorillonite composites: reinforcing effects and toughening mechanisms. Polymer **48**, 7632–7644 (2007)

Karande, V.S., Bharimalla, A.K., Vigneshwaran, N., Kadam, P.G., Mhaske, S.T.: Cotton linter nano-fibers as the potential reinforcing agent for guar gum. Iran. Polym. J. **23**, 869–879 (2014)

Lee, J., Sun, Q., Deng, Y.: Nanocomposites from regenerated cellulose and nanoclay. J. Biobased Mater. Bio. **2**, 162–168 (2008)

Liu, M., Zhang, Y., Zhou, C.: Nanocomposites of halloysite and polylactide. Appl. Clay Sci. **75–76**, 52–59 (2013a)

Liu, R., Luo, S., Cao, J., Peng, Y.: Characterization of organo-montmorillonite (OMMT) modified wood flour and properties of its composites with poly (lactic acid). Compos. A **51**, 33–42 (2013b)

Lu, J., Wang, T., Drzal, L.T.: Preparation and properties of microfibrillated cellulose polyvinyl alcohol composite materials. Compos. A **39**, 738–746 (2008)

Mathew, A.P., Oksman, K., Sain, M.: Mechanical properties of biodegradable composites from polylactic acid (PLA) and microcrystalline cellulose (MCC). J. Appl. Polym. Sci. **97**, 2014–2025 (2005)

Oksman, K., Mathew, A.P., Bondeson, D., Kvien, I.: Manufacturing process of cellulose whiskers/polylactic acid nanocomposites. Compos. Sci. Technol. **66**, 2776–2784 (2006)

Pandey, J.K., Chu, W.S., Kim, C.S., Lee, C.S., Ahn, S.H.M.: Bio-nano reinforcement of environmentally degradable polymer matrix by cellulose whiskers from grass. Compos. B **40**, 676–680 (2009)

Pereda, M., Amica, G., Rácz, I., Marcovich, N.E.: Structure and properties of nanocomposite films based on sodium caseinate and nanocellulose fibers. J. Food Eng. **103**, 76–83 (2011)

Petersson, L., Oksman, K.: Biopolymer based nanocomposites: comparing layered silicates and microcrystalline cellulose as nanoreinforcement. Compos. Sci. Technol. **66**, 2187–2196 (2006)

Petersson, L., Kvien, I., Oksman, K.: Structure and thermal properties of poly (lactic acid)/cellulose whiskers nanocomposite materials. Compos. Sci. Technol. **67**, 2535–2544 (2007)

Qu, P., Gao, Y., Wu, G., Zhang, L.: Nanocomposite of poly(lactid acid) reinforced with cellulose nanofibrils. BioResources **5**, 1811–1823 (2010)

Sanchez-Garcia, M.D., Lagaron, J.M.: On the use of plant cellulose nanowhiskers to enhance the barrier properties of polylactic acid. Cellulose **17**, 987–1004 (2010)

Satyamurthy, P., Jain, P., Balasubramanya, R.H., Vigneshwaran, N.: Preparation and characterization of cellulose nanowhiskers from cotton fibres by controlled microbial hydrolysis. Carbohydr. Polym. **83**, 122–129 (2011)

Shayan, M., Azizi, H., Ghasemi, I., Karrabi, M.: Effect of modified starch and nanoclay particles on biodegradability and mechanical properties of cross-linked poly lactic acid. Carbohydr. Polym. **124**, 237–244 (2015)

Shyang, C.W., Kuen, L.S.: Flexural, morphological and thermal properties of poly(lactic acid)/organo-montmorillonite nanocomposites. Polym. Polym. Compos. **16**, 263–270 (2008)

Sinha Ray, S., Okamoto, M.: Polymer/layered silicate nanocomposites: a review from preparation to processing. Prog. Polym. Sci. **28**, 1539–1641 (2003)

Thellen, C., Orroth, C., Froio, D., Ziegler, D., Lucciarini, J., Farrell, R., Ratto, J.A.: Influence of montmorillonite layered silicate on plasticized poly (L-lactide) blown films. Polymer **46**, 11716–11727 (2005)

Vroman, I., Tighzert, L.: Biodegradable polymers. Materials **2**, 307–344 (2009)

Wang, X., Qu, P., Zhang, L.: Thermal and structure properties of biobased cellulose nanowhiskers/poly (lactic acid) nanocomposites. Fiber Polym. **15**, 302–306 (2014)

Wang, Y., Cao, X., Zhang, L.: Effects of cellulose whiskers on properties of soy protein thermoplastics. Macromol. Biosci. **6**, 524–531 (2006)

Effect of Nanoclay on the Dielectric, Transport, Thermal and Fire Properties of Coconut Sheath/MMT Clay Polyester Hybrid Composites

N. Rajini, J.T. Winowlin Jappes, S. Karthikeyan and A. Varada Rajulu

Abstract The emission of the inorganic pollutant and the non-biodegradable compounds to the atmosphere has been a threatening aspect for the increase of the environmental deterioration and posing a human health hazard. Therefore, the developing of the research in the field of science and engineering is mainly focused to create the eco-friendly environment. It is the fact that this research investigated the use of the natural fiber as reinforcement for the possibility of producing composite laminates as an alternative to the synthetic fibers in the industrial application ranging from building construction, existing house holding components like the roofs and the door panels, automotive, and the food packaging, where a good fire performance and the mechanical properties are essential. Therefore, it is necessary to design an eco-friendly material with all beneficial at the same time without compromising the basic mechanical properties. Hence in this chapter, the conductivity and the barrier properties of CS/PE composites were experimented with the effect of chemically treated CS and oMMT nanoclay addition. Thermo-gravimetric analysis has also been performed to study the effect of varying the weight of the nanoclay on the thermal stability of the CS/PE composites.

Keywords Clay · Conductivity properties · Barrier properties · Thermal properties

N. Rajini (✉) · J.T. Winowlin Jappes · S. Karthikeyan
Department of Mechanical Engineering, Kalasalingam University, Anand Nagar, Krishnankoil 626126, Tamil Nadu, India
e-mail: rajiniklu@gmail.com

A. Varada Rajulu
International Research Center, Center for Composite Materials, Kalasalingam University, Anand Nagar, Krishnankoil 626126, Tamil Nadu, India

M. Jawaid et al. (eds.), *Nanoclay Reinforced Polymer Composites*, Engineering Materials, DOI 10.1007/978-981-10-0950-1_6

1 Introduction

Recently, the importance of blended nanoclay in polymer matrix composites explored significantly by many commercial applications such as fire retardant, food packaging, electronic circuit boards and automotive industries (El-Khodary et al. 2008; Sengwa and Sankhla 2008; Zheng and Wong 2003). The addition of nanoclay in polymer matrix can enhance the functional properties of nanocomposites due to the uniform distribution of particle with increasing surface area. Indeed, to achieve the higher mechanical properties, the nanocomposite requires the addition of reinforcing element. Based on the demand of application, the reinforcement can be in the form of particulate, fiber or woven fabric. Nowadays due to the advancement of biocomposites, there are lots of plants fibers have been used as primary reinforcement element in dielectric applications. Moreover, the conductive behavior of natural fibers due to the presence of water molecules can influence the dielectric characteristics of composites. Similarly, the addition of clay in polymer as particulate reinforcement can act as either insulator or conductor depending upon the type (from the source where it has been extracted) of clay. Accordingly, some of the researchers have studied the performance of dielectric properties of natural fibers in the applications such as stabilizer cover, bushings, sleeves and electronic circuit boards (Mukherjee et al. 1978).

By considering the effect of natural fibers and nanoclay particulate as primary and secondary reinforcement in polymer matrix on dielectric and other transport properties were found to be studied only limited. Apart from all other advantages of eco friendly circumstances, the natural fibers also have some demerits. Normally, the natural fibers are hydrophilic in nature and it's have a major drawback of moisture absorption capability due to the presence of chemical constituents such as hemicelluloses, lignin and pectin. These components were found to be the responsible factors for providing strong polarized hydroxyl groups in plant fibers (Hepworth et al. 2000; John and Thomas 2008). So, the presence of moisture content on the fiber surface does not exhibit the better compatibility with hydrophobic polymer matrix. It will create a poor interfacial adhesion between fiber and matrix, which in turn, reduce the performance of the composite in all the aspects. Hence, it requires some topography changes on fiber surface by the chemical treatment so that the enhancement of fiber and matrix adhesion will get improved. Moreover, the morphological changes on fiber surface can create a strong fiber/matrix adhesion which can enhance the characteristics and performance of these composites (John and Anandjiwala 2008; Edeerozey et al. 2007; Alvarez et al. 2003).

Having all said, the factors like fiber length, fiber content, addition of particulate, manufacturing method, treatment of fibers were found to be influenced the dielectric properties composites(Lai et al. 2005; Kumar et al. 2011). Some of the published work related to impedance spectroscopic using natural fiber as reinforcement in thermoset and thermoplastic polymer matrix has been described as below.

The dielectric behavior of Low density polyethylene/sisal fiber composite has been studied and reported as the enhanced dielectric constant was observed with the increasing amount of fiber content in matrix (Paul et al. 1997).

The parameters such as fiber orientation and varying frequency were found to be influence the dielectric strength of the composites (Navin et al. 2005). In addition, the effect of chemical treatment of natural fiber composites on dielectric properties has also been studied with regards to change in magnitude of interfacial adhesion. Paul et al. (1997) studied the dielectric properties with the improved interfacial bonding between sisal and low density polyethylene with imposing different kind of chemical modifications. The dielectric properties of alkali treated Roystonea/epoxy matrix exhibits lower conductivity value than that of unmodified one, and found to be increased with frequency (Gould and Rao 2011). Similarly, the effect of hybridization associated with the dielectric properties has also been studied using the combination of two different natural fibers such as short alfa fiber and wool fibers for varying temperature (Triki et al. 2011). However, the dielectric property of composites with the influence of nanoclay as secondary reinforcement in untreated and chemically treated natural fiber composites was found to be scanty. Therefore, there are lot scope is available for studying the importance of fiber reinforced composites with the addition of nanoclay particulate on dielectric properties. So that the better suitable dielectric composite can be identified for different industrial application without sacrificing the strength of composites.

Moreover, the enhanced barrier behavior of nanocomposite motivates material researchers to spotlight the influence of nano-particle in polymer matrix (Messersmith and Giannelis 1995; Linchan et al. 2011; Kim et al. 2005). Lee et al. 2005 have carried out CO_2 gas permeability studies in polymer system and found that the permeability can be significantly reduced due to the uniformly distributed clay content. Permeability test was carried out for LDPE/clay nanocomposites in O_2 gas environment and determined that the increasing amount of clay content can decrease the rate of permeability, however it was applicable only for certain wt% beyond that the permeability was found to be getting increased (Arunvisut et al. 2007).

Among the different types of properties of composites it is essential to study the characteristics of molecular barrier properties, such as, gas permeability, thermal conductivity, and surface and volume resistivity of a composite. Generally, these transport properties of composites are significantly affected by the fiber size, fiber content, fiber orientation, and chemical modification of fiber surface and the addition of the filler. The dielectric parameters, such as the AC conductivity (σ_{ac}), dielectric constant (ε_r) and dielectric loss tangent (tan δ), were measured for CS/MMT hybrid composites in the frequency range of 42 Hz to 1 MHz at room temperature. Besides the behavior of the molecular transport properties such as thermal conductivity, gas permeability, volume and surface resistivity has been studied. Scanning Electron Microscopic investigation (SEM) has been carried out to find the silica rich area on the coconut sheath fiber.

2 Materials and Methods

2.1 Fabrication of Hybrid Polyester Nanocomposites

The fabrication process of the composite is one of the vital criteria, which decides the strength of the composite. Among all the fabrication processes, the compression molding method provides better cohesion in the prepared composite structure. In this work the fabrication of the coconut sheath fiber/MMT nanoclay reinforced hybrid polymer composites was carried out in two steps. In the first step, the proportionate wt% of nanoclay was taken according to the wt% of the unsaturated isophthalic polyester resin, and the above mixture is stirred by the high speed mechanical shear mixer at 500 rpm for about 2 h, with a radial turbine blade. In the second step, the CS/MMT clay/PE hybrid composites were fabricated using the compression molding machine by the following procedure.

The final product collected from the high shear mixer was allowed for the degassing process. The mats were impregnated with the clay-polyester mixture to which 1.5 vol% of cobalt naphthenate and MEKP were added. The middle plate is placed over the bottom plate with the help of circular pins to maintain the position fixed, as shown in Fig. 1a. Then the clay-resin mixture with the curing agent was spread over the CS fiber mat with a steel brush in each layer, which was placed

Fig. 1 Steps involved in fabrication process. **a** Positioning of mold. **b** Fiber arrangement. **c** Closed mold. **d** Mold placed in the compression machine

inside the mold cavity of size 300 × 125 × 10 mm^3 (see Fig. 1b) by keeping one over another up to six layers (48 ± 2 wt%). The stacked CS fiber mats were rolled with the help of a roller in such a way as to remove the air bubbles. The top plate was placed over the middle plate in such a way as to cover the CS fiber mats and positioned by the circular pins at the corners, as shown in Fig. 1c. The mold was closed completely by applying 150 kgf/cm^2 pressure in order to obtain the 3 mm thickness plate using the compression molding machine, as shown in Fig. 1d. The compressed mold was allowed for room temperature curing, for another 24 h. Finally, the split mold was separated by removing the pins at the corners. The rectangular composite plate of size 300 × 125 mm^2 and thickness of 3 mm was carefully removed through the cavity of the middle plate. Then, the test specimens of the required size were cut from the fabricated composite sheet.

2.2 *Impedance Measurement*

For the dielectric analysis, the samples were placed between two parallel plate electrodes. A sinusoidal voltage creating an alternating electric field was applied to these electrodes thus producing polarization in the samples, which stimulates oscillations at the same frequency as the electric field, but has a shift in phase angle. This phase angle shift was measured with respect to the applied voltage and the measured current, which in turn was separated into capacitive and conductive components. Measurements of capacitance and conductance were used to calculate the dielectric parameters. Figure 2a shows the computer controlled LCR meter for the measurement of conductivity (frequency range 42 Hz–1 MHz), at room temperature condition. The calibration was done for the range of ±5 % error before the experimentation of dielectric measurements. Figure 2b shows the connections to the sample with the holder.

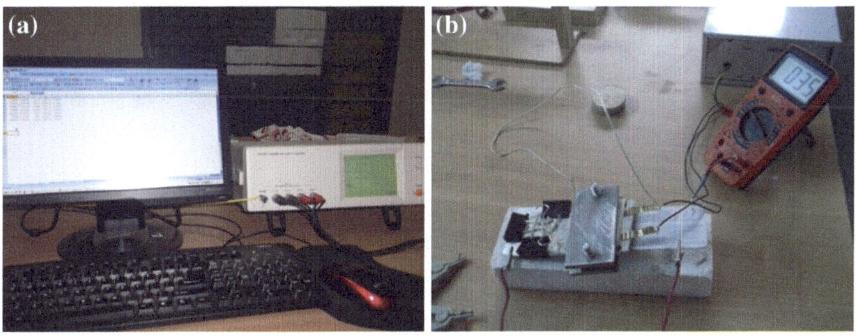

Fig. 2 **a** LCZ meter. **b** Connections to the samples with the holder

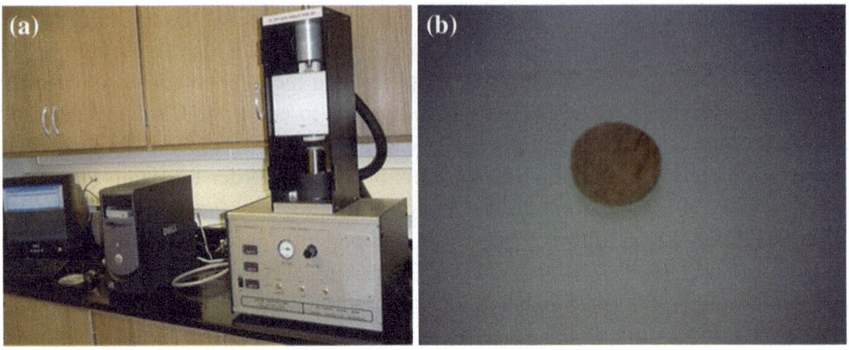

Fig. 3 Thermal conductivity. **a** Measuring apparatus. **b** Sample

2.3 *Thermal Conductivity*

As per the ASTM E-1530 standard the thermal conductivity property of CS/MMT hybrid composite was recorded. The measurements were taken using the conductivity meter, Unitherm model 2022 from ANTER Corp., Pittsburgh, PA, as shown in Fig. 3a. The 50 mm diameter sample was used for the testing (Fig. 3b). In this work, the thermal conductivity was directly measured by the guarded heat flow meter method, which is a computerized thermal conductivity measuring system.

2.4 *Gas Permeability*

The Lyssy Gas Permeability equipment is used to measure the permeability of any common gas oragas mixture through thin films and laminates. The permeability measuring equipment is shown in Fig. 4 (Elixir Technologies, Bangalore, India). In this work, it was used to measure the permeability of CO_2 gas through the composites. ASTM D14382 was followed for the measurement. The samples of coconut sheath/polyester composite laminate were prepared for the size of $110 \times 110 \times 3$ mm^3. Figure 5a, b presents the photographs of uncoated and Teflon coated composite samples. Before the Teflon coating the surface was smoothened using 600-grid SiC emery sheet and cleaned with acetone.

2.5 *Volume and Surface Resistivity*

Volume resistivity can be used as an important parameter while designing an insulator for specific applications. Surface resistivity is the resistance to leakage current along the surface of an insulating material. Surface conductivity helps in assessing the surface contamination of the material.

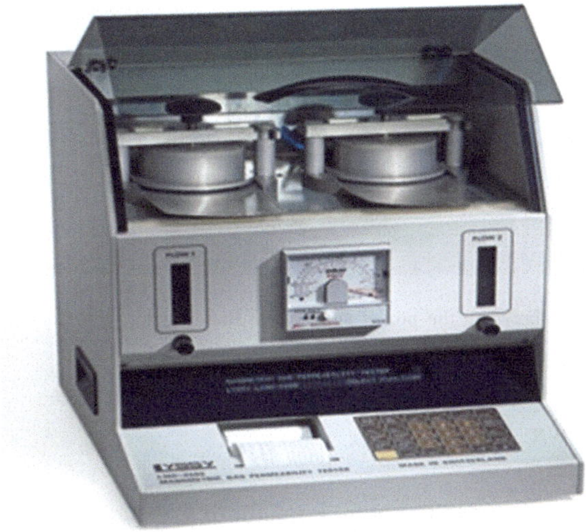

Fig. 4 Gas permeability measuring unit

(a) (b)

Fig. 5 Gas permeability samples, **a** as—fabricated. **b** Teflon coated

ASTM standard of D257 was used for the measurement of volume and surface resistivity. The Keithley resistivity adaptor, model 6105 was used to record the volume and surface resistivity (see Fig. 6a). The resistance across the sample was measured with a Knick model Teraohmmeter (see Fig. 6b). Disk shaped samples having a 100 mm diameter (see Fig. 6c) was used for the measurement.

The volume and surface resistance values were calculated based on the standard relations.

$$\left.\begin{aligned} \rho_V = A_e R_V/t \quad A_e = \pi(D_1+g)^2/4 \\ \rho_S = P\,Rs/g\,(\text{here } P = \pi D_o) \end{aligned}\right\} \tag{1}$$

Fig. 6 **a** Connections between the power supply and the ohmeter. **b** Resistivity adapter. **c** Disk shaped sample

2.6 Breakdown Voltage

Breakdown voltage is the ability of an insulator to withstand the maximum voltage difference that can be applied across the material, before the insulator collapses and starts conducting. The ASTM D149 method was employed to find out the dielectric strength of the sample using the breakdown voltage. Figure 7a shows the high voltage electrical connections to the hydraulic oil tank. In order to withstand the

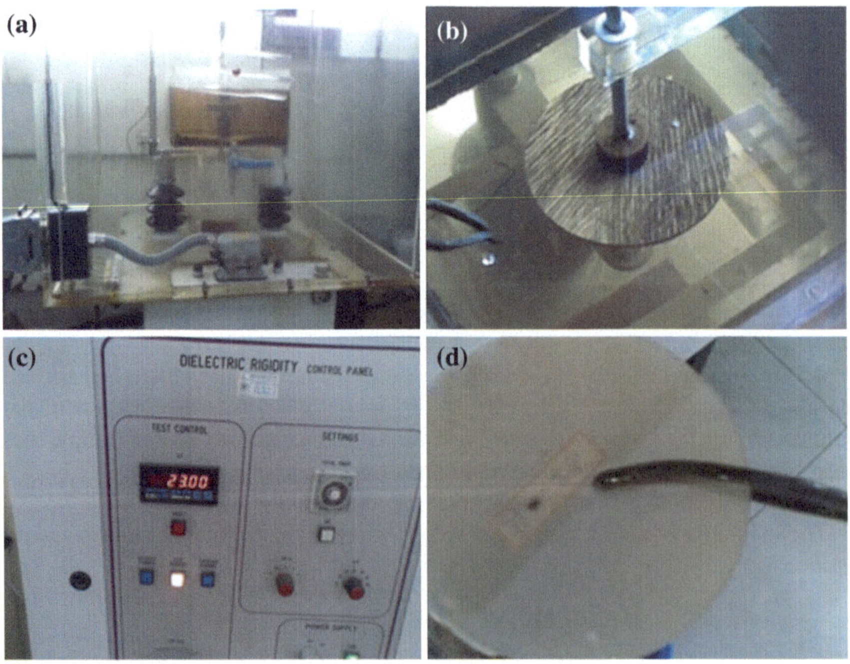

Fig. 7 **a** Hydraulic oil tank. **b** Sample inside the oil tank. **c** Breakdown voltage measuring setup. **d** Break down sample

applied high voltage, the sample of the size 100 mm in diameter was placed inside the hydraulic oil, as shown in Fig. 7b.

The dielectric strength was calculated using the following standard relation given in Eq. 2.

$$D_S = B_y/h \tag{2}$$

where, D_S is the dielectric strength (kV/mm), B_V is the breakdown voltage (kV) and h is the thickness (mm) of the specimen.

In this work, the breakdown voltage of the composites was measured, using an insulation breakdown tester supplied by CEAST (Italy) as shown in Fig. 7c. The B_V is directly measured from the setup given in Fig. 7c. The black mark in the sample, as shown in Fig. 7d, indicates the failure of the materials due to the applied electric field.

2.7 *Thermo-Gravimetric Analysis*

The thermal stability of the polyester nanocomposites is assessed using SII EXSTAR TG 6000 module. About 3–6 mg powder of sample is subjected to dynamic TGA scans at an increased heating rate of 5 °C/min in the range of ambient temperature to 800 °C in N_2 atmosphere. The TG curves are analyzed as the percentage of weight loss as a function of temperature. The average of three powder samples are reported in each cases and that are obtained from the three different locations of same specimen using diamond cutter.

2.8 *Cone Calorimetric Analysis*

Fire resistant properties of coconut sheath reinforced polyester composites have been carried out on a cone calorimeter in accordance with ISO 5006-1. The test specimen of size 100 × 100 × 10 mm^3 has been taken for the measurements. All the data collected from the cone calorimeter test are available in engineering units.

Conditions of the test:

- Specimen position: horizontal
- Heat flux: 50 kW/m^2
- Spark igniter was used
- Scan: 1800 s.

3 Results and Discussions

3.1 Characterization Studies

The XRD spectrum of pure organically modified oMMT clay and the polyester reinforced with particulate nanoclay composites with 1, 2, 3, and 5 wt% are given in Rajini et al. (2012). The Pure oMMT clay exhibits a diffraction peak at 4.5° and it indicate the d-spacing of 1.963 nm. This high intensity peak at around 4.5° indicates the crystallinity behavior of nanoclay. In the other cases composites there was no peak was identified at the same two theta angle. It represents that the distance between the nanoclay particles exceeds the value of 1.963 nm ($2\theta < 3°$) or it could be well dispersed in the matrix. At the same time, the inclusion of higher weight percentage of nanoclay(5 wt%) found to be shifting diffraction peaks towards lower angle side, which could be the reason for the decreasing distance between the clay platelets. The uneven curing process due to the addition of clay particle creates the possibility of intercalated nanocomposites. From this observation, it was found that the separation of clayplatelets has also depends on the wt% of clay content. Since, higher the wt% of clay content will not allow the particles to segregate easily due to the constraint of surrounding fillers. The similar kind of observation has also been reported by Yasmin et al. (2003).

Based on our previous work, it was found that the distribution of the nano particle in polyester matrix can be seen clearly like zebra pattern, which indicates that the platelets were separated from the stack due to the shearing force by mechanical stirring (Rajini et al. 2013). Moreover, the aggregate was found to be inevitable in the case of higher amount of clay content. Accordingly, anything less than 5 wt% of clay content was observed to produce better performance. In addition, the FTIR and chemical analysis were performed on untreated and chemically treated coconut sheath fiber in regards to understand the variation in chemical proportions after the removal of weak and unwanted contaminants. The FTIR spectrum as well as the varying proportion of chemical constituents obtained from the conventional chemical analysis was already reported in our previous work (Rajini et al. 2013).

3.2 Conductivity Spectra Analysis at Room Temperature

The frequency dependence of the AC conductivity (σ_{ac}) of CS/MMT clay reinforced hybrid composites at room temperature, in a wide range of frequencies was measured. It is clearly found that the σ_{ac} of the hybrid composite is higher in the alkali treated condition, and that may be due to the incorporation of more polar molecules (hydroxyl groups) present in the fiber.

Rajini et al. (2013) studied the variation of the A.C. conductivity of the composites with UTC, ATC and STC coconut sheaths at room temperatures. The spectra of the two specimens (ATC, STC) exhibit "two well—defined regions". From their observation it is clear, that the A.C. conductivity depends strongly on the treatment of the coconut sheath. Moreover, they also showed the variation of the AC conductivity of the addition of different weight percentages of nanoclay with UTC+ x wt%, ATC+ x wt% and STC+ x wt% coconut sheaths at room temperatures. The addition of the MMT nanoclay can influence the A.C. conductivity in the ATC and STC reinforced composites more than in the UTC composites (Rajini et al. 2013).

All the data obtained from the experimentation is found to obey Jonscher's power law (Joseph and Thomas 2008). Jonscher's power law equation may be written as

$$\sigma(\omega) = \sigma_{dc} + A\omega^{n} \tag{3}$$

where n is the frequency exponent in the range $0 < n < 1$. Both σ_{dc} and A are constants. The frequency-independent conductivity is observed in the low frequency region, which indicates the dc conductivity of the material. In the high frequency region, the power law feature σ (ω) α $A\omega^{n}$ is observed.

3.3 *Effect of Temperature on Dielectric Properties of CS/MMT Hybrid Composites*

The AC Conductivity versus frequency is drawn and shown in Fig. 8a–f for the untreated and chemically treated CS reinforced nanocomposite with lower and higher temperatures. The two selected temperatures are 50 and 90 °C. This is because of molecular mobility due to the increasing frequency. From the Fig. 8a, it was found the conductivity was increased linearly with the increase of frequency. At the same time there is no variation was found for the addition of varying % of nanoclay addition. The similar trend was also observed at the high temperature condition also (see Fig. 8b). But, the entirely different type of ac conductivity curve was observed in the case of for the alkali and silane treated composites in Fig. 8c–f. A two stage curve was observed throughout the entire frequency region for the alkali and silane treated composites. A significant change in ac conductivity value was observed for all the wt% of clay addition. In low frequency region, the difference was found to be large and then it merged into single line at high frequency domain. The conductivity increases as the temperature increases for a frequency at a particular composition. This effect is created because of the ionic motion of composite atoms. As the temperature increases the atoms gets their energy and gets ionize. The plots were also drawn for all the combination to identify the behavior of

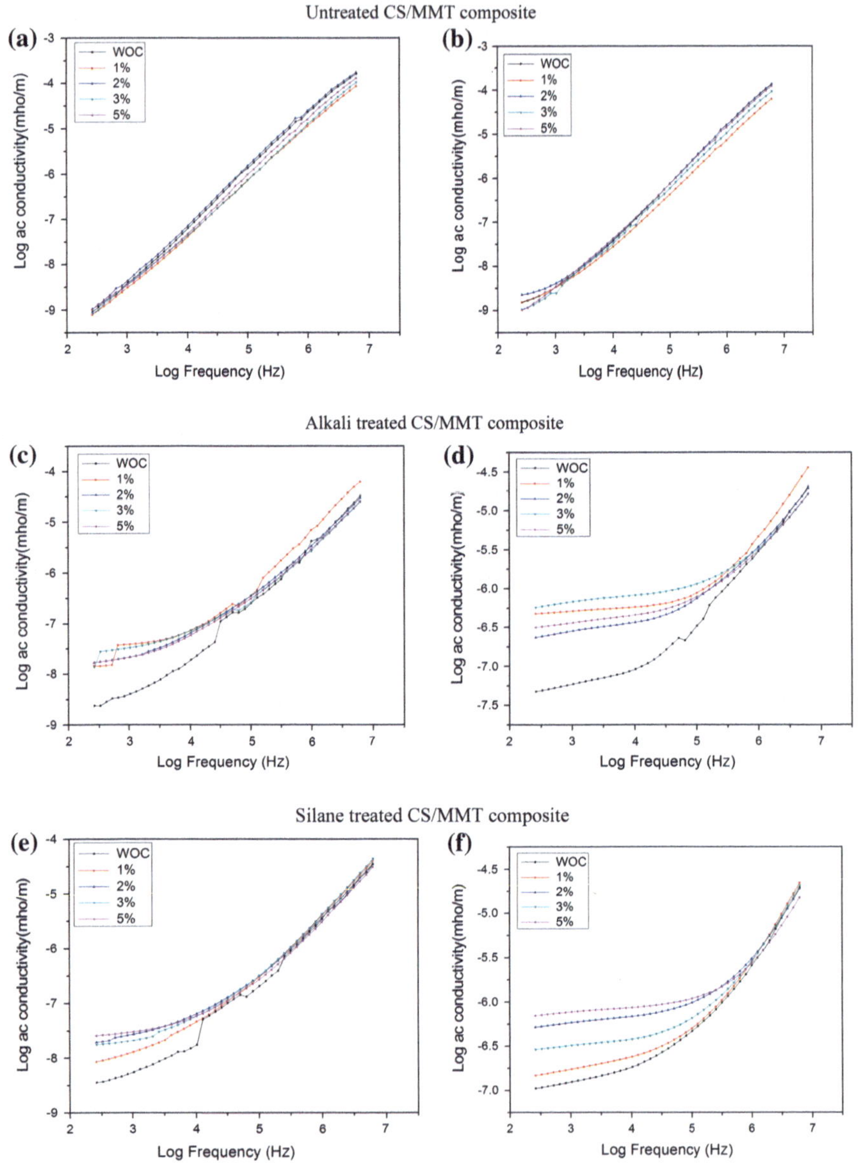

Fig. 8 Effect of temperatures on the variation of the AC conductivity for clay filled UTC, ATC and STC coconut sheath composite

dielectric constant and dielectric loss factor for same two temperatures (Figs. 9 and 10). Figure 9 clearly shows that the dielectric constant decreases with the increase in frequency. The decreasing rate is high at lower frequencies and very low or nearly same at higher frequencies. This is because of the absence of water dipole.

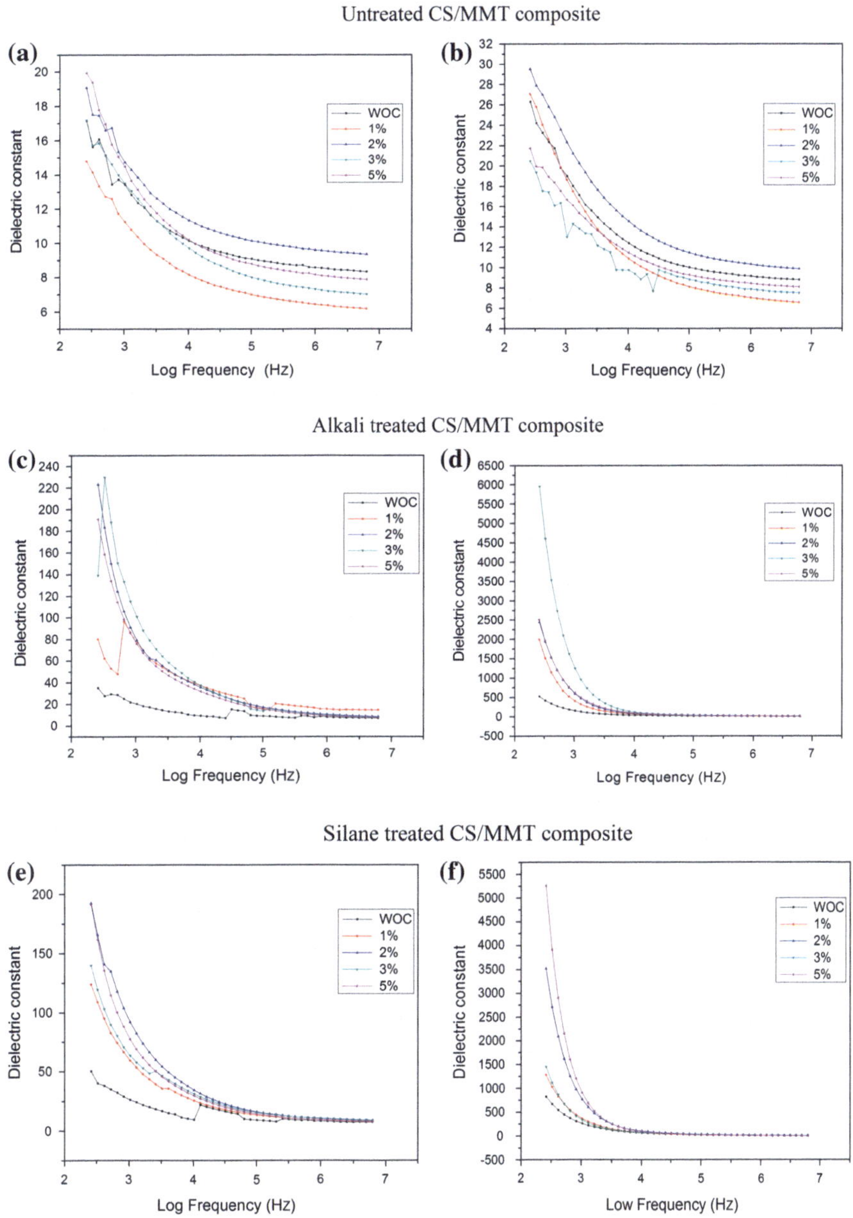

Fig. 9 Effect of temperatures on the variation of the dielectric constant for clay filled UTC, ATC and STC coconut sheath composite

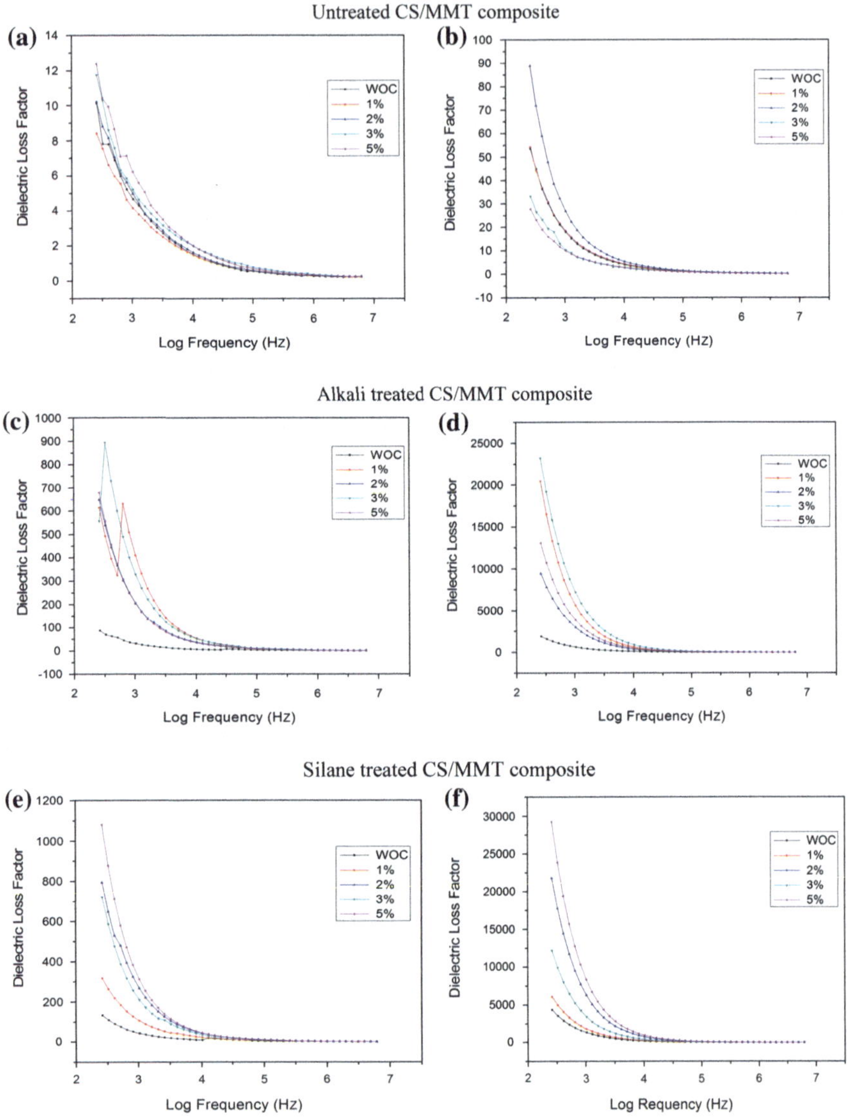

Fig. 10 Effect of temperatures on the variation of the dissipation factor for clay filled UTC, ATC and STC coconut sheath composite

As temperature increases, the dielectric constant increases due to the enhanced conductivity of the composite at higher temperatures, because when temperature increases the mobility of molecular dipoles increases. Similarly, the presence of cellulose content in the natural fibers is highly related to the resistivity of the fiber, more the cellulose content can lower the resistivity.

From the Fig. 10a–f, one can be interfered that for untreated samples dissipation factor decreases as the frequency increases and the values coincide as the frequency is further increased whatever the temperature. But in case of silane treated and alkali treated samples the values first increases, reaches a peak and then decrease at higher frequency irrespective of the temperature. Thus the values tend to become similar. This may be accounted due to ionic migration conduction at low frequencies. The peak is associated with molecular dipole rotation and occurs when molecular rotation can just keep with the alteration of electric field, a type of molecular resonance condition. Irrespective of the chemical treatment, the addition of clay content with the varying wt% found to be influence the dielectric strength and dielectric loss factor. However, there is no specific trend was observed with the increasing percentage of nanoclay.

3.4 Thermal Conductivity Behavior of CS/MMT Hybrid Composite

The behavior of thermal conductivity by the effect of nanoclay addition and chemical treatment of the CS/MMT hybrid composite is shown in Fig. 11. Form Fig. 11, it can be clearly seen that the thermal conductivity decrease with the addition of nanoclay with the varying wt%. The presence of nanoclay is distributed randomly throughout the polymer matrix and acts as a block for the heat source motion. Hence, there is a decrease in the thermal conductivity was noticed. But it was not the case for the addition of higher wt% of clay. A slight increase in thermal conductivity is observed, perhaps it could be due to the clay agglomeration, but still it is was found to be lower than pure polymer. The increasing trend of thermal conductivity was observed in the alkali treated CS composites with the effect of varying wt% of clay. A decrease in the thermal conductivity was noticed for the

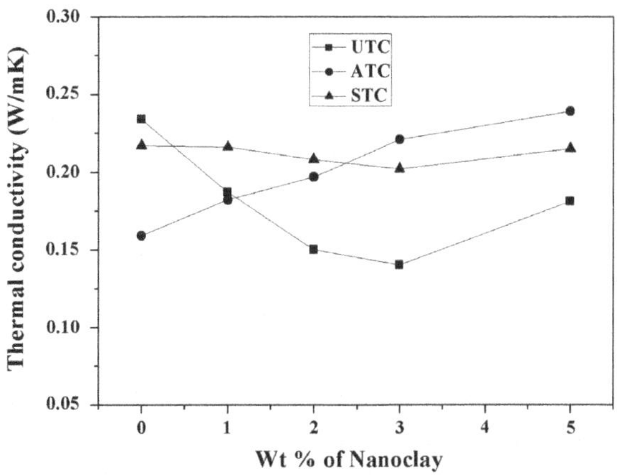

Fig. 11 Effect of nanoclay addition and chemical treatment of coconut sheath on thermal conductivity

silane treated CS due to the presence of nanoclay. In can happen due to the heat insulated behavior of clay which tries to decrease the thermal conductivity, except in the NaOH treated condition. The silane coating over the fiber surface can reduce the direct exposure of coconut sheath fiber surface. The reduction of surface area decreases the thermal conductivity, since the flow of heat is directly proportional to the available surface area. It is further confirmed by the following equation (Kreith and Bohn 2001):

$$Q = KA\,\Delta T/h \tag{5}$$

where Q is the heat flow (W), k is the thermal conductivity (W/mK), A is the surface area (m^2), ΔT is the temperature difference between two ends of composite and h is the thickness (m).

3.5 *Gas Permeability Properties of CS/MMT Clay Hybrid Composites*

In this section, the effect of parameters such as clay addition and chemically treatment CS fiber is studied on the CO_2 gas permeability in polyester matrix. Figure 12 shows the experimentation results of CS/polyester composites with the combination of clay and chemical treatment of CS fibers. From Fig. 12, the maximum drop in gas permeability was clearly observed in the alkali (NaOH) treated composites compared to the UTC/polyester composites. It can happen due to removal of unwanted contaminants at the surface after the alkali treatment of CS fiber. The flow of gas was restricted since the reduction of crack and voids due to the formation of strong interfacial adhesion. Moreover, the removal of contaminants leads to the way for better compatibility between the fiber and matrix which in turn increases the surface to area contact between fiber and matrix. Hence, the closely

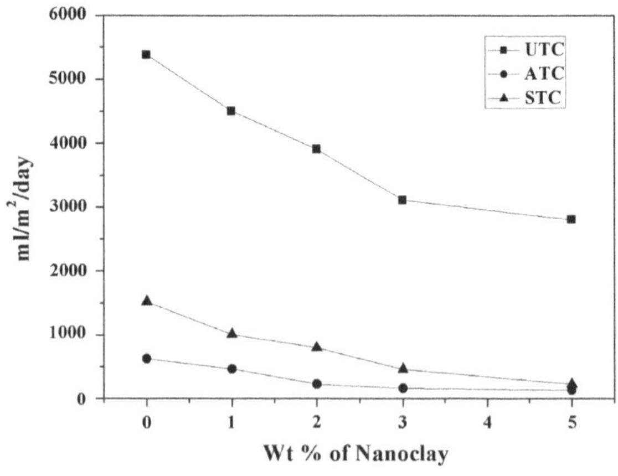

Fig. 12 Gas permeability results of the CS/PE composites with the effect of clay addition and chemical modifications

packed materials will arrest the options of gas travel in random direction. Besides, the interlocking of nano fillers inside the pores of fibers and dispersion with the polymer matrix can create the complicated path for the flow of gas. But the reduction of gas permeability is decreased by 72 % in the STC composites; this can be due to the presence of chemical coupling between the silane and the matrix. It may lead the way to a leakage of gas flow along the fiber region.

3.6 Thermo-Gravimetric Analysis (TGA)

The thermal behavior of the CS/polyester composite samples were investigated at the heating rate of 5 °C/min from ambient temperature to 800 °C under nitrogen atmosphere at a flow rate of 60 ml/min. The TGA curves of the untreated, alkali and silane treated coconut sheath without nanoclay and varying wt% percentage of nanoclay loaded polyester hybrid composite are represented in Fig. 13a–d. It is evident from the Fig. 13a that the initial degradation (10 wt%) was started at

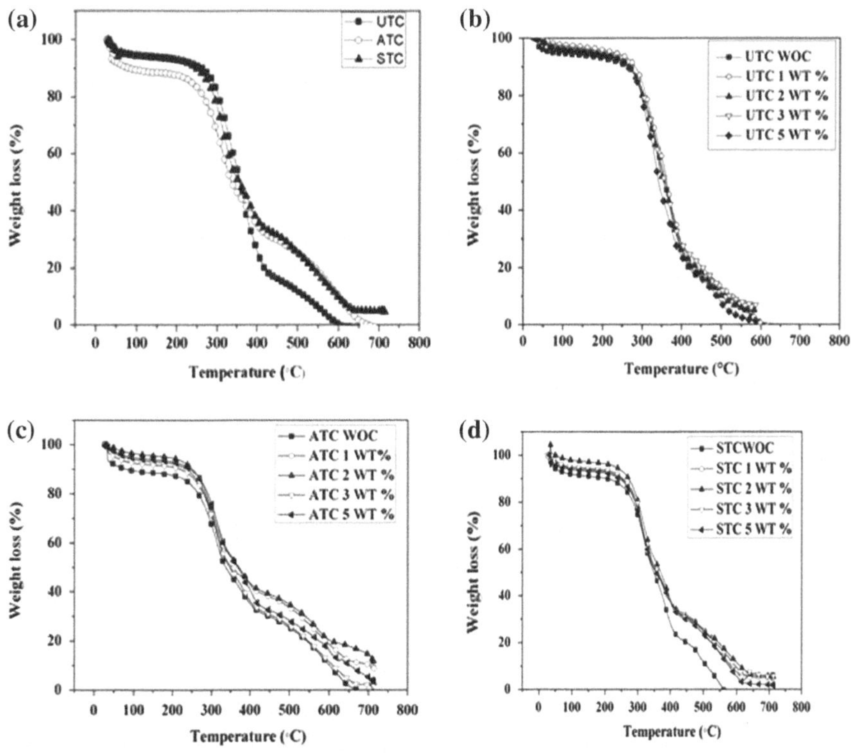

Fig. 13 The TGA curves for untreated, alkali and silane treated coconut sheath, **a** without nanoclay and **b–d** with varying wt% percentage of nanoclay

254.4 °C and 100 % degradation was noticed at 490 °C for untreated composites. At the same time, the initial degradation temperature of the alkali treated composite was decreased to 120 °C, whereas silane treated composites raises it to 238.7 °C. This may be due to the changes in the rate of dehydration of the fibers after the chemical treatment. However, as for as the major weight loss (100 %) is concerned, there was a substantial enhancement in the thermal stability noticed for both alkali and silane treated composites. The alkali treated composites raise the final degradation temperature to 608.2 °C giving a charred residue of 4.8 %, whereas the silane treated raises it to 713.4 °C giving a charred residue of 3.2 %. This could happen due to the better interfacial bonding and the breakage of decomposition products (cellulose, hemicelluloses, lignin and waxes) at the second stage lead to the formation of charred residue.

Figure 13a also reveals that the untreated coconut sheath covered by the waxes, greasy layer and the surface impurities restricts the rate of heat flow along the fiber orientation and causes the slow decomposition process which is observed from the lesser slope in TGA curve. From the curve, the first range of weight loss which begins from 30–110 °C is due to the loss of water content by the fiber. At about 200–350 °C, the degradation occurred which is associated with the thermal decomposition of the lignin and the hemicelluloses. The final stage of weight loss occurs at temperature of about 350–400 °C, which indicates the degradation of α-cellulose and other non-cellulosic compounds (Pracella et al. 2006). During alkalization, it was observed that the coconut sheath fiber started degrading at 200 °C and displayed a lower thermal stability and it might be due to the less dehydration by the appearance of the amorphous cellulose and the lignin content (Rajini et al. 2013).

The thermal degradation behavior of the clay unfilled CS/polyester composites (untreated, alkali and silane) and the CS/polyester nanocomposites with various percentage of clay loading is shown in Fig. 13b–d. The TGA of all the composites follow a two-stage decomposition behavior. Figure 13b shows the TGA for the clay filled UTC/polyester hybrid nanocomposites. Initial thermal decomposition (10 wt %) of the UTC/polyester composites without the addition of clay starts at 254.4 °C. On the addition of the clay in the UTC/polyester composites, the decomposition starts at higher temperature. For UTC with 1 wt% clay, the decomposition starts at 269 °C, for 2 wt% clay, the decomposition starts at 261 °C and for 3 wt% clay it starts at 276 °C. This suggests that the addition of clay increases the thermal degradation temperature of the UTC/polyester composites and it reaches a maximum value for 3 wt% clay. The addition of 5 wt% clay in the UTC composites does not show any improvement in decomposition and in fact, it decreases the initial degradation temperature to 236 °C. The formation of the agglomeration with 5 wt% of clay content could not penetrate into the surface of UTC due to the coverage of the inorganic contaminates (waxes and dusts). It could make a weak interface in the UTC composite and lead to the increase in weight loss. However, for the ATC and the STC samples, the enhanced thermal stability was observed even for the addition of 5 wt% clay (Fig.13c, d). A significant shift in the degradation temperature for the initial and the final decomposition was noticed for the alkali and silane treated composites with the varying wt% of nanoclay. The increase in the thermal properties of the hybrid nanocomposites is predominant at higher temperature (>400 °C) with the increase in the clay content but in an anomalous manner This encouraging thermal behavior of clay blended polyester composite can happen because of varying interaction between the distributed clay and polyester matrix to large surface area and attributed to the strong rigid interface. Among all the composites, the final residue in the CS/polyester clay filled nanocomposites is comparatively higher than the unfilled clay composites.

3.7 Fire Properties

The fire properties of untreated and treated CS fiber reinforced composites have been studied with the aid of cone calorimeter experiment. Figure 14 shows the HRR curves for the polyester (PE), untreated coconut sheath (UTC), silane treated coconut sheath (STC) and the alkali treated coconut sheath (ATC) reinforced composites in without clay (WOC) condition that were obtained from cone calorimeter, as a function of time. Only one peak in HRR curve of polyester matrix was observed in Fig. 14 and it can be due to the greater rate of combustion of matrix by quicker thermal decomposition. In all CS reinforced composites, the UTC is one which shows a higher HRR value compared to the other two composites. It should be due to the poor interfacial adhesion between the fiber and the matrix by the presence of the waxy layer and the inorganic contaminates over the fiber surface. This weak interface can separate the two phases (reinforcement and polymer) apart in case of fire.

However, as for as HRR is concerned, the mean (average) heat release rate value for ATC sample is 122.512 kW/m^2 whereas it rises into 152.59 kW/m^2 for UTC. The minimum value of mean HRR is found to be 85.67 kW/m^2 in the case of silane

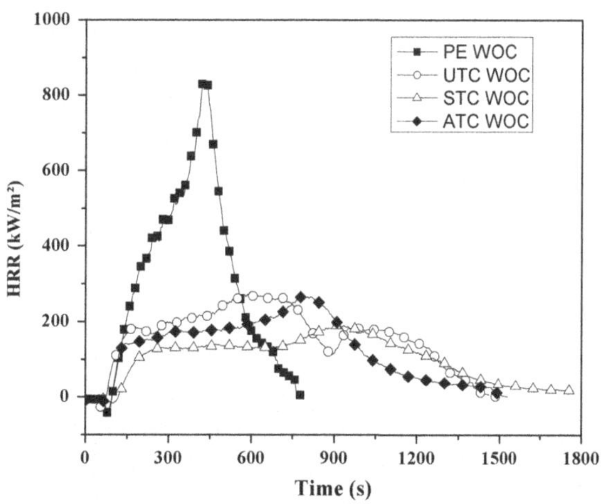

Fig. 14 HRR curves for CS fiber reinforced composite with unfilled clay

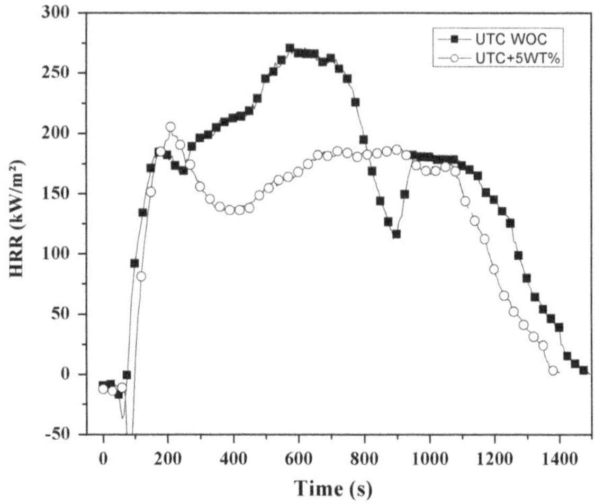

Fig. 15 HRR curve for UTC with the addition of 5 wt% of nanoclay

treated composite. In alkali treated condition, the removal of organic impurities can form the openings between the thin fibers which form spider-web like porous structure. The deposition of silane forms a chemical bond between the fiber surface and the matrix through a siloxane bridge. Thus, during the fire, the polymer layer present at the surface tends to burn first and the siloxane layer which act as a barrier improves the flame-out time of the composite. After the siloxane layer is burnt it results in the formation of char which acts as an insulating barrier at the interface of the composite thus resulting in the decreasing of HRR. Figure 15 shows the variation of HRR without and with the addition of 5 wt% of nanoclay to UTC reinforced composites. Nanoclay addition was found to have much influence on the conventional composite in all the conditions of the coconut sheath up to the time period of 600 s beyond that it was not valid for silane treated composite. It promotes the formation of the charred residue like a porous structure, which covers on the fiber surface and acts as a protection barrier against the heat conduction and the mass transport, leading to enhanced flame retardancy. It can also change the melt flow index of the polymer and the increase in the viscosity resulting in a reduced flame spread. Even the clay agglomeration in higher wt% of nanoclay in the matrix could create the torturous path of fire advance and it can increase the flame out time of composites.

3.8 CO and CO_2 Yield

The Fig. 16 shows the average values of (a) CO yield, and (b) CO_2 yield where PE (C1), PE + 5 wt% (C2), UTC WOC (C3), UTC + 5 wt% (C4),STC WOC (C5), STC + 5 wt% (C6), ATC WOC (C7) and ATC + 5 wt% (C8). From the above Fig. 16a, it is inferred that the pristine polyester emits more CO than the other fiber

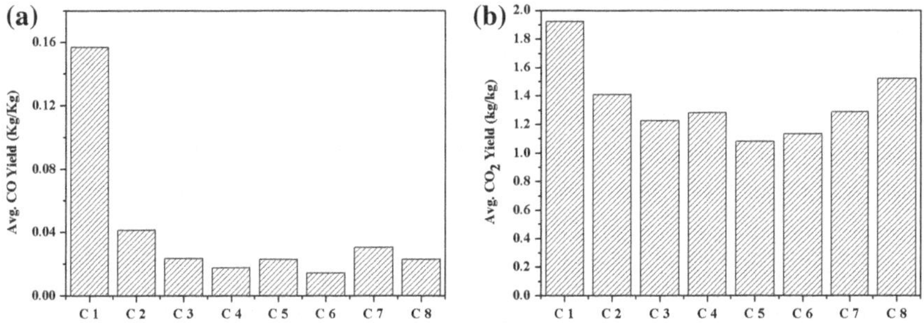

Fig. 16 Comparison of average value of **a** CO yield, **b** CO_2 yield for PE, UTC, STC and ATC reinforced composites with the addition of 5 wt% of nanoclay

reinforced composites. An increase in the heat release rate is attributed to increase in the yield of CO gas. When the clay is incorporated into the pristine polyester, the CO yield is reduced significantly. This shows clearly that the addition of clay to the polymer reduces the toxic CO emission. This is because the clay is naturally a fire retardant and thus the clay particles present in the polymer improves the flame out time of the polyester composite and thus results in the reduction of CO emission. However, with the addition of combustible CS fiber, the amount of CO gets reduced when compared to PE. The lowest value of CO yield was observed in silane treated CS composites which indicated that the complete combustion had taken place. The higher amount of CO_2 was evolved in the polyester matrix as shown in Fig. 16b and it may be due to the breakdown of the ester bonds within the resin. In the case of the clay unfilled CS fiber reinforced composites, the lower value of CO_2 was observed for UTC composites. However, unlike CO yield the controlling mechanism of CO_2 yield is complex and is dependent on factors other than just the heat release rate. With the addition of clay, the CO_2 yield follows the same trend of CO yield for the ATC and the STC fiber reinforced composites except the UTC.

4 Conclusions

This chapter is mainly focused on the fabrication of polyester/clay nanocomposites and coconut sheath fiber reinforced hybrid composites. The effects of chemical treatment (NaOH, silane) on the coconut sheath fiber are studied, with the changes in the physical and chemical properties. The addition of organoclay as a secondary reinforcement in the CS/PE composites can further enhance the properties of the hybrid composites to a certain absolute level. This chapter focuses on the research findings on the dielectric, transport, and the thermo-gravimetric analysis, properties of CS/MMT clay reinforced hybrid polyester composites. A detailed report, on the summary of the findings in the chapter is given below:

- The dielectric behavior was found to be influencing the CS polyester composites with the effect of chemical treatment and addition of nanoclay. The alkali and silane treated composites have shown significant variation dielectric behavior due to the changes in surface morphology and chemical composition. However, the increasing percentage of clay content does not show any specific trend in dielectric spectra.
- The temperature is found to be one of the most influencing parameter affects the dielectric behavior of natural fiber reinforced composites. The increasing temperature can increase the behavior of dielectric characteristics due to the hindering effect of ionization on fiber surface.
- The thermal conductivity of CS/polyester is as follow as per the ascending order below: UTC > STC > ATC. It implies that the increase in the moisture content due to the alkali treatment of CS fiber could be the reason for this lower thermal conductivity value.
- A significant reduction of gas permeability was noticed in the CS/MMT composites with the addition of nanoclay content. The randomly distributed clay in polymer matrix can reduce the micro voids and which can create the torturous path. So that the complication was found in the flow thus decreases a rate of flow considerably. Moreover, the improved adhesion between the fiber and matrix after the removal of contaminates should also be the reason for the decrease in gas permeability.
- According to the TGA results, the thermal stability of the alkali and silane treated condition gets reduced, because the presence of more water content after the removal of globular protrusion seeks to absorb more OH groups.
- But with the addition of nanoclay, the degradation shifted to a higher temperature for the fiber reinforced composites in all conditions, because of the resistance to the heat flow along the tortuous path produced by the dispersed nanoclay.
- The effect of chemical treatments on CS fiber exhibits good fire resistance properties. Similarly, encouraging results were also found in the UTC with the addition of 5 wt% clay content.

References

Alvarez, V.A., Ruscekaite, R.A., Vazquez, A.: Mechanical properties and water absorption behavior of composites made from a biodegradable matrix and alkaline-treated sisal fibers. J. Compos. Mater. **37**, 1575–1588 (2003)

Arunvisut, S., Phummanee, S., Somwangthanaroj, A.: Effect of clay on mechanical and gas barrier properties of blown film LDPE/clay nanocomposites. J. Appl. Polym. Sci. **106**, 2210–2217 (2007)

Chand, N., Jain, D.: Effect of sisal fibre orientation on electrical properties of sisal fibre reinforced epoxy composites. Compos. Part A-Appl. Sci. **36**, 594–602 (2005)

Chan, M.L., Lau, K.-T., Wong, T.-T., Ho, M.-P., Hui, D.: Mechanism of reinforcement in a nanoclay/polymer composite. Compos. Part B-Eng. **42**, 1708–1712 (2011)

Edeerozey, A.M.M, Akil, H.M., Azhar, A.D., Ariffin, M.I.Z.: Chemical modification of kenaf fibers. Mater. Lett. **61**, 2023–2025 (2007)

El-Khodary, A., Oraby, A.H., Abdelnaby, M.M.: Characterization, electrical and magnetic properties of PVA films filled with $FeCl_3$–$MnCl_2$ mixed fillers. J. Magn. Magn. Mater. **320**, 1739–1746 (2008)

Gould, G., Rao R.N.: The effect of alkali treatment on dielectric properties of Roystonea regia/epoxy composites. Int. J. Polym. Anal. Charact. **16**, 239–250 (2011)

Hepworth, D.G., Hobson, R.N., Bruce, D.M., Farrent, J.W.: The use of unretted hemp fibre in composite manufacture. Compos. Part A-Appl. Sci. **31,** 1279–1283 (2000)

John, M.J., Anandjiwala, R.D.: Recent developments in chemical modification and characterization of natural fiber-reinforced composites. Polym. Compos. **29**, 187–205 (2008)

John, M.J., Thomas, S.: Review biofibres and biocomposites. Carbohydr. Polym. **71**, 343–364 (2008)

Joseph, S., Thomas, S.: Electrical Properties of Banana Fiber-Reinforced Phenol Formaldehyde Composites. J. Appl. Polym. Sci. **109**, 256–263 (2008)

Kim, J.K., Hu, C., Woo, R.S.C., Sham, M.L.: Moisture barrier characteristics of organoclay–epoxy nanocomposites. Compos. Sci. Technol. **65**, 805–813 (2005)

Kreith, F., Bohn, M.S.: Principle of Heat Transfer, 6th edn, pp. 6–7. Cengage Learning Publishers, New Delhi (2001)

Kumar, V., Kushwaha, P.K., Kumar, R.: Impedance-spectroscopy analysis of oriented and mercerized bamboo fiber-reinforced epoxy composite. J. Mater. Sci. **46**, 3445–3451 (2011)

Lai, C.Y., Sapuan, S.M., Ahmad, M., yahya, N., Dahlan, K.Z.H.M.: Mechanical and electrical properties of coconut coir fiber-reinforced polypropylene composites. Polym. Plast. Technol. Eng. **44**, 619–632 (2005)

Lee, J.H., Jung, D., Hong, C.E., Rheeand, K.Y., Advani, S.G.: Properties of polyethylene-layered silicate nanocomposites prepared by melt intercalation with a PP–g–MA compatibilizer. Compos. Sci. Technol. **65**, 1996–2002 (2005)

Messersmith, P.B., Giannelis, E.P.: Synthesis and barrier properties of poly(E-caprolactone)-layered silicate nanocomposites. J. Polym. Sci. A. **33**, 1047 (1995)

Mukherjee, P.S., De, A.K., Battcharjee, S.: Electrical anisotropy of asbestos: a fibrous tremolite. J. Mater. Sci. **13**, 1824–1827 (1978)

Paul, A., Thomas, S.: Electrical properties of natural-fiber-reinforced low density polyethylene composites: a comparison with carbon black and glass-fiber-filled low density polyethylene composites. J. Appl. Polym. Sci. **63**, 247–266 (1997)

Paul, A., Joseph, K., Thomas, S.: Effect of surface treatments on the electrical properties of low-density polyethylene composites reinforced with short sisal fibers. Compos. Sci. Technol. **57**, 67–79 (1997)

Pracella, M., Chionna, D., Anguillesi, I., Kulinski, Z., Piorkowska, E.: Functionalization, compatibilization and properties of polypropylene composites with Hemp fibers. Compos. Sci Technol. **66**, 2218–2230 (2006)

Rajini, N., Winowlin Jappes, J.T., Rajakarunakaran, S., Jeyaraj, P.: Mechanical and free vibration properties of MMT clay dispersed with naturally woven coconut sheath composite. J. Reinf. Plast. Compos. **30**, 1364–1376 (2012)

Rajini, N., Winowlin Jappes, J.T., Rajakarunakaran, S., Bennet, C.: Effects of chemical modifications and MMT nanoclay addition on transport phenomena of naturally woven coconut sheath/polyester nanocomposites. Chinese J. Appl. Polym. Sci. **31**, 1074–1086 (2013)

Rajini, N., Jappes, J.T.W., Rajakarunakaran, S., Prabu, M., Bennet, C.: Influence of MMT nanoclay on impedance spectroscopy analysis of naturally woven coconut sheath/polyester hybrid composite. J. Polym. Sci., Appl (2013). doi:10.1002/APP.39149

Sengwa, R.J., Sankhla, S.: Characterization of ionic in ethylene glycol oligomers. Polym. Bull. **60**, 689–700 (2008)

Triki, A., Guicha, M., Hassen, M.B., Arous, M., Fakhfakh, Z.: Studies of dielectric relaxation in natural fibres reinforced unsaturated polyester. J. Mater. Sci. **46**, 3698–3707 (2011)

Zheng, W., Wong, S.-C.: Electrical conductivity and dielectric properties of PMMA/expanded graphite composites. Compos. Sci. Technol. **63**, 225–235

Yasmin, A., Abot, J.L., Daniel, I.: Processing of clay/epoxy nanocomposites by shear mixing. Scripta. Mater. **49**, 81–86 (2003)

A Review on Thermal Performance of Hybrid Natural Fiber/Nanoclay Polymer Composites

M. Norkhairunnisa and D.L. Majid

Abstract This chapter review on the thermal analysis on the performance of hybrid natural fiber/nanoclay in polymer composite. During the last decade, there has been an interest in using natural fiber as an alternative to synthetic fiber such as glass fiber. The major attraction in using natural fiber as part of reinforcing materials in nanocomposite are renewable materials, biodegradable, low material price, and weight saving. Some efforts have been made on improving the thermal properties of natural fiber reinforced polymer (NFRP) composite by inclusion of thermally stable layered silicate of nanoclay. The layered silicate structure of nanoclay can act as a barrier towards heat by forming layer of char residue after burn at high temperature. This characteristic delays the degradation process and further improves the thermal stability of the composite. In addition, the flammability of the hybrid natural fiber/nanoclays polymer composite was found to decrease with presence of exfoliated layered structure of nanoclay.

Keywords Nanoclay · Natural fiber · Thermal stability · Flammability · Crystallization temperature

M. Norkhairunnisa (✉)
Institute of Tropical Forestry and Forest Products (INTROP),
Universiti Putra Malaysia, 43400 Serdang,
Selangor Darul Ehsan, Malaysia
e-mail: norkhairunnisa@upm.edu.my

M. Norkhairunnisa · D.L. Majid
Department of Aerospace Engineering, Faculty of Engineering,
Universiti Putra Malaysia, 43400 Serdang,
Selangor Darul Ehsan, Malaysia

D.L. Majid
Aerospace Manufacturing Research Centre (AMRC),
Universiti Putra Malaysia, 43400 Selangor,
Serdang Darul Ehsan, Malaysia

M. Jawaid et al. (eds.), *Nanoclay Reinforced Polymer Composites*,
Engineering Materials, DOI 10.1007/978-981-10-0950-1_7

1 Introduction

Over the past few years, natural fibers have attracted much attention from researcher worldwide due to its unique properties. Common natural fibers used in composite include empty fruit bunch, kenaf, bagasse, bamboo, flax, jute, hemp, pineapple, animal fibers and fibers from wood. The attractive features of natural fibers are green in nature which is eco-friendly, biodegradable, reproducible as the origin comes from renewable resources, processing advantages at low cost. Natural fibers are alternative materials used to replace conventional fibers as reinforcing material in polymer composite especially in producing low weight composite. Processing of composite with natural fiber possess some benefit such as less machine wear and the fiber itself will bend rather than fracture during processing. Unlike synthetic fiber such as glass fiber, it breaks easily when intensively mixed with polymer resin. However, due to poor compatibility of natural fiber with hydrophobic type of polymer matrix, temperature difference of compounding process and flammability characteristics of natural fibers, limit the use of biocomposite in certain applications especially for aerospace and automotive areas.

Clay materials have been used progressively in heat resistant and flame retardant applications. Layered silicate or known as nanoclays are the most common nano-fillers which have been used in polymer composite to improve the thermal characteristic of the polymer matrix (Duquesne et al. 2003; Haurie et al. 2007; Olewnik et al. 2010). Common type of nanoclays are montmorillonite, bentonite, mica, saponite and hectorite. Size reduction of macro/-microclay to nanoclay is expected to improve the thermal barrier property of the composite due to the increase in surface area. The improvement of the properties for composite embedded with nanoclays are well beyond than what can generally be achieved from pure polymer or conventional composites. Nanoclays have gained acceptance as possible reinforcing materials due to its unique properties such resistance to fire, excellent heat distortion temperature, and have good permeability towards gas and water. Dispersion of low amount of nanoclay within a polymer matrix tend to enhance the properties of the pure polymer or polymer reinforced with fiber composite (Rajasekar et al. 2009; Alhuthali et al. 2012). Polymer filled with nanoclay also have better toughening properties due to its plate structure, high surface to volume and strength to weight ratio (Liu et al. 2005; Liqun and James 2005; Subramaniyan and Sun 2007).

Recently, development of hybrid nanocomposite consist of polymer, natural fiber and nanoclay, often exhibit remarkable improvement in mechanical, thermal and various other properties compared to conventional composite (Reddy et al. 2010; Rajini et al. 2012; Zahedi et al. 2013; Eng et al. 2014). Inclusion of nanoclay in natural fiber/polymer composite (NFRP) system to overcome the drawback of the natural fiber due to it hydrophilicity in nature, low weathering resistance, and delamination of the composite. Thus, addition of nanoclay tend to improve the fiber/polymer composite strength, improve the thermal stability of the composite, function as a bridge between fiber and resin, and reduce gas and water permeability in the composite (Kumar et al. 2012). Moreover, the layered silicate structure of nanoclay can delay the gasification process through formation of carbonaceous-silicate char by providing a protective layer on the nanocomposite surface. Thus, thermal stability of the fiber nanocomposite can be improve and reduced the flame spread.

Combination of low cost natural fiber with nanoclay as an excellent thermal insulator, can be an ideal alternatives to the available synthetic fiber which is commonly used in composite for insulation applications. In this chapter, a review on thermal analysis of hybrid natural fiber/nanoclay reinforced polymer composite based on recent research findings. Various thermal analysis such as thermal stability, flammability, glass transition temperature, etc. were carried out to investigate the effect of nanoclay and natural fiber the thermal properties of the polymer composite.

2 Studies on the Thermal Stability of Hybrid Nanocomposite

Determination of thermal stability and weight loss of polymer composite as a function of temperature under controlled atmosphere are generally performed by using thermogravimetric analysis (TGA) instruments. In general, it have been reported that thermal stability of natural fiber reinforced polymer composite are thermally less stable than virgin polymer (Lee and Wang 2006; Lei et al. 2007a; Arrakhiz et al. 2013). Plant fiber primarily consist of cellulose, hemicellulose and lignin. There are four principle steps in degradation of natural fibers which are (i) removal of water uptake by fiber at region between 50–150 °C; (ii) depolymerisation of hemicellulose and cleavage of glycosidic linkage at region between 250–370 °C; (iii) degradation of lignin ranging from 200–500 °C, and (iv) decomposition of cellulose at region more than 430 °C (Dorez et al. 2013; Manfredi et al. 2006). Figure 1a shows the example on the degradation behaviour of natural fibres.

Variation in initial decomposition temperature for different type of natural fibers was due to content of the chemical composition of the fibers (Manfredi et al. 2006). Table 1 shows different chemical composition of some natural fibers which tend to affect the thermal stability of the fibers. Presence of high lignin content tend to increase the thermal decomposition temperature which is due to its aromatics ring structure consist of various branches (Liu et al. 2004).

Figure 1b, c shows the thermogravimetric curves of natural fiber reinforced polymer and hybrid natural fiber/nanoclay reinforced polymer composite, respectively. Incompatible surface nature between natural fiber and polymer matrix, tend to reduce the thermal stability of the natural fiber/polymer composite as shown in Fig. 1b. Work done by Eng et al. (2014), shows improvement of thermal stability of the OPMF/blend polymer composite with inclusion of nanoclay (Fig. 1c). Therefore, from Fig. 1, it can be concluded that the degradation trends of natural fiber reinforced polymer nanocomposite change with presence of different phases which are natural fiber, nanoclays and the polymer matrix itself. Inclusion of clay into polymer system promote flame retardancy of the pure polymer. Formation of carbonaceous residue over the molten polymer at high temperature acts as

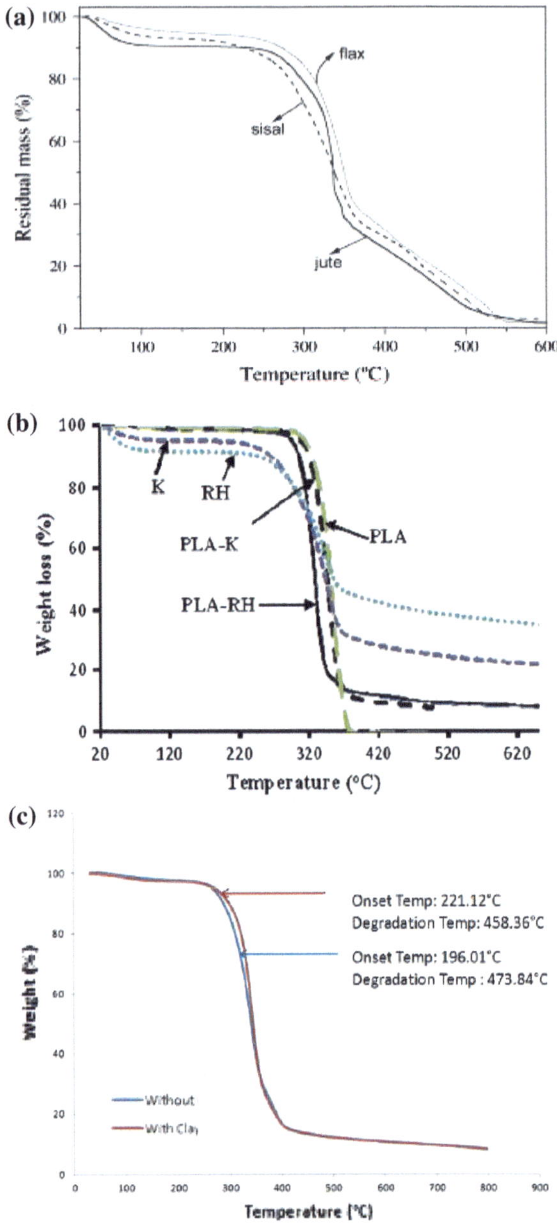

Fig. 1 Trend on the thermogravimetric curves of **a** natural fibers (Manfredi et al. 2006); **b** kenaf, rice husk, PLA/kenaf and PLA/rice husk (Yussuf et al. 2010); and **c** hybrid oil palm mesocarp fiber (OPMF)/nanoclay in polymer blend of PLA/PCL (Eng et al. 2014)

protective barrier to hinder transfer of volatile products and heat transfer between flame and polymer (Zanetti et al. 2000; Sánchez-Jiménez et al. 2012). The carbonaceous residue consist of high concentration of silicate layers from nanoclay which function as an excellent insulator (Manias et al. 2001). On top of that, increase in thermal decomposition temperature was attributed to the hindered

Table 1 Chemical composition of some natural fibers (% of total)

Type of fiber	Extractive	Holocellulose	α-cellulose	Lignin	Ash	Reference
Oil palm frond	4.5	83.5	49.8	20.5	2.4	Abdul Khalil et al. (2007)
Coconut	6.4	56.3	44.2	32.8	2.2	
Pineapple leaf	5.5	80.5	73.4	10.5	2.0	
Banana stem	0.2–8.5	60–80	30–60	21–37	<1	
Kenaf core	4.7	87.2	49	19.2	1.9	Abdul Khalil et al. (2010)
Kenaf bast	5.5	86.8	55.0	14.7	5.4	

diffusion of volatile decomposition products in the composite, together with physic-chemical adsorption of volatile degradation products on the silicate surface of nanoclay (Iman and Maji 2012). Presence of acidic sites on the clay layered structure can increase the thermal-oxidative degradation too (Qin et al. 2005). Moreover, tortuous path created by silicates layer, delay the volatilization of carbon-carbon scission to produce degradation products (Qin et al. 2004; Deka and Maji 2011).

Reinforcing of nanoclay in fiber/polymer composite tend to increase the decomposition temperature. Work done by Shuvo et al. (2015), observed increment of maximum thermal decomposition temperature of jute/polyester composite with inclusion of 1 wt% of nanoclay for about 3 %. The authors suggested that heat distribution between fibres and nanoclay are uniform and more energy required to thermally degrade the composite structure. Similar finding by Kumar and Kumar (2012), in which they observed increased in thermal stability after adding 1 wt% of nanoclay in bamboo/epoxy composite. Good dispersion of nanoclay combined with good interfacial adhesion of nanoclay in polymer matrix restrict the thermal motion on epoxy molecules which subsequently improve the thermal stability of the composite. Thus, less surface area of nanoclay being contact with polymer and further reduce the interfacial adhesion between agglomerate nanoclay and polymer resin. However, adding nanoclay content up to 10 wt%, does not improve the thermal stability of the bamboo/epoxy composite. High content of nanoclay may lead to the formation of agglomerates, in which high particle to particle interactions than particle to polymer interactions in the nanocomposite (Zainuddin et al. 2010). Compare to hemp/PLA nanocomposite, 10 wt% of nanoclay is needed to increase the thermal decomposition temperature. Low content of nanoclay does not contribute significant thermal performance due to less amount of clay platelet function as barrier towards heat (Hapuarachchi and Peijs 2010).

Figure 2 shows the micrographs on the composite upon TGA test conducted by Kovacevic et al. (2015). Composite contain nanoclay (MMT) showed the firm and intact structure of the composite surface (Fig. 2b), while for composite without nanoclay was found to be totally damaged (Fig. 2a). Authors suggested that movement of nanofiller from the inner part of the composite to the surface, form char residue thus prevent degradation on the composite structure. It was observed that

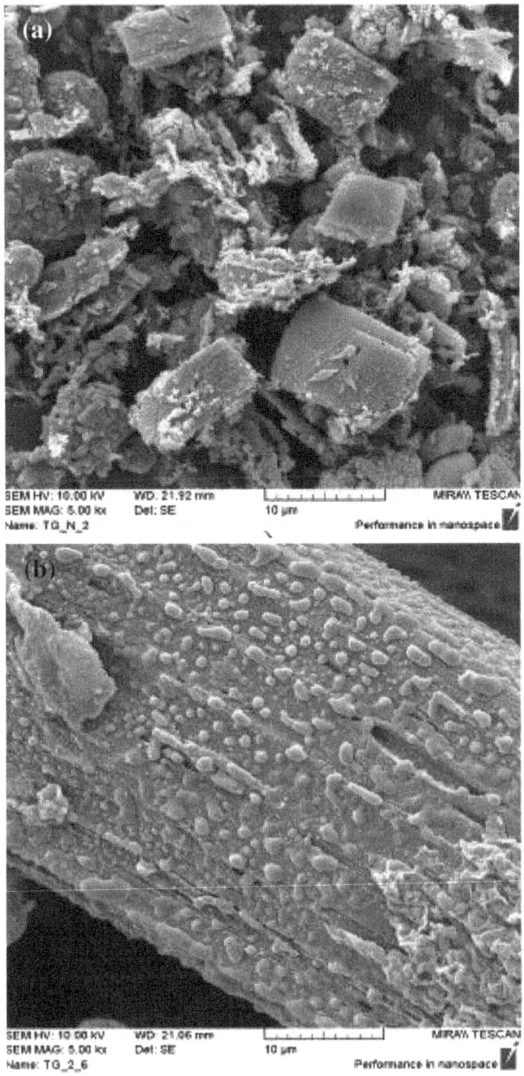

Fig. 2 Scanning electron micrograph of chars after TGA for: **a** *Spartium junceum* L. fiber/PLA composite and, **b** *Spartium junceum* L. fiber/PLA/MMT nanocomposite (adapted from Kovacevic et al. 2015)

char residue increased for about 12 % as the clay content increase to 5 wt% for 0.75 MDF/UF nanocomposite (Zahedsheijani et al. 2012). It was said that increasing amount of nanoclay as flame retardant materials will subsequently increase the char residue of composite at relatively low temperatures and thus, improved the thermal insulation property.

One of the issue addressing the stability and flammability of clay composite is the behaviour of the layered silicate structure in polymer. Morgan (2006) reviewed that both intercalate and exfoliate of nanoclay structures shown reductions in heat release. He highlighted that the most important is the dispersion of nanoclays in the polymer matrix. Poor dispersibility or agglomerations of nanoclays will reduced the heat transfer efficiency. If poor interfacial adhesion between clay and polymer matrix, clay will no longer function as insulator and tend to increase the flammability property of the composite (Biswal et al. 2012). The dispersibility of the nanoclays can be enhance by modified the nanoclay surface. The initial onset degradation temperature for composite filled with modified nanoclay shows low decomposition temperature compared to unfilled polymer at earlier stage of degradation. The early stage of degradation temperature was attributed to the loss of water and solvent from the fiber, and due to the hydrophobicity nature of the plant fiber (da Silva Santos et al. 2010). In addition, this scenario occurred due to the Hofmann elimination reaction, in which clay function as catalyst to promote the degradation of the matrix (Madaleno et al. 2010; Hapuarachchi and Peijs 2010). However, when reached to the second stage of decomposition, composite filled with nanoclay shows higher thermal decomposition temperature than unfilled polymer. Similar case can be observed from polymer nanocomposite reinforced with natural fibers (Alamri et al. 2012). High temperature of thermal decomposition was due to the properties of the nanoclay which hinder the diffusion of volatile products from the nanocomposite.

Use of modified or unmodified nanoclay give an effect to the thermal properties of the composite as well. Unmodified nanoclay have high tendency to clump together and form large agglomerations of layered structure. Alongside that, no intercalation or exfoliation of the clay structure appear in the polymer composite and the composite can be classified as microcomposite. Modified nanoclay provide good contact or adhesion with polymer resin. In addition, surface modified nanoclay showed high tendency to change the organized layered structure into exfoliate during mixing with polymer resin.

Effects of different type of nanoclay on the thermal properties of natural fiber reinforced polymer nanocomposite was studied by Delhom et al. (2010). They prepared cotton fiber nanocomposite with different type of nanoclay which are montmorillonite and cloisite. Generally, can be observed that adding nanoclay increase the degradation temperature of the cotton fiber nanocomposite. However, different type of nanoclay produce different degradation temperature peak. The cotton fiber nanocomposite filled with montmorillonite demonstrated the high thermal degradation temperature peak value. This is because different intercalant chemistry can effects the degradation property of the polymer (Lei et al. 2006; Rahimi-Razin et al. 2012; Molinaro et al. 2013).

Saw (2015) prepared nanocomposite made from modified nanoclay (organo modified montmorillonite, OMMT) with coir fiber in blend of epoxy novolac resin (ENR) and diglicidyl ether of bisphenol A (DGEBA) based epoxy resin. TGA analysis shows that incorporation of 30 wt% of coir fiber and 2 wt% OMMT in blend epoxy (ENR:DGEBA) have better thermal stability than in nanocomposite with single epoxy component. The polymer chains of blend epoxy is interact well with the OMMT and coir fiber surfaces through hydrogen bonds. In addition, the catalytic effect of OMMT enhance the curing rate of the epoxy blends. Therefore, the crosslink density of the epoxy blends increase and high temperature need to break the polymer chain. On top of that, the char residue was found high at 800 °C which is 28.19 % which support good adhesion and ordered structure of the nanocomposite. However, increase the ENR loading to 65 %, a decrease in thermal degradation temperature of the nanocomposite was observed. This attributes to the high intensity of polymer chains, thus form less ordered structure and lower the dispersibility of the coir fiber and OMMT in the epoxy blend.

Mohan and Kanny (2015) fabricate short banana fiber/nanoclay reinforced epoxy composites by resin casting method. Banana fibers and nanoclay were initially treated in alkaline solution. High shear force was applied in the solution mixture in order to ease the infusion of nanoclay particles into the fibers. From tensile analysis, it shows improvement in modulus and strength of treated fiber/nanoclay in epoxy matrix. Strengthening mechanism of nanoclay in fiber greatly retard the deformation process of the fiber/epoxy composite. In addition, fiber infused with nanoclay can effectively bear load applied on it due to good interfacial adhesion with epoxy matrix. As shown in Fig. 3a, untreated fiber easily being pull out from epoxy matrix. While for treated fiber, the pull out process is difficult due to high resistance from the nanoclay inside of the fiber (Fig. 3b). Authors add that the thermal stability of treated fiber/nanoclay/epoxy composite was higher than epoxy filled with untreated fiber. Interestingly, the mass loss of treated fiber/nanoclay/epoxy composite was low due to reinforcing ability of nanoclay and it delays the degradation process by form a protective layer on the composite surface. A summary of the decomposition temperature of some natural fiber reinforced polymer is given in Table 2.

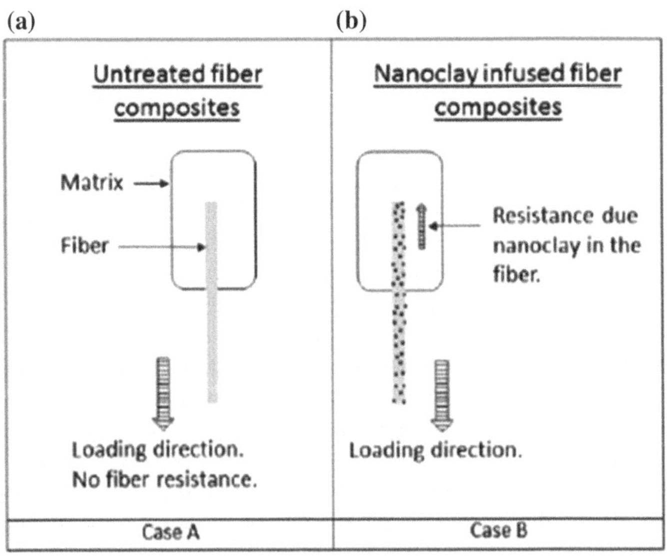

Fig. 3 Schematic illustration on the fiber matrix interface: **a** untreated fiber composite, and **b** nanoclay infused fiber composite (adapted from Mohan and Kanny 2015)

Table 2 Decomposition temperature of some natural fiber reinforced polymer nanocomposites

Sample	Initial decomposition temperature (°C)	Maximum degradation temperature (°C)	Final decomposition temperature	References
Polymer blend/5phr PE–co–GMA/40 phr wood	255	451	At 600 °C = 7.2 %	Deka and Maji (2011)
Polymer blend/5 phr PE–co–GMA/40 phr wood/3 phr nanoclay/1 phr ZnO	277	503	At 600 °C = 10.8 %	
Polymer blend/5 phr PE–co–GMA/40 phr wood/3 phr nanoclay/3 phr ZnO	289	517	At 600 °C = 14.5 %	
Polymer blend/5 phr PE–co–GMA/40 phr wood/3 phr nanoclay/3 phr ZnO	281	510	At 600 °C = 12.7 %	
Wood/polymetylmethacrylate (PMMA)/glycidyl methacrylate (GMA)	215	411	At 600 °C = 10.6	Hazarika et al. (2012)
Wood/PMMA/GMA/0.5 % nanoclay	229	418	At 600 °C = 13.5	
Wood/PMMA/GMA/1.0 % nanoclay	245	425	At 600 °C = 16.4	
Wood/PMMA/GMA/1.5 % nanoclay	272	439	At 600 °C = 23.1	
Polymer blends (PB)	251	407	At 600 °C = 6.1	Deka et al. (2011)
40 wt% Nals (W40)/ PB/5 wt% PE–co–GMA (G5)	257	455	At 600 °C = 7.2	
W40/ PB/G5/1 wt% nanoclay (N1)	260	466	At 600 °C = 8.4	
W40/ PB/G5/N3	270	496	At 600 °C = 11.7	
W40/ PB/G5/N5	263	480	At 600 °C = 10.2	
Spartium junceum L. fiber/PLA/MMT	328	364.8	At 800 °C = 4.9	Kovacevic et al. (2015)
Jute/polyester	–	385.30	At 700 °C = 1.591 %	Shuvo et al. (2015)
Jute/polyester/1 wt% graded nanoclay (GNC)	–	396.97	At 700 °C = 1.746 %	
Arboform®/lignin	–	343	–	Guigo et al. (2009)
Arboform®/lignin/2 % sepiolite	–	344	–	
Arboform®/lignin/5 % sepiolite	–	348	–	
Arboform®/lignin/2 % OMMT	–	341	–	
Arboform®/lignin/5 % OMMT	–	353	–	

(continued)

Table 2 (continued)

Sample	Initial decomposition temperature (°C)	Maximum degradation temperature (°C)	Final decomposition temperature	References
PP	–	398	–	Biswal et al. (2011b)
PP/nanoclay (OPP)	–	452	–	
Mercerized banana fiber (MBF)/OPP	–	475	–	
PLA	–	364	At 800 °C = 1 %	Hapuarachchi and Peijs (2010)
Hemp/PLA/1 wt% sepiolite	–	359	At 800 °C = 2 %	
Hemp/PLA/10 wt% sepiolite	–	364	At 800 °C = 7 %	
0.65 medium density fiberboard (MDF)/urea-formaldehyde (UF)/2.5 % NaMMT	–	332	At 500 °C = 14.77	Zahedsheijani et al. (2012)
0.65 MDF/UF/5 % NaMMT	–	339	At 500 °C = 15	
0.75 MDF/UF/2.5 % NaMMT	–	339	At 500 °C = 12.98	
0.75 MDF/UF/5 % NaMMT	–	336	At 500 °C = 14.8	
Cotton	–	282	–	Delhom et al. (2010)
Cotton/montmorillonite	–	327	–	
Cotton/cloisite	–	311	–	
Ramie	–	290	–	
Ramie/montmorillonite	–	335	–	
Ramie/cloisite	–	313	–	
Kenaf	–	283	–	
Kenaf/montmorillonite	–	321	–	
Kenaf/cloisite	–	305	–	
Coir/DGEBA/nanoclay	223.4	371.9	At 800 °C = 0	Saw (2015)
Coir/ENR/nanoclay	235.9	393.8	At 800 °C = 0	
Coir/ENR:DGEBA(50:50)/nanoclay	294.7	485.7	At 800 °C = 28.19	
Coir/ENR:DGEBA(65:35)/nanoclay	274.4	461.6	At 800 °C = 22.78	

3 Studies on the Flammability of Hybrid Nanocomposite

Flammability is one of the crucial parameters that limit the properties of the composite for certain area of applications. It described the reactions of particles in producing fire and can be classified into four different phases such as ignitability, flame spread, heat release and the resulting products from the fire process (Walters and Lyon 2008).

Thermal conductivity of natural fiber was 0.29–0.32 W/mK, which is very low and make it suitable for thermal barrier application. However, due to lack of oxidation resistance from the aromatic rings of lignin, the thermal stability of the fiber would be low. In addition, poor fire resistance of natural fiber, limit its application especially in aerospace and automotive industries.

Therefore, in order to improve flammability of the fiber reinforced polymer composite (NFRP), a flame retardant additives such as nanoclay is suitable to be embedded into the composite system. Another point of view is, during combustion process, nanoclay particle inhibit formation of vigorous bubbling by develop a large lateral surface crack layer (Pandey et al. 2011). Recently, there are few number of researchers looking into the flammability performance of nanoclay in NFRP (Zhao et al. 2006). The flammability analysis can be carried out using cone calorimeter according to ASTM E1354-92 and ISO/DIS 13927 standards (Zanetti et al. 2000). From this instrument, a number of parameters such as char residue (CR), total time to ignite (TTI), time to peak heat release rate (TPHRR), smoke production rate (SPR) and total smoke production (TSP) data can be evaluated. Another way to measure the flammability of the nanocomposite is by limiting oxygen index (LOI) test. This test can be done in accordance to ASTM D-2863 method. LOI test is conduct to measure the tendency of material to sustain flame by applying oxygen and nitrogen gaseous mixture over the burning specimen.

Chigwada et al. (2005) carried out research on the effect of fire behavior on ABS, PE and epoxy with incorporation of nanoclays through cone calorimeter. This research works proves that addition of nanoclays into polymer matrix decreases maximum heat release rate and increases the ignition time. Similar findings by Zhao et al. (2006), based on a study on flammability of wood flour/polyvinyl chloride (PVC) nanocomposite with varying loading of organo modified montmorillonite (OMMT). From cone calorimeter analysis, it was observed that at OMMR loading of 4.5 %, the TPHRR was approximately 63 % higher than composite without OMMT. The TTI of the fiber nanocomposite also improve from 25 s for wood flour/PVC composite to 78 s for wood flour/PVC nanocomposite filled with 4.5 % OMMT. Improvement in TPHRR and TTI indicate that presence of OMMT delayed the formation of ignition and heat release process in the fiber nanocomposite. There is a small peak (peak A) at 25 s shows in Fig. 4 indicates decomposition of some PVC and wood flour molecules. While the following peaks correspond to the latter ignition of the fiber nanocomposite samples. According to the study, homogeneously distribution of OMMT in the PVC matrix tend to form an inorganic shell-like structure until the ignition start to occur by heating.

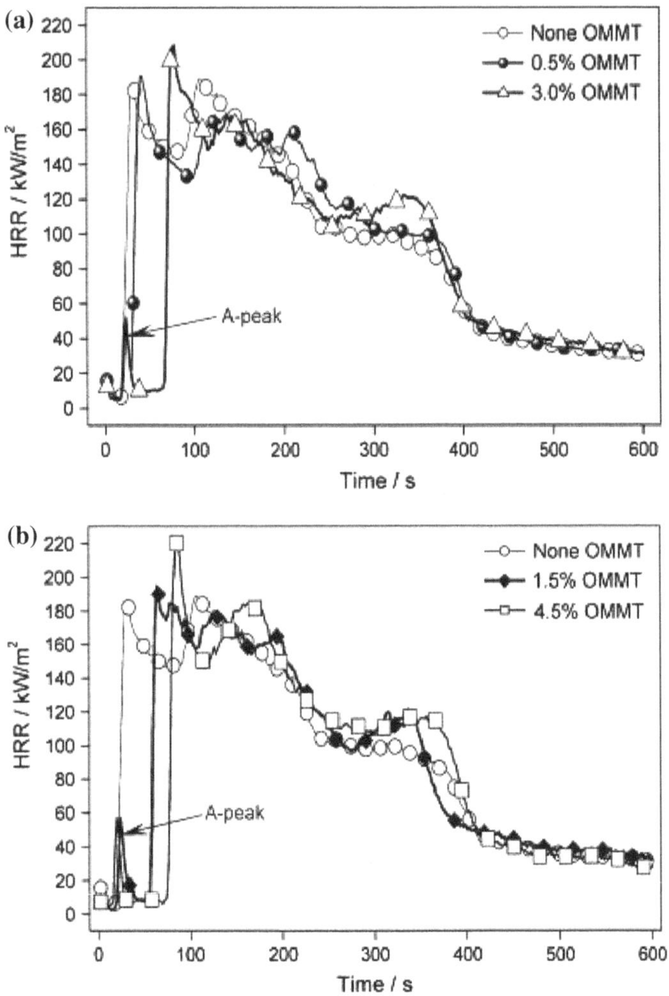

Fig. 4 Heat release rate (HRR) versus time relations for different loading of OMMT in wood flour/PVC nanocomposites (adapted from Zhao et al. 2006)

Moreover, high degree of nanoclay exfoliation have high oxidation protection against heat (Camino et al. 2005). On top of that, good interfacial adhesion between PVC, wood flour and OMMT delay the time for fire to start.

The burning behaviour of the fiber nanocomposite is largely determined by the synergistic effect on the adhesion between matrix and surface of the fiber or nanoclay according to Guo et al. (2007), in a study on the flammability of wood fiber/HDPE/nanoclay composites with and without coupling agent. Different concentration of maleic anhydride (MAH) was added into the composite to improve the interfacial adhesion between wood flour and HDPE. As reported by Zhang et al. (2004), blending of coupling agent with polymer system tend to improve the interfacial adhesion between polymer resin and fillers. As reported by Guo et al. (2007), they added 0–50 % of MAH coupling agent in wood fiber/HDPE nanocomposites. As shown in Fig. 5a, it is observed under TEM view that for wood fiber/nanocomposite without MAH, the layered silicate structure did not expand and distribute well in the HDPE resin. With addition of 5 % of MAH, the size of the nanoclay agglomeration decrease and orientation of the layered silicate change to intercalate structure (Fig. 5b). Thus, by increasing the MAH content up to 50 %, the layered silicate structure of nanoclay have been fully exfoliated in the HDPE matrix (Fig. 5c). Therefore, owing to the good dispersion of exfoliated nanoclay in fiber nanocomposite with 50 % of MAH, the flammability analysis shows low burning rate of the nanocomposite which is 18 % less than fiber nanocomposite without nanoclay (Fig. 6). Thus, sufficient amount of coupling agent to be blend with polymer resin is important to facilitate the diffusion of polymer chains into the silicate layers of nanoclay (Mittal 2013). Lee et al. (2010) reported that in order to

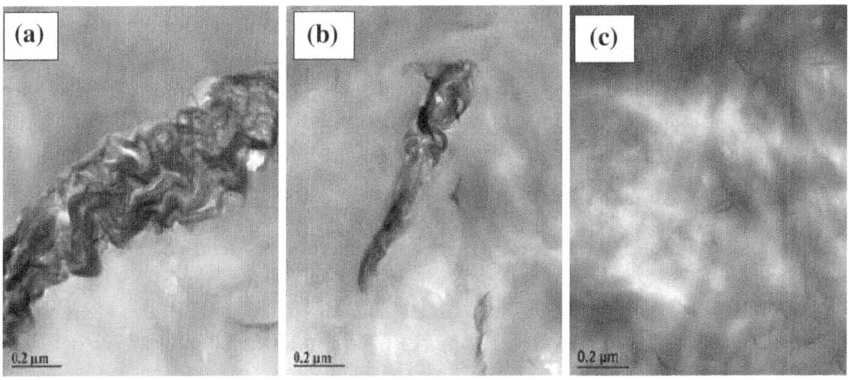

Fig. 5 Transmission electron micrographs (TEM) for different nanoclay orientation in fiber/HDPE nanocomposites; **a** Conventional, **b** intercalated, and **c** exfoliated (adapted from Guo et al. 2007)

Fig. 6 Burning rate versus clay content for fiber nanocomposites with different type of structural orientation layered of nanoclays (adapted from Guo et al. 2007)

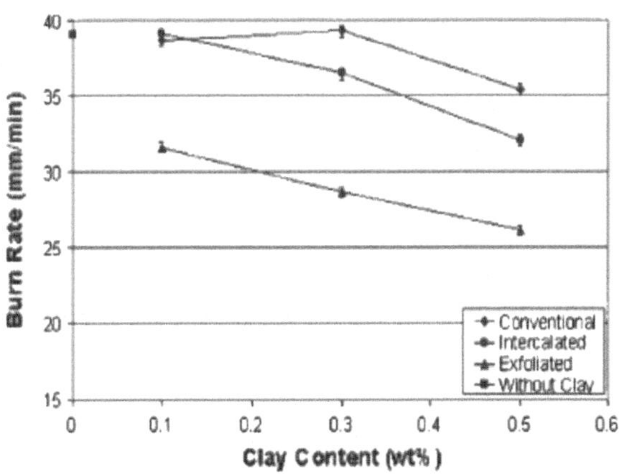

improve the flame retardant properties of the fiber/polymer nanocomposite with low loading of nanoclay, it is crucial to have a higher degree of exfoliation of nanoclay.

Paluvai et al. (2015), investigated on the LOI of unsaturated polyester (UP) toughened epoxy nanocomposites filled with alkaline silane treated sisal fiber (ASTF) and cloisite 30B type nanoclays (C30B). The high pressure mixing (HPM) method was used to disperse the C30B into epoxy/UP blend matrix. Good exfoliation of silicate layer nanoclay structure have been obtained due to the interactions between carboxylic ester group from polymer and OH group from the nanoclay. The LOI analysis increase to 27 % with addition of C30B into the ASTF/UP/epoxy composite system. Reason to the improvement in LOI analysis is due to the strong covalent bonding between Si–O–Si linkages of the ASTF surface with alkyl ammonium ion on C30B.

Biswal et al. (2012) prepared mercerized banana fiber (MBF) reinforced polypropylene nanocomposites (OPP). The banana fiber was soaked into 5 % NaOH solution to remove the presence of impurities before being compound with organically modified nanoclay in PP/MA-g-PP matrix. Different loading of banana fiber in the composite system was prepared to investigate effects of fiber loading. By applying 50 kW/m^2 radiant heat flux from cone calorimeter analysis, it was observed a decreasing of TTI value from 30 s for blank PP to 36 s for PP filled with nanoclay (OPP). Then, the trend shows reduction of TTI value to 26 for OPP filled with 30 wt% of MBF. The inconsistent TTI value was due to the addition of MBF into the nanocomposite system. Authors mentioned during pyrolysis process, the levoglucosan from lignin and hemicellulose was found increase. Thus, lead to the formation of flammable fluid tar which subsequently reduced the TTI value as shown in Table 3. On the other hand, the HRR value was found to be increased with addition of MBF. With addition of 10 wt% of MBF in OPP, the HRR value increase approximately 26 % over the OPP. Poor interfacial interaction between

Table 3 Cone calorimeter information of some natural fiber reinforced polymer nanocomposite

Sample	Total time to ignite (TTI) (s)	Peak heat release rate (HRR), kW/m^2	References
Epoxy/UP	61	829.2	Paluvai et al. (2015)
Epoxy/UP/1 wt% C30B	66	647.2	
Epoxy/UP/ASTF	65	610.9	
Epoxy/UP/1 wt% C30B/ASTF	64	583.3	
PP	30	2498.2	Biswal et al. (2012)
OPP	36	1164.5	
OPP/10 wt% MBF	28	1581.9	
OPP/20 wt% MBF	27	1746.5	
OPP/30 wt% MBF	26	1612.7	

MBF and OPP, explained the increment in HRR. It was suggested by author that in this case, probably PP, MBF and nanoclay does not compliment with each other. PP matrix burn vigorously due to poor adhesion with fillers, lignocellulosic characteristic of banana fiber which is highly flammable and nanoclay does not act as insulator to protect the composite against heat. However, the HRR value for fiber reinforced nanocomposite for OPP/MBF was still lower than blank PP. This indicate that presence of banana fiber and nanoclay gives variation in reaction to fire parameters for the fiber reinforced nanocomposite.

In summary, the addition of nanoclay can improve the flammability of natural fiber reinforced polymer composites through various flame retardant mechanisms which are (i) formation of barrier against heat and volatiles due to the migration of nanoclay towards the surface, formation of char between the surface and the flame, and inhibit formation of bubbles during combustion process.

4 Studies on the Melting and Crystallization Behaviour of Hybrid Nanocomposite

Dynamic Scanning Calorimetry (DSC) is one of thermoanalytical technique to measure the isothermal crystallization behaviour of the composite. Some reactions may occur without any weight loss, therefore in this instance differential scanning calorimetry (DSC) can detect these reactions. The crystallization and melting curves of the nanocomposite can be obtained from cooling and heating process of non-isothermal in DSC measurements.

Two phases of crystallization process of semicrystalline polymer which are the macroscopic expansion of the degree of crystallinity and, crystallization of lateral and interfibrillar chains (Bouafif et al. 2009). Degree of crystallinity depends on several processing parameters such as crystallization temperature, nucleation density, cooling rate type of fiber and annealing time (Wang and Liu 1999). Contact between solid surface and semicrystalline polymers during crystallization process of the polymer matrix induces heterogeneous nucleation. Buoafif et al. (2009) reviewed that presence of high heterogeneous nucleation activity, lead to the formation of transcrystalline layer on the fiber matrix interface as to improve bonding between fiber and matrix. However, presence of nanoclay tend to alter the crystalline structure of the fiber too. This was observed by Delhom et al. (2010) in which the melting temperature of cotton fibre composite drop from 186 to 172 °C after adding with nanoclay.

Crystallization temperature of composite filled with nanoclay was found to be higher than unfilled composite. As shown in Fig. 7a, crystallization temperature for composite contained treated kraft fiber (high density polyethylene-maleic anhydride polyethylene, HDPE-MAPE) and nanoclay was higher than composite without presence of nanoclay. However, comparing with Fig. 7b, presence of clay in treated

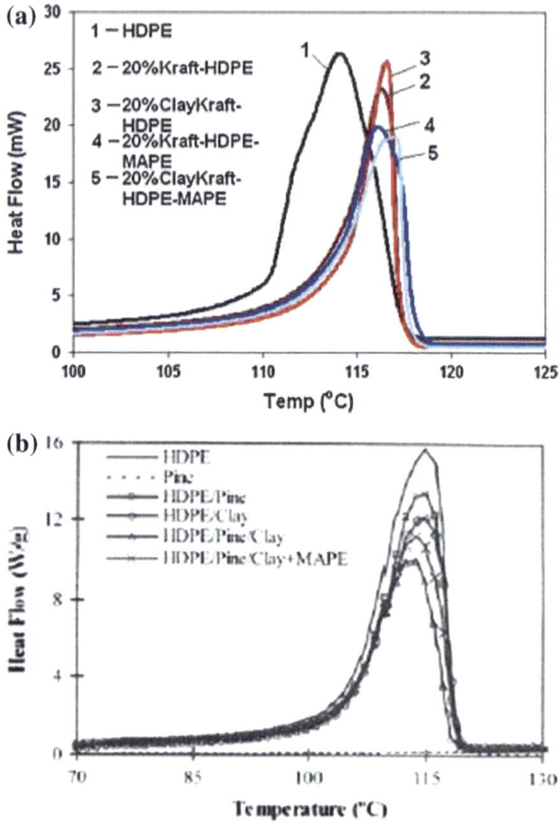

Fig. 7 DSC curves of crystallization temperature of **a** 20 wt% treated nanoclay or untreated kraft fiber-HDPE and HDPE-MAPE composites including blank HDPE (adapted from Chen and Yan 2013a) and, **b** pine, HDPE, and HDPE blend with pine and/or clay (adapted from Lei et al. 2007b)

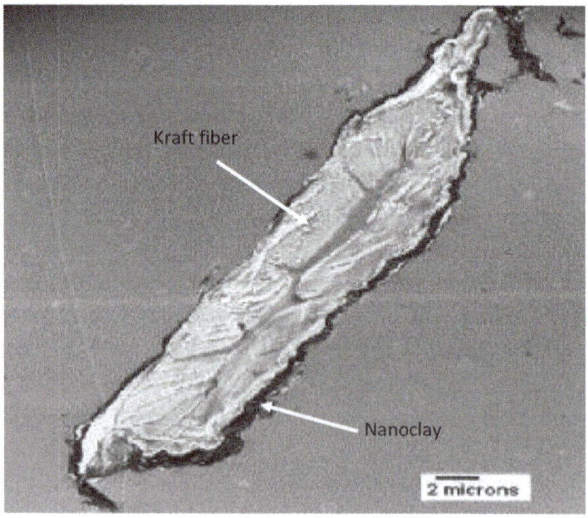

Fig. 8 TEM image of cross section on nanoclay treated kraft fiber (adapted from Chen and Yan 2013b)

pine fiber/HDPE/MAPE nanocomposite does not give significant effect on the crystallization temperature. Basically there are two reasons explaining on this phenomenon. First is the adhesion between nanoclay and fiber. By looking at Fig. 8, kenaf fiber has been coated with nanoclay. According to Chen and Yan (2013b) in their experiment, the exfoliated nanoclay was adsorbed onto the kraft fiber surface to change the surface nature to hydrophobic. It is expected that surface modification of kraft fiber with nanoclay is important to well blend with HDPE. However, it turns out that there is poor distribution of treated nanoclay-kraft in HDPE matrix. Thus, kraft/HDPE nanocomposite have low crystallinity temperature than kraft/HDPE composite. This occurred due to the low interfacial interaction between kraft fiber surface, nanoclay and HDPE matrix. It was explained that presence of nanoclay reduce the chain mobility of the HDPE polymer matrix. Addition of MAPE as compatibilizer improved the adhesion property between kraft fiber, nanoclay and HDPE. Therefore, MAPE was added to improve chain mobility of HDPE matrix with presence of nanoclay. Moreover, the nucleating site at the fiber surface was found to be increase and good adhesion was observed between HDPE-nanoclay-kraft fibers which subsequently increased the crystallinity temperature. Presence of nanoclay treated kraft with addition of MAPE act as an efficient nucleation agent and facilitate the crystallization of HDPE in which the crystallization peak is shift to high temperature. Another factor effects the crystallinity temperature of fiber/nanoclay composite was the particle size of the fiber itself. Nucleating ability of natural fibers depends on the fiber surface topography, wettability, fiber chemistry, surface energy and fiber length. Lei et al. (2007b) explained in their work that addition of nanoclay in pine/HDPE does not give significant effect to the crystalline temperature (Fig. 7b). In fact, addition of 2 % of nanoclay and 30 % of pine fiber flour drop the crystalline temperature and crystallization rate due to reductions of HDPE chain mobility. Thus, adding of nanoclay does not improve the chain mobility of HDPE matrix. This situation can be seen clearly as shown in Fig. 7a, crystallization temperature of blank HDPE is lower than HDPE composites while in Fig. 6b shows the crystallization temperature of blank HDPE is higher than HDPE composites. Therefore, Chen and Yan (2013a) concluded that fiber length is one of the factor that affect the nucleating activity in the polymer matrix. Table 4 summarizes the crystalline peak temperature, crystalline enthalpy and crystalline level of some natural fiber reinforced polymer nanocomposites.

Incorporation of pineapple leaf fiber (PALF) and nanoclay in polypropylene (PP)/maleic anhydride grafted polypropylene (MA–g–PP) shows increased in crystallization temperature from 104.43 °C for blank PP to 110.67 °C for PP/PALF/MA–g–PP/organoclay nanocomposite. MA–g–PP was added into the composite mixture to function as a compatibilizer between nanoclay surface, banana fiber and PP matrix. Observation under optical microscope as shown in Fig. 9 revealed that addition of PALF and nanoclay can act as nucleation agents in PP matrix due to the strong bonding between PALF, nanoclay and PP matrix. Biswal et al. (2009) explained that a number of nucleus were produce in limited

Table 4 Crystallization peak temperature and crystalline level of some natural fiber reinforced polymer nanocomposite

Sample	Crystalline peaks (°C)	Crystalline enthalpy, ΔH (J/g)	Crystalline level (%)	References
10 wt% bamboo/PP	165.3	227.2	55.0	Rahman et al. (2015)
10 wt% bamboo/PP/5 % nanoclay	161.4	220.3	50.2	
10 wt% bamboo/PP/5 % treated nanoclay	162.2	215.4	48.1	
Pine/HDPE/clay	113.1	158.3	54	Lei et al. (2007b)
Pine/HDPE/MAPE/clay	113.6	146.4	50	
PP/PALF/MA–g–PP/C20A	110.67	108.65	45.17	Biswal et al. (2009)
OPP	116.72	65.38	26.68	Biswal et al. (2012)
OPP/10 wt% MBF	118.08	147.41	60.16	
OPP/20 wt% MBF	118.81	189.36	77.28	
OPP/30 wt% MBF	120.99	172.96	70.59	
HDPE	113	192	66	Lee et al. (2010)
Wood polymer composite (WPC)	117.5	112.6	54.9	
5 % intercalated filled WPC	118.2	98.9	51.9	
5 % exfoliated filled WPC	118.6	100.9	52.97	

space with addition of PALF/MA–g–PP/PP, which lead to the evolution of small size crystal of spherulite (Fig. 9b). The crystal growth initially start from fiber itself. However, every neighboring fiber growth together in perpendicular direction and they are competing each other in order to act as a nucleation sites. While, inclusion of nanoclay in PP/PALF/MA–g–PP composite diminish the spherulite crystal with increasing the crystallinity level from 40.60 % for PP/PALF/MA–g–PP composite to 45.17 % for PP/PALF/MA–g–PP/nanoclay nanocomposite (Fig. 8c).

From DSC analysis observed by Biswal et al. (2012), increasing the banana weight fraction up to 30 wt%, does increase the crystallization temperature peak (Table 4). The interaction between anhydride group from MA–g–PP was found to form ester linkage with—OH group from banana fibers. Thus, MA–g–PP are able to diffuse into the PP polymeric network and lead to the formation of interchain entanglements, resulting in improved of interfacial adhesion between fiber, nanoclay and matrix (Bikiaris et al. 2001). The highest crystalline level was observed on the nanocomposite with loading of 20 wt% banana fiber. Presence of banana fibers and nanoclay act as nucleating sites in PP matrix which subsequently accelerated the crystallization rate and increase the crystalline level.

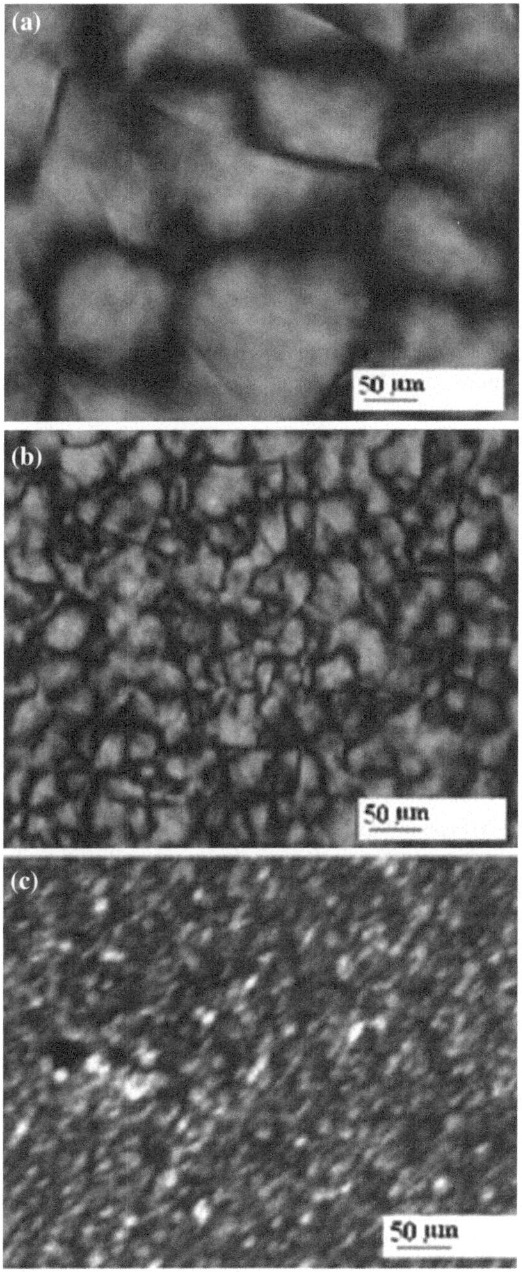

Fig. 9 Optical microscopy of (a) PP, (b) PP/PALF/MA-g-PP, and (c) PP/PALF/MA-g-PP/C20A (adapted from Biswal et al., 2009)

5 Glass Transition Temperature, T_g

The glass transition temperature, T_g of polymer information provide understanding of the polymer properties and this characteristic can be altered by addition of fibers and fillers into the polymer system. Increased or reduction on the T_g were attributed to surface chemistry of the nanoclay, nanoclay/fiber content or loading, dispersibility of nanoclay/fiber and presence of compatibilizer agent. Paluvai et al. (2015) blend surface modified nanoclay with alkaline-silane treated sisal fiber in epoxy/unsaturated polyester (UP) blend matrix. The T_g of the epoxy/UP filled with 1 % modified nanoclay and alkaline silane treated sisal increased for about 41 °C over the epoxy/UP polymer blend. Three factors lead to the improvement in T_g which are: (1) presence of alkyl ammonium on the surface of modified clay functioned as catalyst for amine cured epoxy monomer; (2) modification on the fiber surface via alkaline treatment to increase the surface roughness and amount of polar groups to interact with polymer matrix and (3) presence of silane as compatibilizer agent improve the mobility chain in the epoxy/PU polymer blend.

Significant effect to the shift of T_g, due to the interfacial adhesion effects between fibers and nanosize fillers. The T_g effect of adding nanoclay in treated banana fiber/PP has been investigated by Biswal et al. (2011a). They revealed that the T_g peak shift about 5 °C for the hybrid banana fiber/clay in PP matrix. Addition of treated banana fiber in PP nanocomposite was found to decrease the mobility chains in the PP matrix. Similar observation have been obtained for natural fibers/nanoclays/lignin nanocomposite (Guigo et al. 2009). Inclusion of 5 wt% of nanoclay increase the T_g the due to good layered silicate dispersibility in lignin medium and presence of organophilic cations on the nanoclay surface improve the interfacial adhesion with lignin matrix. Besides that, presence of fibers can alters the response on the polymer chain as well. This effect was investigated by Lin and Renneckar (2011) for wood fiber/nanoclay in PP nanocomposite. Besides nanoclay, fiber can act as nucleating agent which then can accelerate the crystallization process of PP by alter the mobility chain inhomogeneously to amorphous zone. However, fast crystallization rate resulted in lower the T_g value (Nuńez et al. 2004).

6 Conclusions

Inclusion of nanoclay in natural fiber reinforced polymer composite was found to improve the thermal performance of the composite. From thermal stability analysis, layered silicate structure of nanoclay delay the degradation temperature of the natural fiber and polymer in the composite system. During degradation process, the nanoclay move to the surface of the composite and created tortuous to hinder the diffusion of volatiles matter in the composite. In addition, strong interfacial adhesion between nanoclay and natural fiber surfaces with polymer matrix, improve the thermal barrier property of the natural fiber nanocomposite. Moreover, addition of compatibilizer to improve the interfacial adhesion between nanoclay and polymer matrix, help in accelerate the crystallization rate and increase the crystallization temperature peak of the polymer composite. Combination with nanoclay particles and short natural fibers can act as nucleating agent in polymer matrix. Interestingly, T_g characteristics of polymer matrix tend to increase with strong adhesion of nanoclay in fiber reinforced polymer composite.

Acknowledgments Authors would like to thank Universiti Putra Malaysia for funding this research study under Research Universiti (RUGS) Grant (9415401 & 9405400).

References

Abdul Khalil, H.P.S., Siti Alwani, M., Mohd Omar, A.K.: Chemical composition, anatomy, lignin distribution, and cell wall structure of Malaysian plant waste fibers. BioResources **1**(2), 220–232 (2007)

Abdul Khalil, H.P.S., Ireana Yusra, A.F., Jawaid, B.M.: Cell wall ultrastructure, anatomy, lignin distribution, and chemical composition of Malaysian cultivated kenaf fiber. Ind. Crops Prod. **31** (1), 113–121 (2010)

Alamri, H., Low, I.M., Alothman, Z.: Mechanical, thermal and microstructural characteristics of cellulose fibre reinforced epoxy/organoclay nanocomposites. Compos. Part B **43**(7), 2762–2771 (2012)

Alhuthali, A., Low, I.M., Dong, C.: Characterisation of the water absorption, mechanical and thermal properties of recycled cellulose fibre reinforced vinyl-ester eco-nanocomposites. Compos. Part B **43**(7), 2772–2781 (2012)

Arrakhiz, F.Z., El, Achaby M., Malha, M., Bensalah, M.O., Fassi-Fehri, O., Bouhfid, R., Benmoussa, K., Qaiss, A.: Mechanical and thermal properties of natural fibers reinforced polymer composites: doum/low density polyethylene. Mater. Design **43**, 200–205 (2013)

Bikiaris, D., Matzinos, P., Prinos, J., Flaris, V., Larena, A., Panayiotou, C.: Use of silanes and copolymers as adhesion promoters in glass fiber/polyethylene composites. J. Appl. Polym. Sci. **80**(14), 2877–2888 (2001)

Biswal, M., Mohanty, S., Nayak, K.: Influence of organically modified nanoclay on the performance leaf fiber-reinforced polypropylene nanocomposites. J. Appl. Polym. Sci. **114**(6), 4091–4103 (2009)

Biswal, M., Mohanty, S., Nayak, S.K.: Mechanical, thermal and dynamic-mechanical behavior of banana fiber reinforced polypropylene nanocomposites. Polym. Compos. **32**(8), 1190–1201 (2011a)

Biswal, M., Mohanty, S., Nayak, S.K.: Banana fiber-reinforced polypropylene nanocomposites. Effect of fiber treatment on mechanical, thermal and dynamic-mechanical properties. J. Thermoplast. Compos. Mater. **25**(6), 765–790 (2011b)

Biswal, M., Mohanty, S., Nayak, S.K.: Thermal stability and flammability of banana-fiber-reinforced polypropylene nanocomposites. J. Appl. Polym. Sci. **125**(S2), E432–443 (2012)

Buoafif, H., Koubaa, A., Perré, P., Cloutier, A., Rield, B.: Wood particle/high density polyethylene composites: thermal sensitivity and nucleating ability of wood particles. J. Appl. Polym. Sci. **113**(1), 593–600 (2009)

Camino, G., Tartaglione, G., Frache, A., Manferti, C., Costa, G.: Thermal and combustion behaviour of layered silicate-epoxy nanocomposites. Polym. Degrad. Stab. **90**(2), 354–362 (2005)

Chen, J., Yan, N.: Crystallization behavior of organo-nanoclay treated and untreated kraft fiber-HDPE composites. Compos. Part B **54**, 180–187 (2013a)

Chen, J., Yan, N.: Mechanical properties and dimensional stability of organo-nanoclay modified biofiber polymer composites. Compos. Part B **47**, 248–254 (2013b)

Chigwada, G., Jash, P., Jiang, D., Wilkie, C.: Fire retardancy of vinyl ester nanocomposites: synergy with phosphorus-based fire retardants. Polym. Degrad. Stab. **89**(1), 85–100 (2005)

da Silva, Santos R., de Souza, A.A., De Paoli, M.A., de Souza, C.M.L.: Cardonal-formaldehyde thermoset composites reinforced with buriti fibers: preparation and characterization. Compos. Part A **41**(9), 1123–1129 (2010)

Deka, B.K., Maji, T.K.: Effect of nanoclay and ZnO on the physical and chemical properties of wood polymer nanocomposite. J. Appl. Polym. Sci. **124**(4), 2919–2929 (2011)

Deka, B.K., Maji, T.K., Mandal, M.: Study on properties of nanocomposites based on HDPE, LDPE, PP, PVC, wood and clay. Polym. Bull. **67**(9), 1875–1892 (2011)

Delhom, C.D., White-Ghoorahoo, L.A., Pang, S.S.: Development and characterization of cellulose/clay nanocomposites. Compos. Part B **41**(6), 475–481 (2010)

Dorez, G., Taguet, A., Ferry, L., Lopez-Cuesta, J.M.: Thermal and fire behavior of natural fiber/PBS nanocomposite. Polym. Degrad. Stab. **98**(1), 87–95 (2013)

Duquesne, S., Jama, C., Blas, M.L., Delobel, R., Recourt, P., Gloaguen, J.M.: Elaboration of EVA-nanoclay systems-characterization, thermal behaviour and fire performance. Compos. Sci. Technol. **63**(8), 1141–1148 (2003)

Eng, C.C., Ibrahim, N.A., Zainuddin, N., Ariffin, H., Yunus, W.M.Z.W., Then, Y.Y.: Effect of hydrophilic nanoclay on morphology, thermal and mechanical properties of polylactic acid/polycaprolactone/oil palm mesocarp fiber biocomposites. Int. J. Inst. Mater. Malays. **1**(1), 51–70 (2014)

Guigo, N., Vincent, L., Mija, A., Naegele, H., Sbirrazzuoli, N.: Innovative green nanocomposites based on silicate clays/lignin/natural fibres. Compos. Sci. Technol. **69**(11), 1979–1984 (2009)

Guo, G., Park, C.B., Lee, Y.H., Kim, Y.S., Sain, M.: Flame retarding effects of nanoclay on wood-fiber composites. Polym. Eng. Sci. **47**(3), 330–336 (2007)

Hapuarachchi, T.D., Peijs, T.: Multiwalled carbon nanotubes and sepiolite nanoclays as flame retardants for polylactide and its natural fibre reinforced composites. Compos. Part A **41**(8), 954–963 (2010)

Haurie, L., Fernández, A.I., Velasco, J.I., Chimenos, J.M., Cuesta, J.-M.L., Espiell, F.: Thermal stability and flame retardancy of LDPE/PVA blends filled with synthetic hydromagnesite/aluminium hydroxide/montmorillonite and magnesium hydroxide/aluminium hydroxide/montmorillonite mixtures. Polym. Degrad. Stab. **92**(6), 1082–1087 (2007)

Hazarika, A., Devi, R.R., Maji, T.K.: Studies on properties of softwood (Ficus hispida)/PMMA nanocomposites reinforced with polymerizable surfactant-modified nanoclay. Polym. Bull. **68** (7), 1989–2008 (2012)

Iman, M., Maji, T.K.: Effect of crosslinker and nanoclay on starch and jute fabric based green nanocomposites. Carbohydr. Polym. **89**(1), 290–297 (2012)

Kovacevic, Z., Bischof, S., Fan, M.: The influence of Spartium junceum L. fibres modified with montmorillonite nanoclay on the thermal properties of PLA biocomposites. Compos. Part B **78**, 122–130 (2015)

Kumar, V., Kumar, R.: Improved mechanical and thermal properties of bamboo-epoxy nanocomposites. Polym. Compos. **33**(3), 362–370 (2012)

Kumar, V., Sharma, N.K., Kumar, R.: Dielectric, mechanical, and thermal properties of bamboo-polylactic acid bionanocomposites. J. Rein. Plast. Comp. (2012). doi:10.177/0731684412461129O

Lee, S.-H., Wang, S.: Biodegradable polymers/bamboo fiber biocomposite with bio-based coupling agent. Compos. Part A **37**(1), 80–91 (2006)

Lee, Y.H., Kuboki, T., Park, C.B., Sain, M., Kontopoulou, M.: The effects of clay dispersion on the mechanical, physical, and flame-retarding properties of wood fiber/polyethylene/clay nanocomposites. J. Appl. Polym. Sci. **118**(1), 452–461 (2010)

Lei, S.G., Hoa, S.V., Ton-That, M.-T.: Effect of clay types on the processing and properties of polypropylene nanocomposites. Compos. Sci. Technol. **66**(10), 1274–1279 (2006)

Lei, Y., Wu, Q., Yao, F., Xu, Y.: Preparation and properties of recycled HDPE/natural fiber composites. Compos. Part A **38**(7), 1664–1674 (2007a)

Lei, Y., Wu, Q., Clemons, C.M., Yao, F., Xu, Y.: Influence of nanoclay on properties of HDPE/wood composites. J. Appl. Sci. **106**(6), 3958–3966 (2007b)

Lin, Z., Renneckar, S.: Nanocomposite-based lignocellulosic fibers 2: layer-by-layer modification of wood fibers for reinforcement in thermoplastic composites. Compos. Part A **42**(1), 84–91 (2011)

Liqun, X., James, L.L.: Kinetic analysis and mechanical properties of nanoclay reinforced unsaturated polyester (UP) resins cured at low temperatures. Polym. Eng. Sci. **45**(4), 496–509 (2005)

Liu, W., Mohanty, A.K., Drzal, L.T., Askel, P., Misra, M.: Effects of alkali treatment on the structure, morphology and thermal properties of native grass fibers as reinforcements for polymer matrix composites. J. Mater. Sci. **39**(3), 1051–1054 (2004)

Liu, W., Hoa, S.V., Pugh, M.: Fracture toughness and water uptake of high-performance epoxy/nanoclay nanocomposites. Compos. Sci. Technol. **65**(15), 2364–2373 (2005)

Madaleno, L., Schjødt-Thomsen, J., Pinto, J.C.: Morphology, thermal, and mechanical properties of PVC/MMT nanocomposites prepared by solution blending and solution blending + melt compounding. Compos. Sci. Technol. **70**(5), 804–814 (2010)

Manfredi, L.B., Rodríguez, E.S., Wladyka-Przbylak, M., Vázquez, A.: Thermal degradation and fire resistance of unsaturated polyester, modified acrylic resins and their composites with natural fibres. Polym. Degrad. Stab. **91**(2), 255–261 (2006)

Manias, E., Touny, A., Wu, L., Strawhecker, K., Lu, B., Chung, T.C. Polypropylene/montmorillonite nanocomposites. Review of the synthetic routes and materials properties. Chem. Mater. **13**(10), 3516–3523 (2001)

Mittal, V.: Polymer Nanocomposite Foams, p. 63. CRC Press (2013)

Mohan, T.P., Kanny, K.: Nanoclay infused banana fiber and its effects on mechanical and thermal properties of composites. J. Compos. Mater. (2015). doi:10.1177/0021998315590265

Molinaro, S., Romero, M.C., Boaro, M., Sensidoni, A., Lagazio, C., Morris, M., Kerry, J.: Effect of nanoclay-type and PLA optical purity on the characteristics of PLA-based nanocomposite films. J. Food Eng. **117**(1), 113–123 (2013)

Morgan, A.B.: Flame retarded polymer layered silicate nanocomposites: A review of commercial and open literature systems. Polym. Advan. Technol. **17**(4), 206–217 (2006)

Nuńez, A.J., Kenny, J.M., Reboredo, M.M., Aranguren, M.I., Marcovich, N.E.: Thermal and dynamic mechanical characterization of polypropylene-woodflour composites. Polym. Eng. Sci. **42**(4), 733–742 (2004)

Olewnik, E., Garman, K., Czerwiński, W.: Thermal properties of new composites based on nanoclay, polyethylene and polypropylene. J. Therm. Anal. Calorim. **101**(1), 323–329 (2010)

Paluvai, N.R., Mohanty, S., Nayak, S.K.: Effect of cloisite 30B clay and sisal fiber on dynamic mechanical and fracture behaviour of unsaturated polyester toughened epoxy network. Polym. Compos. (2015). doi:10.1002/pc.23480

Pandey, P., Dayanidhi, A., Mohanty, S., Nayak, S.K.: Effect of clay loading on flammability of poly(methyl methacrylate)/clay nanocomposites. J. Thermoplast. Compos. **26**(5), 663–679 (2011)

Qin, H., Zhang, S., Zhao, C., Feng, M., Yang, M., Shu, Z., Yang, S.: Thermal stability and flammability of polypropylene/montmorillonite composites. Polym. Degrad. Stab. **85**(2), 807–813 (2004)

Qin, H., Zhang, S., Zhao, C., Hu, G., Yang, M.: Flame retardant mechanism of polymer/clay nanocomposites based on polypropylene. Polymer **46**(19), 8386–8395 (2005)

Rahimi-Razin, S., -Kalajasi, M.S., Asl, V.H., Mamaqani, H.R.: (2012) Effect of different modified nanoclays on the kinetics of preparation and properties of polymer-based nanocomposites. J. Polym. Res. **19**(9), 1–16

Rahman, M., Hamdan, S., Hashim, D.M.A., Islam, M., Takagi, H.: Bamboo fiber polypropylene composites: Effect of fiber treatment and nano clay on mechanical and thermal properties. J. Vinyl Addit. Techn. **21**(4), 253–258 (2015)

Rajasekar, R., Pal, K., Heinrich, G., Das, A., Das, C.K.: Development of nitrole butadiene rubber-nanoclay composites with epoxidized natural rubber as compatibilizer. Mater. Design **30**(9), 3839–3845 (2009)

Rajini, N., Winowlin Jappes, J.T., Rajakarunakaran, S., Jeyaraj, P.: Mechanical and free vibration properties of montmorillonite clay dispersed with naturally woven coconut sheath composite. J. Reinf. Plast. Comp. **31**(20), 1364–1376 (2012)

Reddy, C.R., Sardhasti, A.P., Simon, L.C.: Preparation and characterization of polypropylene-wheat straw-clay composites. Compos. Sci. Technol. **70**(12), 1674–1680 (2010)

Sánchez-Jiménez, P.E., Pérez-Maqueda, L.A., Perejón, A., Criado, J.M.: Nanoclay nucleation effect in the thermal stabilization of a polymer nanocomposite: a kinetic mechanism change. J. Phys. Chem. C **116**(21), 11797–11807 (2012)

Saw, S.K.: Static and dynamic mechanical analysis of coir fiber/montmorillonite nanoclay-filled novolac/epoxy hybrid nanocomposites. In: Book Eco-friendly Polymer Nanocomposites. Advanced Structured Materials, pp. 137–154. Springer, India (2015)

Shuvo, S.N., Shorowordi, K.M., Islam, M.A.: Effect of nanoclay on jute fiber reinforced polyester composites. Int. J. Adv. Eng. Nano Technol. **2**(8), 20–26 (2015)

Subramaniyan, A.K., Sun, C.T.: Toughening polymeric composites using nanoclay: crack tip scale effect on fracture toughness. Compos. Part A **38**(1), 34–43 (2007)

Walters, R.N., Lyon, R.E.: Flammability of polymer composites. Office of aviation research and development. Federal aviation administration (2008)

Wang, C., Liu, C.-R.: Transcrystallization of polypropylene composites: nucleating ability of fibres. Polymer **40**(2), 289–298 (1999)

Yussuf, A.A., Massoumi, I., Hassan, A.: Comparison of polylactic acid/kenaf and polylactic acid/rice husk composites: the influence of the natural fibers on the mechanical, thermal and biodegradability properties. J. Polym. Environ. **18**(3), 422–429 (2010)

Zahedi, M., Tabarsa, T., Ashori, A., Madhoushi, M., Shakeri, A.: A comparative study on some properties of wood plastic composites using canola stalk, Paulownia, and nanoclay. J. Appl. Polym. Sci. **129**(3), 1491–1498 (2013)

Zahedsheijani, R., Faezipour, M., Tarmian, A., Layeghi, M., Yousefi, H.: The effect of Na+ montmorillonite (NaMMT) nanoclay on thermal properties of medium density fibreboard (MDF). Eur. J. Wood Prod. **70**(5), 565–571 (2012)

Zainuddin, S., Hosur, M.V., Zhou, Y., Narteh, A.T., Kumar, A., Jeelani, S.: Experimental and numerical investigations on flexural and thermal properties of nanoclay-epoxy nanocomposites. Mater Sci. Eng. A-Struct. **527**(29), 7920–7926 (2010)

Zanetti, M., Lomakin, S., Camino, G.: Polymer layered silicate nanocomposites. Macromol. Mater. Eng. **279**(1), 1–9 (2000)

Zhang, H., Rizvi, G.M., Park, C.B.: Development of an extrusion system producing fine-celled HDPE/wood fiber composite foams using CO_2 as a blowing agent. Adv. Polym. Tech. **23**(4), 263–276 (2004)

Zhao, Y., Wang, K., Zhu, F., Xue, P., Jia, M.: Properties of poly(vinyl chloride)/wood flour/montmorillonite composites: effects of coupling agents and layered silicate. Polym. Degrad. Stab. **91**(12), 2874–2883 (2006)

Effect of Nanoclay on Natural Fiber/Polymer Composites

Zhou Chen, Tairong Kuang, Zhaogang Yang and Xiaofeng Ren

Abstract Nanoclays play a significant role to improve composite performance by enhancing their properties such as thermal stability, mechanical strength, and barrier properties. Some of the important parameters contribute most to modify the properties of a variety of composites include the content, shape, size, and the affinity towards matrix material. With their enhanced performance, nanoclay filled polymer matrix based nanocomposites have drawn much attention in the materials industry. In this book chapter, the authors provide an overview of the effect of nanoclay on natural fiber/polymer composites, including the rheological properties, mechanical and thermal properties, morphological and structural properties, modeling of mechanical and rheological properties of nanoclay on natural fiber/polymer composites.

Keywords Nanoclay · Natural fiber · Polymer composites · Mechanical properties · Thermal properties

1 Introduction

Nowadays, fiber reinforced polymer (FRP) composites are widely applied in many areas, such as navigation, aviation, aerospace, building construction, medical industries because of their excellent performance (Chen et al. 2012, 2015a).

Z. Chen · T. Kuang · Z. Yang
Department of Chemical and Biomolecular Engineering,
The Ohio State University, Columbus, OH 43210, USA

Z. Chen
College of Material Science and Technology,
Nanjing University of Aeronautics and Astronautics,
Nanjing 210016, China

X. Ren (✉)
Department of Polymer Engineering, The University of Akron,
Akron, OH 44325, USA
e-mail: chris.uakron@gmail.com

M. Jawaid et al. (eds.), *Nanoclay Reinforced Polymer Composites*,
Engineering Materials, DOI 10.1007/978-981-10-0950-1_8

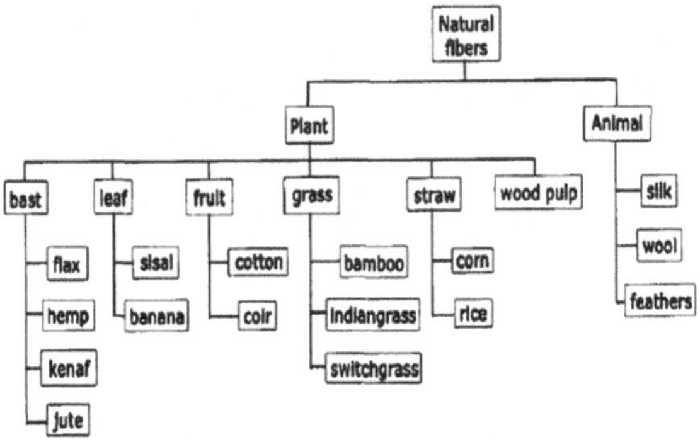

Fig. 1 The classification of natural fibers

However most of synthetic fiber reinforced polymer composites cannot be broken down, thus becomes a burden for the environment nowadays (Bajpai et al. 2013). Natural fiber is a renewable resource, and there are thousands of natural fibers available. Figure 1 shows the classification of natural fibers. Once natural fibers have been added to the biodegradable polymers, it can lead to the production of new biocomposites. Natural fibers have been used as possible alternative to conventional synthetic fibers such as glass fibers, chemical fibers, and carbon fibers (Mohanty et al. 2005). The natural composite structure generally composed of cellulose fibrils and lignin matrix presents high flexural strength. Further, natural fibers possess good electrical resistance as well as thermal and acoustical insulating. Thus it is possible that incorporation of the fibers into low-modulus matrix, e.g. polyester, leads to the wider application due to enhanced properties (Saheb and Jog 1999). Meanwhile, compared to inorganic fibers, they demonstrate many advantages, such as low cost and density (1.25–1.5 g/cm^3), high toughness, good biodegradability, as well as low tool wear. They could be widely used as the potential micro-reinforcements because of the improvement of mechanical properties, thermal, as well as structural properties of biodegradable polymer composites (Leng and Lau 2011).

In many applications, natural fiber reinforced polymer (NFRP) composites are gradually used to replace synthetic fiber reinforced polymer composites due to their being biodegradable and cost effective (Mallick 2008; Bajpai et al. 2012). Research shows that 50,000 MJ of net energy can be saved when 30 % of glass fibers are alternative by hemp fibers, equivalent to cutting 3 tons of carbon dioxide (CO_2) emissions. In addition, the results indicate that, compared to polypropylene (PP) reinforced with glass fiber, 21 % weight was reduced for PP reinforced with hemp. Due to new environmental regulations and depletion of oil resources, attempts have been made to replace the petroleum-based materials by natural fiber reinforced composites. The waste and recycled natural fibers is used to reinforce polymer which is a major step taken in promoting environmental sustainability (Bajpai et al. 2014). However, there are also several limitations of natural fibers, including high moisture absorption, low melting point and relatively narrow incompatibility with the polymer matrix (Taj et al. 2007; Alamri and Low 2012). Because the hydroxyl content of cellulose is very high, natural fibers are prone to absorb water, leading to deduction of the mechanical properties of composite such as flexural strength, flexural modulus and fracture toughness. When the temperature is higher than 200 °C, natural fiber degradation will occur, with hemicellulose degradation first, followed by the degradation of lignin. The degradation will cause odor, discoloration, release of volatiles, and reduction of mechanical properties.

Stress concentration and poor interfacial adhesion between natural fiber and polymer have been considered as a main cause of composites material brittleness. Thus, the technology of fiber surfaces and polymeric matrices modification have been used to improve the interface between natural fibers and polymer matrix (Uma Devi et al. 2004; Cantero et al. 2003; Qiu et al. 2011, 2012; Ren et al. 2012a, b, 2013a, b). Lots of researchers have been working on hygrothermal aging of natural fiber, and those researches have been successfully used in different load-bearing applications. Thermoplastic matrices (polyethylene, polypropylene, polyvinyl alcohol, polystyrene and polylactic acid) have been mixed with natural fibers to fabricate composites to obtain desired attributes (Patel et al. 2005; Rout et al. 2001).

In recent advances in the composites field, nanoparticles (nanoclays, carbon nanotubes or silicates nanoparticles) are added to improve the thermal, mechanical, and electrical properties of composite materials (Chen et al. 2015b). The size of conventional fillers is from 10 μm to 1 cm; however, the size of nanofiller ranges is from 1 to 500 nm. It indicates that the high aspect ratio of the nanoclays leads to a greater interfacial area, enhancing reinforcement properties. Nanoclays are also widely used in different industries (cable coatings, adhesives, inks, pharmaceutical and automotive) (Sattler 2011). Research found that nanoclays in the natural fiber/polymer system can improve the mechanical properties (Nguong et al. 2013). In order to obtain similar results, 3–5 wt% nanoclay is greater than 30 wt% for a conventional filler (Hetzer and De Kee 2008). Since most of the nanoclays are hydrophilic in natural state meanwhile unevenly distributed, they have to be organically improved to avoid agglomeration between the platelets in the dispersion media (Nazare et al. 2006; Bartholmai and Schartel 2004). The degree of exfoliation is increased by this improvement leading to increase the level of surface interaction. This improvement can be done through common dispersion techniques, i.e., exfoliation absorption, in situ polymerization, melt-intercalation or sonication (Wang et al. 2000; Haq et al. 2009). Surface modification of nanoclays have been used extensively in the fields of engineering and technology due to the improvement of properties such as strength (Song et al. 2005), modulus (Hetzer and De Kee 2008), solvent resistance (Mravčáková et al. 2006), biodegradability (Ray et al. 2003), and water vapor transmission resistance (Osman et al. 2004), and all the improvement is based on the uniformity of dispersion of nanoclays in polymer resin.

2 Rheological Properties

Well intellectual of the fluxional phenomenon is compulsive and the liquidity need to be guaranteed so that fibers can be fully processed and prepare composites with a variety of strength and thermal property. Polymer dispersion characteristics and flow characteristics of the resin are influenced by many factors, i.e., molecular morphology, molecular weight, size. The viscosity and flow of the material properties are influenced by the dispersion and deformation characteristics of enhanced nanoparticles. To master the processing performance and organizational structure to the material, it is necessary to know the flow property of polymer materials and nano enhanced phase. The research of liquidity and viscosity of reinforced material also makes analysis of material in melt state easier.

Sepiolite nanoclay has uniform grain size, high specific surface area, porosity, and therefore can be used as reinforced phase to greatly improve the flow performance of the nano reinforced material (Bilotti et al. 2008). Increase of the reinforcing particles can solve the problem of viscosity increase and improve the strength of the materials (Shen et al. 2005).

2.1 *Effect on Viscosity*

Bensadoun et al. (2011) prepared nanoclays modified flax fiber/unsaturated polyester resin (UPR) composites (FF/UPRs) with different forming methods and nanoclays (Tables 1 and 2). Nanoclays and styrene were used as the modified

Table 1 Properties of nanoclays (Bensadoun et al. 2011)

	Cloisite 11B	Cloisite 15A	Cloisite 30B
D-spacing (Å)	18.4	31.5	18.5
Density (g/cm^3)	1.9–2.1	1.66	1.98
Surface treatment	Dimethyl, salts with bentonite	Dimethyl, salt with bentonite	Alkyl quaternary ammonium bentonite

Table 2 Different forming methods of FF/SUPRs

No.	Dispersion method			Add	Remove (stirring + heat)
	Raw material	Modified material	Dispersion method		
0	Resin	/	Ultrasonication	Styrene	Styrene
1	/	Nanoclays + styrene	Ultrasonication	Resin	Styrene
2	Resin	Nanoclays + styrene	Ultrasonication	/	Styrene
3	Resin	Nanoclays	Ultrasonication	/	/
4	Resin	Nanoclays	Ultrasonication	Styrene	Styrene

Fig. 2 Shear viscosity sweep at 23 °C for 3 %wt C30B nanoclay dispersed in R937 resin by different methods (Bensadoun et al. 2011)

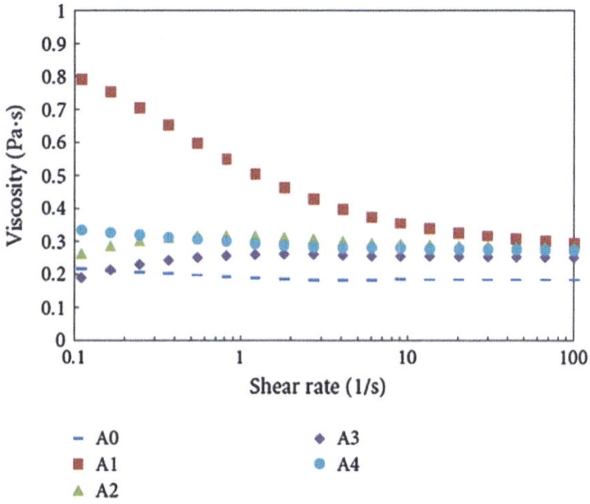

material and dispersion media for resin, respectively. Styrene was either deleted or increased after ultrasonication to assure the same proportion.

As shown in Fig. 2, the viscosity of UPR with 3 wt% nanoclay has greatly improved compared to the UPR without nanoclay (No. 0). However, a non-Newtonian shear-thinning behavior appears in the No. 1 mixture, while No. 0 to the neat resin is similar to mixtures No. 2 to No. 4 in Newtonian behavior. The resin and the nanoparticles are able to a direct result of the connection in the non-Newtonian behavior of mix No. 1. These results show that the exfoliated and well-dispersed structure was able to produce in the only dispersion technique of No. 1. Due to the physical properties and intercalation in the polymer diffusion of MMT nanoclay, an exfoliated structure could lead to a higher interaction of surface (Carreau et al. 1997; Morrison 2001). Continuous shear measurements were conducted in No.1 dispersion method to study the change of orientation structure of MMT nanoclay. In the Fig. 3, during the rotation, the prepreg viscosity of 0.8 Pa s under the 0.1 (1/s) can be turned into 0.3 Pa s under the 100 (1/s). In the rotational axis, this shear force reduction trend is closely related to the orientation structure of strengthening nanoclay. When the test is started for about 1 min in first stage, the nanoclays have the tendency to recombine with the original stochastic construction. Consequently, the shear measurements show that the viscosity with low shearing rate in the first rotation has higher number than the second. It also applies to the third shearing stage, as was neatly demonstrated in Fig. 3. The numerical value of viscosity in three shear measurements is the same in 100 (1/s), which means the same reorganization of nanoclays for the three sweeps.

Duran et al. (2000) showed the similar discipline of MMT suspension in different shearing force testing, and Sinha (Ray and Okamoto 2003) observed that the composition with the addition of the nanoparticles altered for a performance of the linear viscoelastic polymer chains. The dispersant of styrene can facilitate uniform distribution of the nanoclays in polymer solution. The nanoparticles can accomplish

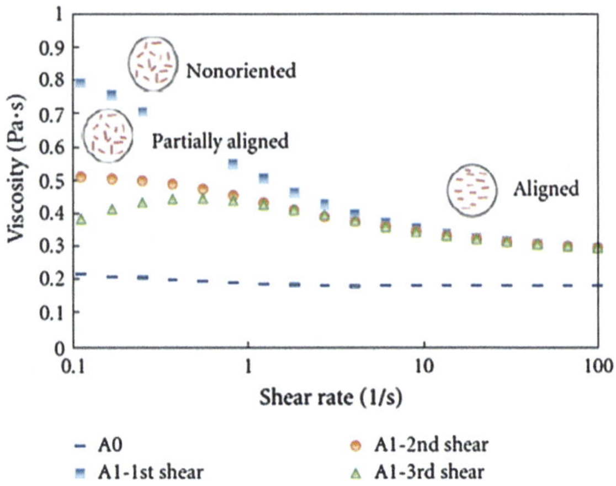

Fig. 3 Successive shear viscosity sweeps at 23 °C for 3 %wt C30B nanoclay dispersed in R937 resin by No. 1 method (Bensadoun et al. 2011)

the super-duper movement because of the low viscosity of styrene (0.762 mPa s). And the special intercalated-exfoliated structure can be kept by appending the polymer.

As shown in Fig. 4, the initial viscosities of UPR with three different nanoclays at 0.1 s^{-1} both increase with the content of nanoclays. However different nanoclays have different viscosity at the same content of nanoclays, the viscosity of C11B and C30B is lower than C15A. Nanoclays and the UPS have a higher surface interaction showing this phenomenon that it can lead to a possible high dispersion about C15A. The chemical d-spacing and the treatment of nanoclay platelets explained this difference. To wipe out sonication energy, the latter can also have an influence on the capacity of the particles. As is showed in Table 1, even though the particles of C15A and C11B have the same chemical treatment, a higher d-spacing belongs to C15A. Higher viscosity and the platelets that have a higher surface interaction improve the diffusion of the polymer macromolecules with a higher d-spacing. Although the chemical treatment of C30B and C11B was very distinct, their

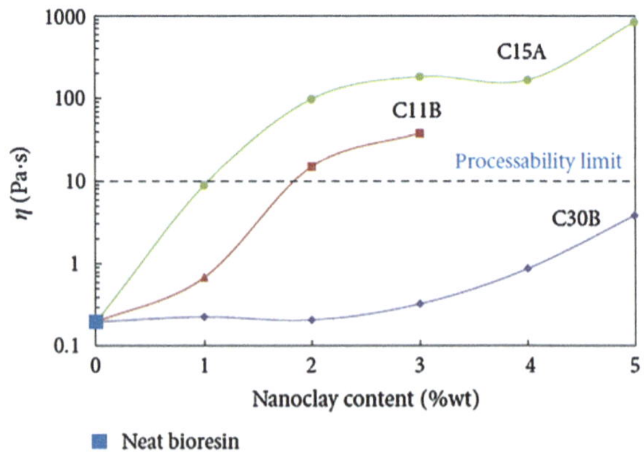

Fig. 4 Viscosity at 23 °C for various types of nanoclays and percentage (Bensadoun et al. 2011)

d-spacing is similar. Therefore, the initial viscosity value will be affected due to the difference, for C30B is 0.28 Pa s and for C11B0.8 Pa s.

2.2 *Modified Nanoclays*

Veronica et al. (Morote-Martínez et al. 2011) improved the performance of UPR-coated natural stone by adding the natural sodium montmorillonite nanoclay (MMT) for modifying the rheology of the prepolymer solution. The changes of viscosity of the different MMT-modified UPR according to the shear rate of polymer solutions were shown in Fig. 5. The viscosity properties of all MMT-modified UPR polymer solutions showed Newtonian behavior, they will not be changed no matter if the shear rate increases. As shown in Fig. 6, the viscosity can be increased by more than three times by adding MMT. And it stabilized with

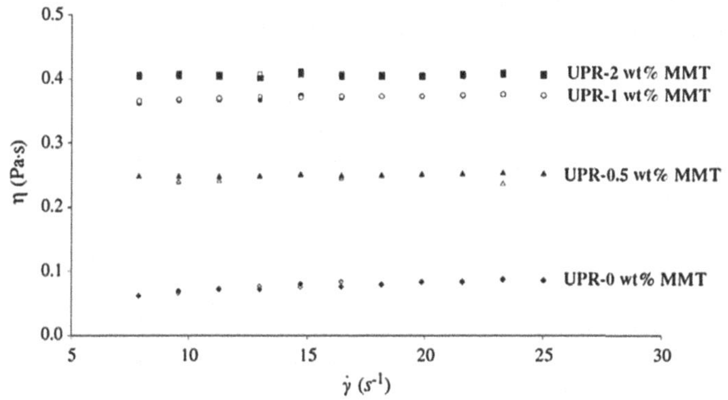

Fig. 5 Variation of the viscosity of the UPR-MMT as a function of the shearing rate (Morote-Martínez et al. 2011)

Fig. 6 Variation of the viscosity of the UPR-MMT as a function of the MMT content (Morote-Martínez et al. 2011)

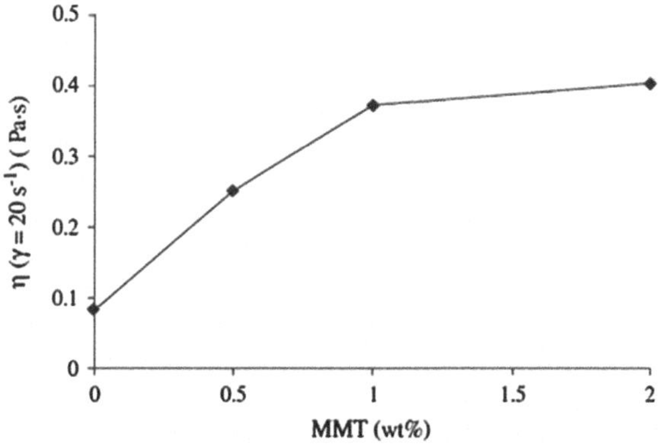

the increase of the addition when the addition of MMT exceeding the 1 wt% limit. The reason of this phenomenon can be described as the unevenness of MMT nanoclay particles in the UPR.

Zhong et al. (2007) incorporated a modified MMT (Cloisite 20A) into wood reinforced polyethylene (PE) composites (MMT-W/PEs) with maleic anhydride grafted polyethylenes (MAPEs). The thermal expansion coefficient and the flexural strength of MMT-W/PEs are reduced by 60 and 24 % for 3 wt% nanoclay loading, respectively. Meanwhile the thermal deformation temperature and the flexural modulus increased by 10 °C and 10 %, respectively. However, the incorporation of the modified MMT had no significant effect on the viscosity of the MMT-W/PEs (Hetzer et al. 2009). Kumar et al. (Pratheep Kumar and Pal Singh 2007) prepared the nanoclay-modificated cellulose fibers/ethylene-propylene (EP) composites (N-M CF/EPs). The elastic modulus of EPs can be significantly improved by adding nanoclays and cellulose fibers from 773 to 1622 MPa, and the water absorption rate of CF/EPs can be reduced by 15 % as the nanoclays are added. The melt temperature of EPs can be increased up to 170 °C with the reinforcement of nanoclays and cellulose fibers. However, the breaking yielding stress, fracture length and elongation at break point are nothing breathtakingly different.

Two sorts of WF/HDPEs composites which comprise intercalated and exfoliated nanoclay were prepared by Lee et al. (2011), WF and HDPE here are the abbreviation of wood fiber and high density polyethylene, respectively. When the contents of wood fiber and nanoclay were different in the foamed WPCs, the die pressure would change with the die temperature differently; the functions between them were shown in Fig. 7. The varying degrees of nanoclay dispersion reflected on the die resistance, when the amounts of clay and wood fiber increased or the exfoliated clay incorporated, the die pressures increased.

Gu et al. (2010) systematically analyzed the effects of different modified nanoclays on wood fiber/linear low density polyethylene (LLDPE) composites (WF/LLDPEs). As shown in Table 3, modified MMT nanoclays can be categorized

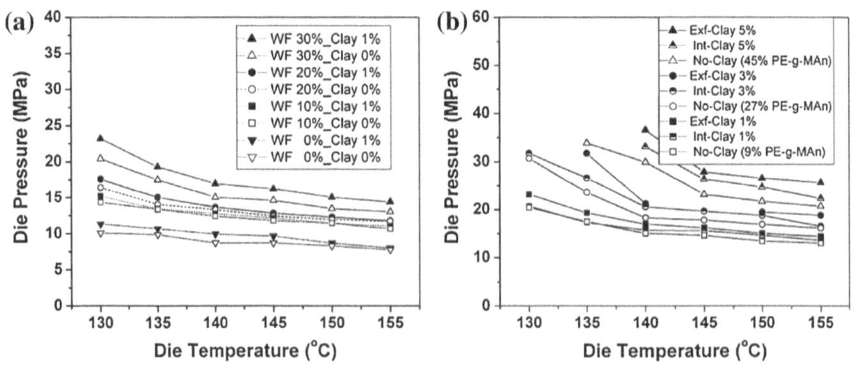

Fig. 7 Die pressure as a function of die temperature **a** with different wood fiber content and **b** with different clay content (30 % wood fiber) (Lee et al. 2011)

Table 3 Different modified nanoclays (Gu et al. 2010)

Type	Physical properties			
	Surfactant[a]	Tails	Particla size (μm)	Density (kg/m^3)
Cloisite® Na$^+$	None	–	2–13	336
Cloisite® 10A	125 meq/100g dimethyl benzyl hydrogenated tallow quaternary ammonium chloride	1	2–13	265
I.28 E	25–30 wt% trimethyl stearyl ammonium	1	8–10	420
I.34 TCN	25–30 % methyl dihydroxyethyl hydrogenated tallow ammonium chloride	2	16–22	300–360
I.44 P	35–45 wt% dimethyl dialkyl (C14–C18) amine	2	≤20	200–500
I.30 E	25–30 wt% octadecyl amine	1	8–10	410
I.31 ps	15–35 wt% octadecyl amine and 0.5–5 wt% aminopropyltriethoxysilane	1	≤20	200–500

[a]Tallow (alkyl) consists of 65 % C18, 30 % C16, and 55 % C14

into five families based on the varying structures of the surfactants except the Cloisite Na$^+$. The first one is the ammonium salts-altered nanoclay with one alkyl tail, such as the Cloisite 10 A and I.28 E. I.34 TCN is a typical example of the second one, the surfactant used here is a kind of amines with double tails. The third one like I.44P is the amines-modified nanoclay with double tails. I. 30E and I.31 PS are the representatives of the fourth family, which is changed by primary amine with one tail. There are hydroxyl groups attaching to either the tallow tail or the ammonium head in the surfactants of the last one like I.34TCN, which can exert an influence on the properties and structure of the composites.

PE nanocomposites can be prepared by a versatile process named melt compounding. At the same time, if the polymer was compatible with the clays, the nanoclays with lamellar structure can be combined with the polymer solution by melt intercalation. However in reality, hydrophilous nanoclays and hydrophobe polymeric resin are incompatible with each other at the virgin states. In order to improve the compatibility between clays and polymers, surface modification of the clay was commonly adopted (Hetzer and De Kee 2008; Fornes et al. 2002), alkyl groups with low molecular mass was introduced to convert the surface of nanoclay from hydrophile to organophile, which also results in the rise of interlaminar d-spacing (Tjong 2006). Figure 8 shows the FTIR spectra of the organo-nanoclays. The typical bands for oxides were found around 1050 cm^{-1}, which belong to the characteristic bands of Si–O–Si or Si=O, and the broad bands at 3600 cm^{-1} are associated with the hydrates including Al–OH and Si–OH (Wu et al. 2002). The bands of the bound H_2O appeared at 1640 cm^{-1} (Frankowski et al. 2007). At 523 and 466 cm^{-1}, two characteristic bands that elastic vibration by Si–O and flexural vibration by Al–O and Si–O of MMT could be found (Wang et al. 2009).

As the symmetrical and asymmetrical chains of CH_2 and CH_3 stretched, two apparent peak values of modified-nanoclays at 2924 and 2846 cm^{-1} can be found

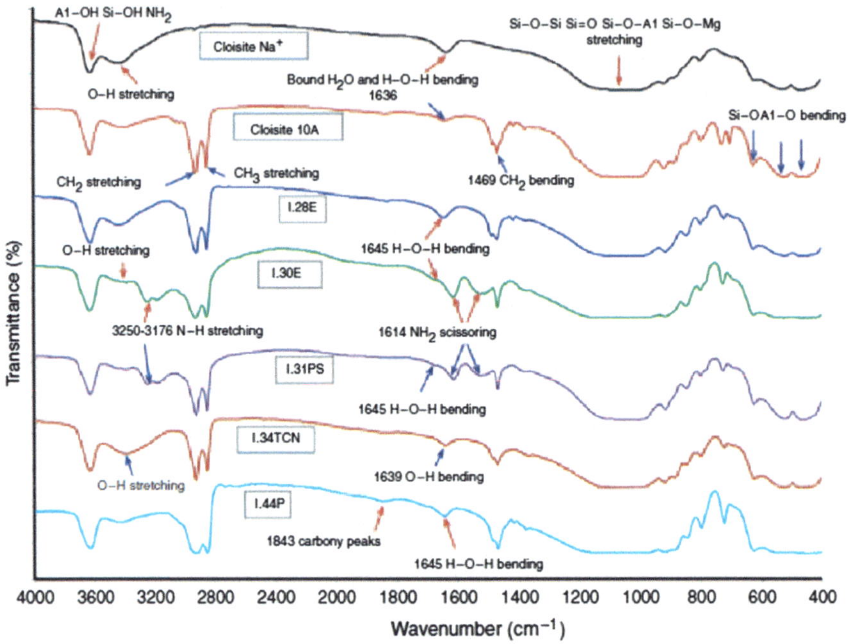

Fig. 8 FTIR spectra of organo-nanoclays with different surfactants (Gu et al. 2010)

(Park et al. 2005). The CH_2 bending corresponds to the band was at 1469 cm^{-1} (Park et al. 2005). The N–H stretching vibrations reflect two bands in the spectra of I.30 E and I.31 PS near 3200 cm^{-1}, which means that the surfactant in the modified nanoclay has amine groups (Wang et al. 2009). The band at 1639 cm^{-1} was further strengthened as indicated in Fig. 8.

Figure 9 shows the impacts of different nanoclays on the tensile property of WF/LLDPEs. The elongation of WF/LLDPEs decreases with the incorporation of nanoclays, and the growth process can be divided into two stages: the growth rate is dramatic changing under 10 wt% of nanoclays, and then the elongation of Cloisite Na^+, I.34 TCN and I.44 P decreases up to 33, 33 and 63 %, respectively

Fig. 9 Effects of nanoclays on elongation of WF/LLDPEs (Gu et al. 2010)

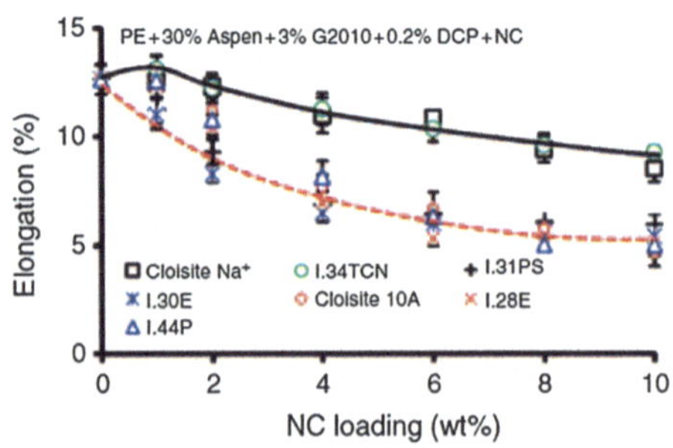

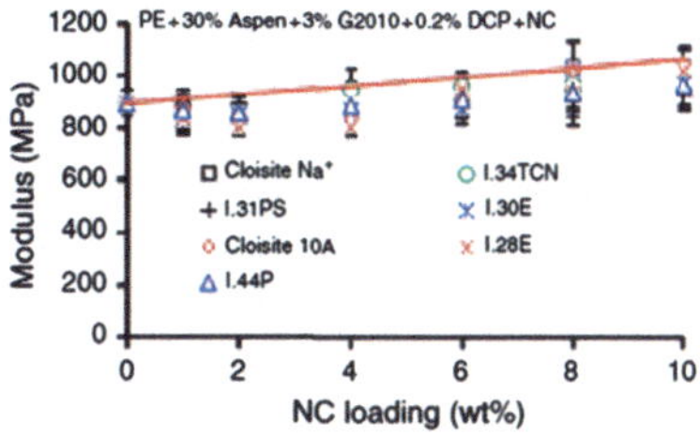

Fig. 10 Effects of nanoclays on modulus of WF/LLDPEs (Gu et al. 2010)

(Lee et al. 2005). This indicates that the surfactants of nanoclays have a critical impact on the interactions between nanoclays and composites. For the WF/LLDPEs including Cloisite 10 A and I.28 E, the elongation at maximum load is about 5 and 9 %, while for the WF/LLDPEs with I.34TCN, (Cloisite Na^+) or without nanoclays, the elongation is up to 12.6 %. The modulus as a function of nanoclays loading is shown in Fig. 10. It is easily to find that the addition of nanoclays make WF/LLDPEs more brittle, corresponding to the increase of modulus in Fig. 10. The stretch modulus show a gradual increase by adding nanoclays, and it peaked at 1000 MPa with 10 wt% nanoclays. This was because the torque and viscosity would be strengthened by the addition of nanoclays, and an increase of the contribution of crystallinity may be caused by the nanoparticles.

2.3 Storage Modulus and Loss Modulus

Rajini et al. (2013a) utilized dynamic mechanical analysis to test the dynamic rheological properties of different contents of nanoclay (0–5 wt% MMT) modified coconut sheath/polyester composite (CS/Ps), such as storage modulus (E′) and loss modulus (E″).

When the temperature is above T_g in the rubbery plateau region, the E′ value of the nanoclay incorporated composites is higher than unfilled composites. It illustrates that the dynamic storage modulus of CS/Ps increased with the addition of nanoclay. The dispersed nanoclay in the state of exfoliated or intercalated would produce efficient stress transition between the fiber and polymer resin at the nanoscale level, resulting in an effect of secondary reinforcement with CS/Ps. It can be observed that the modulus was enhanced because of the large surface-to-contact ratio of nanoclay with polymer in the hybrid nanocomposites, and it can also find the intercalation of nano-multilayers in the chains of PE (Rajini et al. 2012). Due to the strengthening of nano-scale second phase, the thermal stability of nanocomposites would be enhanced with the temperature increasing. At about 100 °C, the E′ of pure CS/Ps is 812 MPa, and the E′ value of CS/Ps with 2 and 3 wt% nanoclay keeps on increasing until more than 1580 MPa. However, evident improvement in E′ was not found with further addition of nanoclay, and it will even decrease the E′ when adding 5 wt% nanoclay. This can be interpreted as the reduction of the efficient stress transition caused by the formation of agglomeration and clustering of nanoclay.

3 Mechanical and Thermal Properties

In this section, the mechanical and thermal properties of composites were comprehensively studied and introduced. Such natural fibers/nanoclay materials are summarized in Table 4.

3.1 Mechanical Properties

Nano-clay reinforced polymer/natural fiber nanocomposites have received much attention due to their excellent characteristics which include improved physical, thermal, and mechanical properties. The addition of nano-clay to the polymer/natural fiber composites resulted in the change of mechanical properties, which providing an effectively way to board the application of polymers in industry area (Table 5).

Table 4 Representative work on nanoclay based natural fiber reinforced polymer composites

Matrix	Filler	Reference
Polypropylene (PP)	Hemp fiber/nanoclay	Kord (2012)
Polypropylene (PP)	Pineapple leaf fiber/nanoclay	Biswal et al. (2009)
Polypropylene (PP)	Banana fiber/nanoclay	Biswal et al. (2011, 2012)
Polypropylene (PP)	Bamboo fiber/nanoclay	Rahman et al. (2014)
Polypropylene (PP)	bagasse fiber/nanoclay	Nourbakhsh and Ashori (2009)
Polyethylene (PE)	Wood fiber/nanoclay	Lei et al. (2007), Lee et al. (2010)
High density polyethylene	Bamboo fiber/nanoclay	Han et al. (2008)
Poly (lactic acid HDPE)	Banana fiber/nanoclay	Kumar et al. (2013)
Polyester	Soybean oil/natural fiber/nanoclay	Haq et al. (2008)
Polyester	Wild cane grass fiber/nanoclay	Prasad et al. (2015)

Table 5 Tensile, flexural and impact properties of biocomposite and its bionanocomposites (Sajna et al. 2014)

Sample	Tensile strength (MPa)	Flexural strength (MPa)	Impact strength (J/m)
PLA + SiB(70/30)	66 ± 3	102 ± 1	150 ± 1.45
PLA + SiB + C30B	67 ± 1	180 ± 3	120 ± 3.45
PLA + NSiB (70/30)	65 ± 5	96 ± 4	143 ± 1.37
PLA + NSiB + C30B	67 ± 5	105 ± 4	119 ± 2.89

NSiB silane-treated fibre; *PLA* poly(lactic acid); *SiB* silane-treated fibre

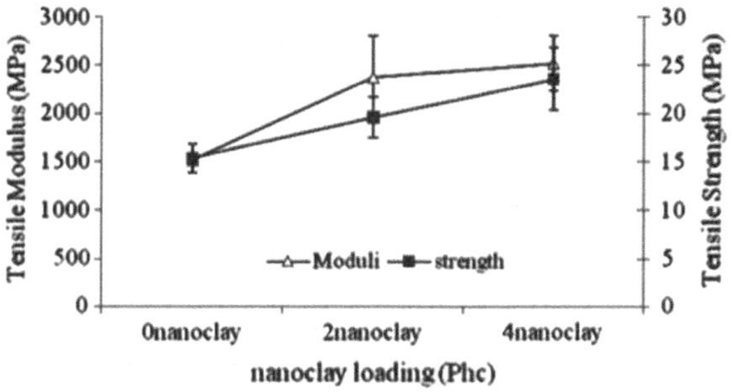

Fig. 11 Effect of nanoclay content on tensile strength and modulus (Najafi et al. 2012)

3.1.1 Tensile Properties

Najafi et al. (2012) found that the incorporation of nanoclay with MAPP as compatibilizer to PP/reed flour fiber composites significantly improved the tensile strength and Young's modulus from 14.6 and 1390 MPa to 28.7 and 2630 MPa, respectively (Fig. 11).

Nayak et al. (Biswal et al. 2009) also reported that the addition of modified nano-clay improves the mechanical strength of PP/pineapple leaf fiber. Also, the presence of MA-g-PP could also contribute to improvement in tensile strength. Similar improvement in the tensile strength was reported by many other works (Faruk and Matuana 2008; Kord and Kiakojouri 2011; Han et al. 2008; Chen and Yan 2013; Alamri et al. 2012). However, Davis D.C. et al. (Withers et al. 2015) reported that at higher nanoclay loadings large aggregates could be formed in the matrix that causes a reduction in mechanical strength. Therefore, the improvement of the tensile strength and modulus is dependent on the dispersion and morphology of nanoclay in the polymer/fiber composites (Fig. 12).

3.1.2 Flexural Properties

Great flexural strength can be achieved in the hybrid composites when nanoclay reinforced polymer/natural fiber was acted as the matrix. The dispersions of nanoclay on the composites are still important for improving the flexural properties. Also, it was found that composites with MAPP treated nanoclay provided significantly higher flexural strength and moduli, compared with untreated nanoclay composites. Ashori et al. (Tabari et al. 2011) indicated that PP/Wood flour fiber composites made with 3 % nanoclay and 5 % MAPP could achieve highest flexural strength and modulus than pure PP and PP/3 % nanoclay with 2.5 % MAPP treated. The enhancement in flexural properties was mainly ascribed to the improved adhesion between polymer composites matrix and nanoclay itself.

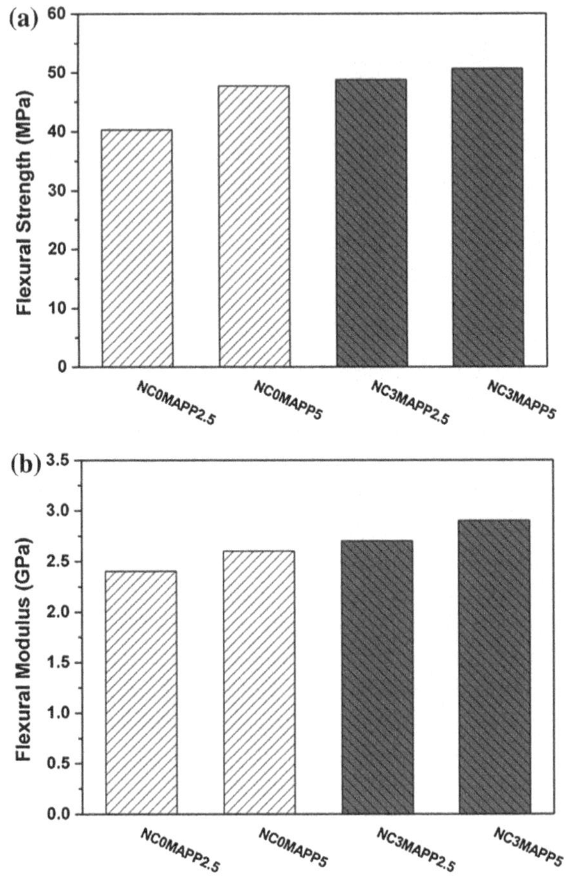

Fig. 12 Comparison of flexural properties of composites as function of the nanoclay and coupling agent loading (Tabari et al. 2011)

3.1.3 Impact Properties

In the presence of nanoclay in the polymer/natural fiber composites, the impact strength would be reduced because the nanocomposites become more brittle, thereby resulted in lower resistance for fracturing. Nayak S.K. et al. (Sajna et al. 2014) studied the addition of nanoclay on the poly (lactic acid) (PLA) with banana fiber composites. They found that the bio-nanocomposites presented maximum tensile and flexural properties than that of pure PLA, whereas the impact strength of composites was reduced due to the inherent brittle of matrix.

3.2 Thermal Properties

In the study of polymers and their practical applications, it is important to understand the thermal properties, which including thermal stability, thermal expansion, glass transition temperature, melting temperature etc., of polymer nanocomposites. In general, the incorporation of nano-clay into polymer matrix could efficiently improve the thermal properties of nanocomposites. Thus, adding nanoclay into polymer/natural fiber composites has attracted more attentions in the past decades.

3.2.1 Thermal Stability

Thermogravimetry was as a way for investigating the thermal stability of polymer composites. The thermal stability is a very important parameter which should be taken into account when choosing polymers for engineering use. Thermal stability of nanocomposites can be investigated using a thermogravimetric analyzer. The test sample was heated from air temperature (room temperature) to desired temperature (above melting temperature) at a speed of desired heating rate (5, 10, 20 °C/min) under nitrogen atmosphere. For polymer/fiber composites, the relatively low thermal stability of these fibers limited the composites application in industry. Thus, it is urgent to find an efficiently method to improve the thermal stability of polymer/ natural fiber composites, thereby boarding its application value. Hezer et al. (Xu and Van Hoa 2008) found that the incorporation of nanoclay to the polymer/wood fiber systems increased their thermal degradation onset temperature, which means the thermal stability of polymer/wood fiber composites was improved. Also, Nayak et al. (Biswal et al. 2011) investigated the thermal degradation behavior of PP/banana fiber/nano-clay, indicated that adding banana fiber into PP/nanoclay nanocomposites occurred a two-stage decomposition behavior (Fig. 13), which mainly due to the dehydration and thermal cleavage at first stage and aromatization involving the dehydration reaction at second stage. Moreover, the addition of nanoclay has been shown to reduce the thermal expansion coefficient of polymer/natural composites.

3.2.2 Thermal Properties

Thermal properties, which including glass transition temperature, melting temperature and crystallization behavior etc., are very important issue when we select which polymer to process and which polymer to use. The thermal properties of nanocomposites were studied by differential scanning calorimetry (DSC). DSC tests were performed at a heating and cooling rate of 5, 10 or 20 °C/min under nitrogen atmosphere protection environment to study the thermal properties of composites. The test samples were first heated to desired temperature (above melting temperature) and kept isothermal at this temperature for 3–5 min to eliminate thermal history of the sample, then cooled to room temperature and subsequently scanned to desired temperature at a heating rate of desired heating rate (5, 10, 20 °C/min).

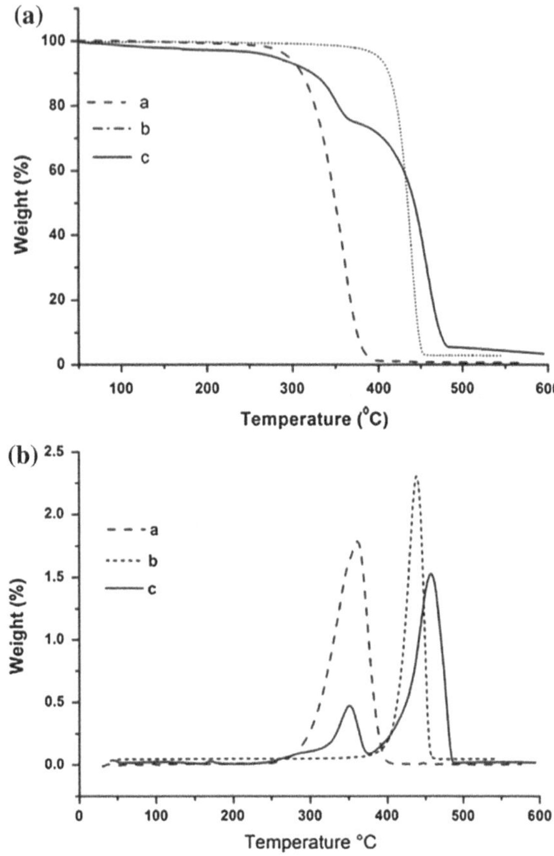

Fig. 13 TGA and DTG curve of PP, OPP, and OPP/MBF nanocomposites (OPP PP/nanoclay composites; OPP/MBF PP/nanoclay/mercerized banana fiber) (Biswal et al. 2011)

The degree of crystallinity (χ_c) of polymer was determined by Eq. (1) (Benhamou et al. 2015)

$$\chi_c = \frac{\Delta H_m - \Delta H_c}{\varphi \times \Delta H_m^0} \times 100\,\% \tag{1}$$

where, ΔH_m^0 is the melting enthalpy and ΔH_c is the crystallization enthalpy obtained by the DSC curves. ΔH_m^0 is the enthalpy of 100 % crystalline polymer sample. Moreover, it is well known that Avarami equation was normally used to analyze the rate of crystallization. The Avarami equation was presented as below (Guo et al. 2015):

$$X_t(t) = 1 - e^{-Kt^n} \tag{2}$$

where, K and n are important parameters for understanding the crystallization mechanism. In general, the addition of fiber into polymer matrix could be reduced the crystallization rate, whereas the addition of nano-clay into polymer/natural fiber composites would be shorten the crystallization time and reached the peak quickly. Also, to investigate the nucleation effect of nano-clay on PP/natural fiber composites, many researchers also studied it by Avarami equation. The values of K and n were calculated by plotting log [$-\ln(1 - X_t)$] versus log(t); n and log (K) are the slope and the intercept values, respectively. They found that there was a remarkable decrease in n, increase in K and the rate of crystallization, which indicates that the nanoclay functions as the nuclearing agent in the composites and thus enhances the degree of crystallinity and rate (Di Maio et al. 2004; Fornes and Paul 2003; Yuan et al. 2006).

The aim of this section was to investigate the benefits of reinforcing polymer/natural fiber composites with nanoclay filler. In most cases, the addition of nanoclay into polymer/natural fiber composites led to a significantly increase in both the mechanical and thermal properties. It will lead to expand the application area of polymer itself due to the most promising and encouraging mechanical and thermal properties. Further research on polymer/natural fiber/nanofiller composites not only used in automotive and industry applications, but also it should be explored further applications on biomedical areas.

4 Morphological and Structural Properties

4.1 Nanoclay-Modified Polymer Matrix

The best way for analyzing morphology of nano composites is transmission electron microscopy (TEM). It can show the size and integrity of nanoparticles from only a few nanometers to hundreds of nanometers. It can describe the decomposition and reduction of agglomerated, non-agglomerated and overlapped structures of nanoclays. Meanwhile it can reflect the discrete degree of the nanoclays in different states in polymer matrix. Figure 14 shows the three different kinds of dispersion of nanoclay in polymer matrix. Referring to immiscibility in Fig. 14, the nanoclay platelets exist in particles which composes of, to some extent, tactoids or aggregates of tactoids, just as they are in the organoclay powder, there is no separation of platelets.

Ali et al. (Ali and Ahmad 2012) focused on the mechanical property and microstructure of MMT-modified empty fruit bunch/polyurethane foam composite (EFB/PUFs). As shown in Fig. 15a, b, the aluminum and silica elements from the MMT distributed uniformly in the composite. MMT is placed at the strut of the cell structures, and with partially exfoliated and intercalated morphologies, MMT platelets were well distributed (Fig. 15c, d). This nanoclay reinforced structure leads to the high compressive modulus and strength of EFB/PUFs.

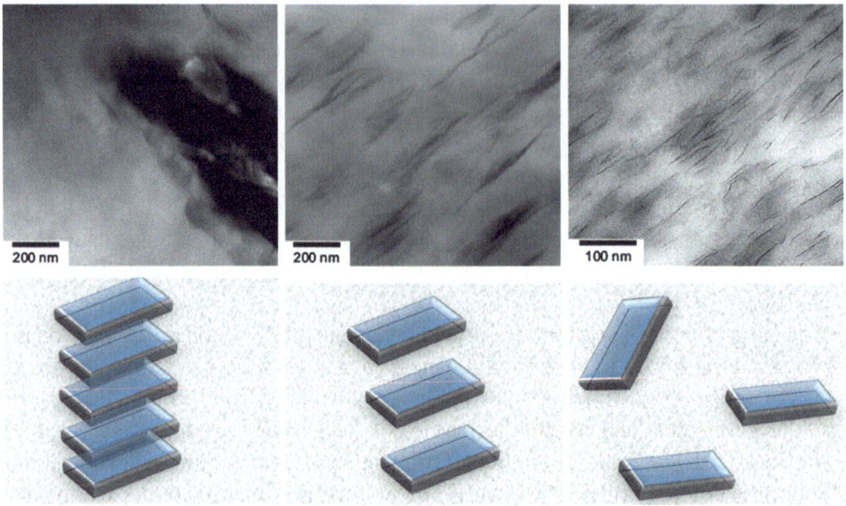

Fig. 14 Three different kinds of dispersion of nanoclay in polymer matrix **a** immiscible dispersion, **b** intercalated dispersion, **c** exfoliated dispersion

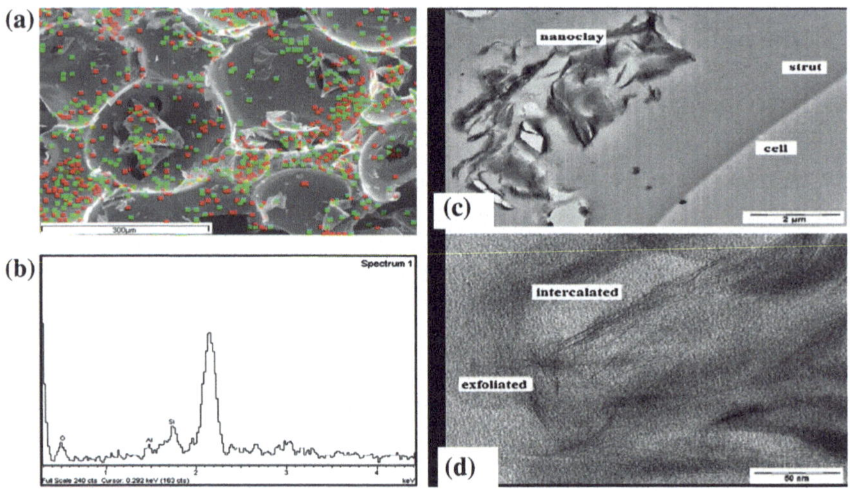

Fig. 15 The micrographs of EFB/PUFs **a** SEM and mapping elements, **b** EDX curves of traced elements, **c** and **d** TEM of EFB/PUFs and nanoclay (Ali and Ahmad 2012)

Islam et al. (2015) studied the tensile fractures morphology of wood fiber and coir reinforced PP composites with MMT nanoclay. As shown in Fig. 16, the wood fiber/PPs, coir fiber/PPs and wood fiber/coir/PP hybrid composite all had rough surface. It demonstrates the interfacial binding force between the natural

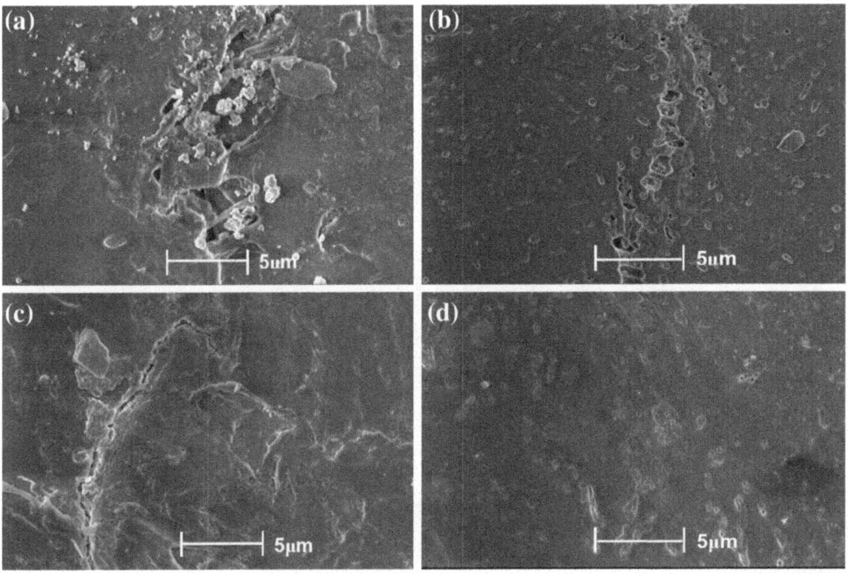

Fig. 16 SEM of the tensile fracture surfaces of the **a** wood fiber/PP composite, **b** coir/PP composite, **c** wood fiber/coir/PP hybrid composite, and **d** MMT-modified wood fiber/coir/PP hybrid composite (Islam et al. 2015)

fibers and during impact, polymeric matrix was weak to resist fiber pull-out, as the energy consumption was small. The MMT-modified wood fiber/coir/PP hybrid composite exhibited a smoother surface than the single fiber composite. As a result, high interfacial adhesion and surface smoothness is obtained by the addition of nanoclay in the hybrid composite, it also increased the mechanical properties of the hybrid composite (Smart et al. 2008).

4.2 Natural Fiber

Kovacevic et al. (2015) researched on the influence of MMT nanoclay modified Spartium junceum L. fiber (SJLF) on thermal properties of SJLF/ PLA composites. Figure 17 shows the SEM picture of pure SJLF, NaOH-treated SJLF, MMT-treated SJLF, and MMT-citric acid (CA)-treated SJLF reinforced PLA composites. The surface of pure SJLF/PLAs was smooth and regular compared with other composites. With the additional treatment of the SJLF, there is increasing roughness on the surface of composites from Fig. 17a to d. With the crosslinker (CA) introduced into the MMT-treated SJLF/PLAs, the size range of clusters of nanoclay became larger.

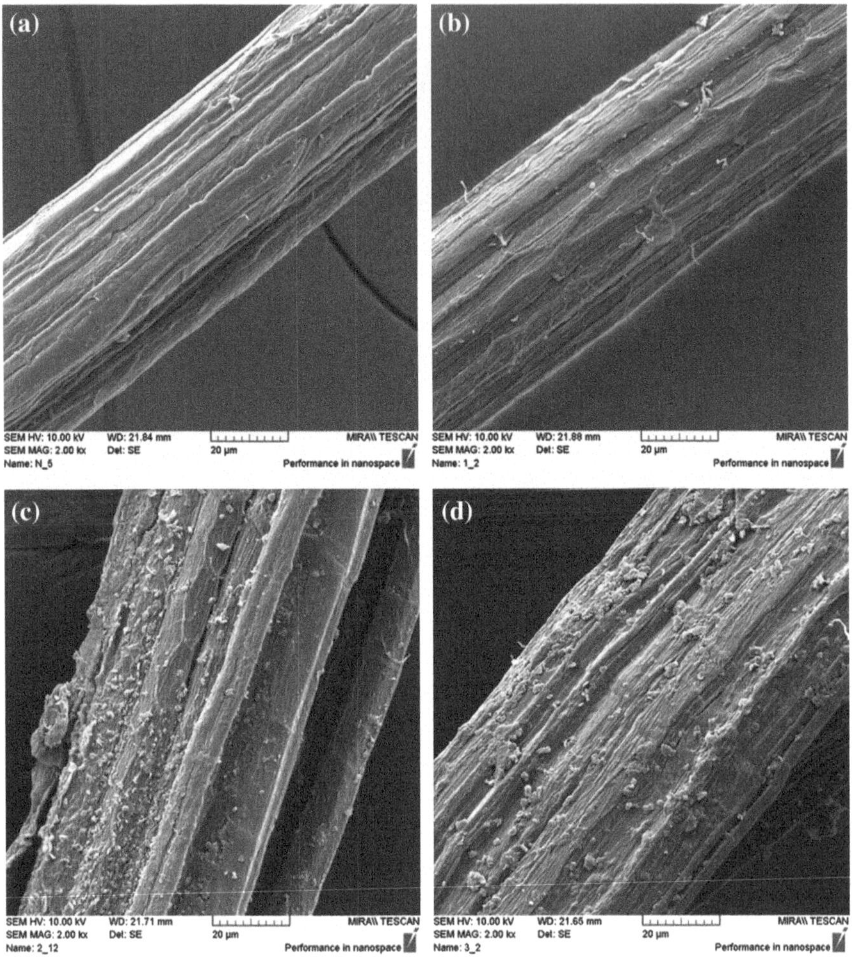

Fig. 17 SEM of **a** pure SJLF, **b** NaOH-treated SJLF, **c** MMT-treated SJLF, and **d** MMT-CA-treated SJLF reinforced PLA composites (Kovacevic et al. 2015)

Figure 18 shows the SEM picture of tensile fracture surface of the four composites above. Interface debonding and fiber pull-out were shown in (a) and (b). The interfacial adhesion between the pure SJLF and matrix was lower (strength of composite is 41.87 MPa) than NaOH-treated SJLF (strength of composite is 42.65 MPa). The MMT-treated SJLF/PLAs had a low strength (19.81 MPa) and fiber-matrix interface separation of the composite was observed (Fig. 18c). The MMT-CA-treated SJLF/PLAs showed smooth fractured surface (Fig. 18d) and good mechanical property (46.67 MPa). It illustrates that the combination of nanoclay and modifier may increase the interaction between polymer and matrix natural fiber.

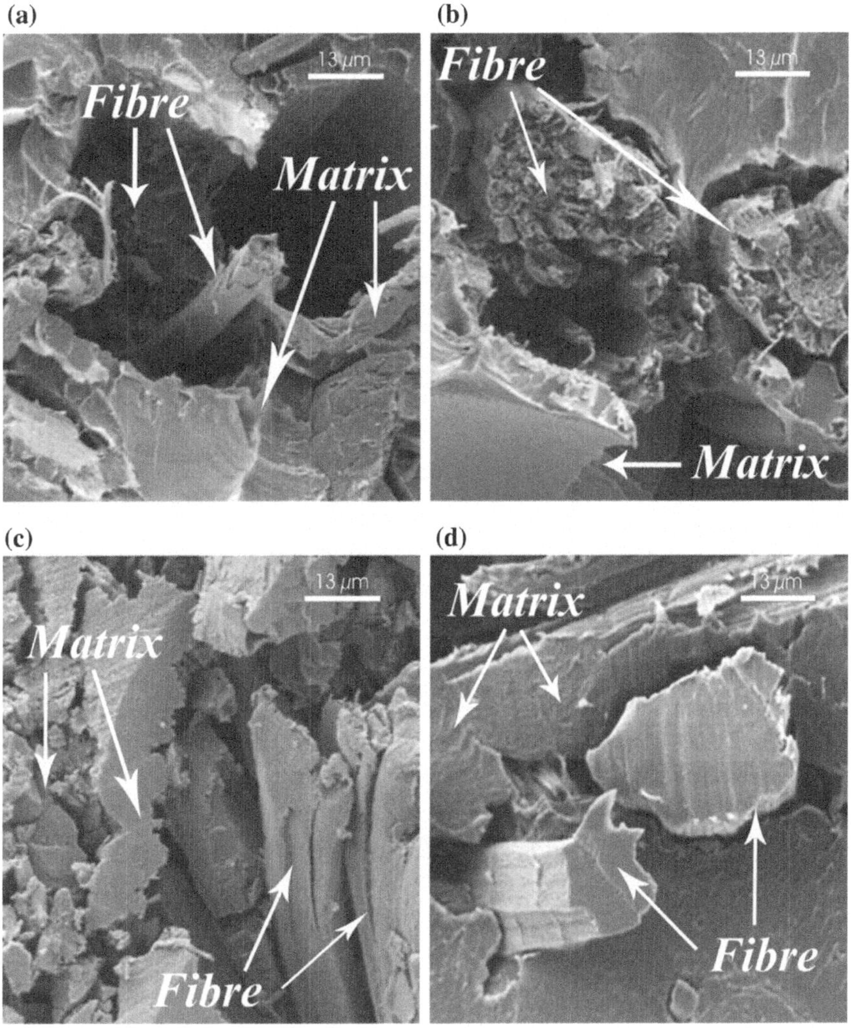

Fig. 18 SEM of tensile fracture surface of **a** pure SJLF, **b** NaOH-treated SJLF, **c** MMT-treated SJLF, and **d** MMT-CA-treated SJLF reinforced PLA composites (Kovacevic et al. 2015)

5 Modeling of Mechanical and Rheological Properties

5.1 Computation Model for Theoretical Modulus

To estimate the composite properties P^*, we can use the micromechanical analysis by the properties of each components according to the rule of mixtures:

$$P^* = P_f \times V_f + P_m \times V_m \tag{3}$$

In Eq. (1), the properties of the fibers (P_f) and matrix (P_m) have influence on the homogenized property P through their volume fractions V_f and V_m. The rule of mixture regards to the interface between phases as perfect condition, which means ignoring the effects of fibers and matrix on each other. Two new coefficients n and g relying on the individual phases and the effective property evaluated by rule of mixture model were put forward by Halpin and Tsai (Halpin and Kardos 1976), which can compensate the effect of the interface condition on the composite effective properties:

$$\xi = \frac{P_f(P^* - P_m) - V_f \times P^*(P_f - P_m)}{P_m(P_f - P^*) - V_m(P_f - P_m)} \tag{4}$$

$$\eta = \frac{P_f - P_m}{P_f + \xi \times P_m} \tag{5}$$

To be suitable for non-perfect interface conditions, the Halpin-Tsai model which estimates the effective properties of unidirectional composite materials could be demonstrated by the following equation:

$$P^* = \frac{P_m(1 + \xi \times \eta \times V_f)}{1 - \eta \times V_f} \tag{6}$$

The composites of flax fiber/soy protein resin (FF/SPRs) modified with nanoclay (Cloisite Na^+) were prepared by Huang et al. (Huang and Netravali 2007). Moreover, they provided the theoretical modulus through the calculation of mixture:

$$E_c = E_y V_y + E_m(1 - V_y) \tag{7}$$

When the fiber volume percentage (V_y) is 48 %, modulus E_y is 12.1 GPa and modulus of resin E_m is 969 MPa, the composite modulus E_c calculated was about 6.3 GPa. Meanwhile, the experimental Young's moduli of FF/SPRs with and without nanoclay were 4.3 and 3.8 GPa, respectively. In the process of fabrication, both the loss of alignment caused by resin shrinking and voids generated in the composites resulted from water vaporization can lead to the difference between the calculated and experimental values. The process can be slower or divided to two or three stages in order to solve these problems.

Yeh et al. (2005) modified the wood fiber/polypropylene (PP) composites (WF/PPs) with layered silicates. MAPEs also used as a compatibilizer. It is demonstrated that the Young's modulus of WF/PPs was increased from 4.2 to 4.58 GPa with a 10 wt% clay loading by the addition of layered silicates. Besides, the adhesion between polymer matrix and wood fibers became poor due to the incorporation of layered silicates to the wood/PP matrix. So it was guessed that the adhesion between the polymer matrix and wood fibers could be improved by the addition of more compatibilizer. The following relation is a modified mixing rule which the modulus of layered silicates reinforced WF/PPs could be expressed:

$$E_c = E_f \xi_1 \xi_0 V_f + E_m (1 - V_f) \quad (8)$$

In the equation, E_c, E_f, and E_m represent the moduli of the composite, fiber and matrix, respectively, V_f means the volume fraction of fiber, $\xi_1 \xi_0$ are length and orientation correction factors, respectively.

5.2 *Dynamic Mechanical Model*

Rajini et al. (2013b) examined the effect of MMT addition and surface treatment of coconut sheath on CS/Ps. To research the impact of nanoclay addition on free shaking characteristic, they have tested the natural frequency and modal loss factor associated with some low frequency bending modes in all cases. Modal damping factor ($\zeta = \Delta\omega / 2\omega_n$) of first three modes are gained by the half power bandwidth method for each case.

The analytical and numerical model is mainly based on the assumption that the CS/HPs are quasi-isotropic. The following equation is to express the angular natural frequencies of first three modes of a cantilever isotropic beam:

$$\omega_n = \beta_n^2 \sqrt{EI / \rho A L^4} \quad (9)$$

where E represents the Young's modulus, I means the area moment of inertia, A is the cross-sectional area, L is the length of the beam, ρ means the density of the beam, and β represents the first n lowest frequency bending modes.

The natural frequencies estimated experimentally are in accordance with analytical and numerical results. As a result of the increased surface area exposed by the nanoclay leading to high internal damping, the fiber and matrix interface will be improved and the frictional resistance and mechanical interlock can be supplied by the incorporation of nanoclay with fiber surface (Ávila et al. 2008). Because the interface between fiber and matrix is weak as a result of the agglomeration and void formation of clay platelets, the damping ratio value becomes reduced as the increase of nanoclay percentage (>3 %).

5.3 *Rheological Properties of Composites*

The amount of the particles in the mixture has great influence on the rheological properties. To explain the effect of filler concentration on the viscosity of suspensions, there are a lot of equations have been put forward. The following relation is one of the equations which describe the effect of volume fraction of the particles φ on the zero-shear viscosity η_0 (Marcovich et al. 2004):

$$\eta_0 = \eta_{0s}(1 - \frac{\varphi}{\varphi_m})^{-b} \tag{10}$$

where η_0 is the zero-shear rate viscosity of the solvent, φ is the volume fraction of the particles, φ_m is the volume fraction at maximum packing and b is an exponent, which is equal to 2 for spherical particles. The Carreau-Yasuda model (Carreau et al. 1997) shown on Eq. (7) is the most common model used for non-Newtonian flows.

$$\frac{\eta - \eta_\infty}{\eta_0 - \eta_\infty} = [1 + (\lambda\gamma)^\alpha]^{(n-1)/a} \tag{11}$$

where η is the shear viscosity, η_0 is the zero-shear rate viscosity, η_∞ is the limiting viscosity at large shear rates, γ is the shear rate, λ is a relaxation time, n is the pseudoplastic parameter (power-law), and a is a parameter which is associated with the curvature in the transition zone between η_0 and the power-law region. Generally, setting $a = 2$ and $\eta_\infty = 0$ can have a good fit which lead to the Carreau model. At last, a simple power for the complex viscosity law expression can represent observed behavior for highly concentrated materials (Marcovich et al. 2004):

$$\eta^* = h(\omega)^{n-1} \tag{12}$$

where η^* represents the complex viscosity, h is a constant, ω is the frequency (rad/s) and n is a power-law index.

The storage modulus (E′), loss modulus (E″) and complex viscosity (η^*) are the most important characteristic parameter in dynamic rheological properties studies (Ying-Chen et al. 2010). The storage modulus (E′) takes as a measure of the energy stored in the materials and recovers from its per capita. It depend on what rearrangements can happen during the period of oscillation, which uses to evaluate the elastic behaviors of the materials. In addition, the energy squandered or lost per capita of sinusoidal deformation is measured by the loss modulus (E″), and it is representative of the viscous behaviors. The viscosity parameter is a measure of the resistance to flow, and the complex viscosity defined as below especially (Shroff and Mavridis 1999):

$$\eta^* = (E'^2 + E''^2)^{1/2}/\omega \tag{13}$$

5.4 Crystallization Kinetics Model

Among the natural fiber plastic composite polymers, there are around 80 % of them semicrystalline polymers. It contains four kinds of crystallization behavior of the polymer matrix. Such behaviors are imperative in determining the final mechanical and thermal properties of the composites (Lenes and Gregersen 2006; Zafeiropoulos et al. 2001).

The relative crystallinity X(t) is the proportion value (crystallized volume/total crystallizable volume) at time t and it was described as following (Zhang et al. 2006):

$$X(t) = \frac{\int_0^t H(t)dt}{\int_0^{t_\infty} H(t)dt} \tag{14}$$

The Avrami equation (Avrami 1939) was defined as following:

$$X(t) = 1 - \exp(-Kt^n) \tag{15}$$

where K denotes crystallization rate constant, and n denotes Avrami exponent related to the form of crystal growth and the mechanism of nucleation (Avrami 1940, 1941).

5.5 Thermal Degradation Kinetics Model

There used to be a variety of kinetic studies been carried out with TGA at different heating rate to determine the rate constants (k), activation energy (E_a), reaction order (n) and Arrhenius pre-exponential factor (A) (Feng et al. 2014). The reaction rate of solid-state reaction ($\frac{d\alpha}{dt}$) can be expressed with the following equation regarding the rate constant k and reaction model f(α):

$$\frac{d\alpha}{dt} = kf(\alpha) \tag{16}$$

The conversion rate α, is derived by weight loss:

$$\alpha = \frac{W_0 - W_t}{W_0 - W_f} \tag{17}$$

where W_0, W_t and W_f are the weight of the samples at different times, respectively represent at initial degradation, the time (t) and final degradation.

Depending on the Arrhenius formula, k can be described as:

$$k = Ae^{(\frac{-E_a}{RT})} \tag{18}$$

Equation 1 becomes:

$$\frac{d\alpha}{dt} = Ae^{(\frac{-E_a}{RT})} f(\alpha) \tag{19}$$

where A stands for the pre-exponential factor (1/s), E_a stands for the activation energy (kJ/mol), R the gas constant (J/mol K) and T the temperature.

Equation 6 shows the relationship between the reaction rate, transformation efficiency and the experimental recording temperature. The heating rate β (°C/min) and change in temperature with time (dT/dt) are adopted to kinetically investigate the TGA, so Eq. 6 can be expressed as follows:

$$\frac{d\alpha}{dT} = \frac{A}{\beta} e^{(\frac{-E_a}{RT})} f(\alpha) \tag{20}$$

Isoconversional methods are widely used for determining the E_a of polymers. The two main isoconversional methods, Friedman method (Friedman 1964) and Doyle's linear approximation (Doyle 1962) applied Flynn-Wall-Ozawa (FWO) method (Ozawa 1965; Flynn 1983; Flynn and Wall 1966), are used to evaluate the accurate value of the E_a at an exact conversion rate. The Friedman method and the Flynn-Wall-Ozawa method follow different procedure (Carrasco et al. 2013).

Friedman equation is as following:

$$\ln \frac{d\alpha}{dt} = A \ln f(\alpha) - \frac{E_a}{RT} \tag{21}$$

The E_a is obtained from the slope of ln $\frac{d\alpha}{dt}$ against 1/T from TGA curves at different heating rates.

Flynn-Wall-Ozawa equation:

$$\log \beta = [\log \frac{AE_a}{R} - \log f(\alpha) - 2.315] - 0.4567 \frac{E_a}{RT} \tag{22}$$

The E_a by FWO method is the slope of log β against 1/T for a variety of degrees of conversion.

Kissinger et al. made use of an equation to describe the E_a. The equation is expressed as follows:

$$\ln \frac{\beta}{T_p^2} = \ln \frac{RA}{E_a} - \frac{E_a}{RT_p} \tag{23}$$

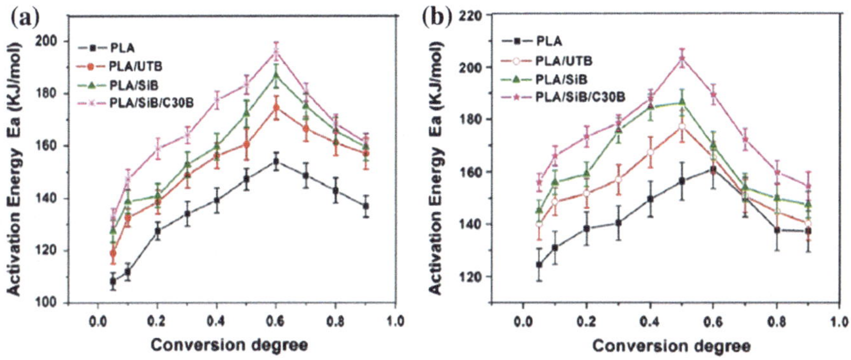

Fig. 19 Activation energy at various conversion rates for PLA, PLA/UTB, PLA/SiB and PLA/SiB/C30B obtained by **a** Friedman method; **b** FWO method (VP et al. 2015)

where T_p is the peak temperature from the corresponding DTG curve. E_a is obtained from the slope of $\ln\frac{\beta}{T_p^2}$ against $1/T_p$ at various heating rates.

VP et al. (2015) studied the reinforcing effect of untreated banana fiber (UTB), silane treated-banana fiber (SiB) and nanoclay (Closite 30B (C30B)) on the thermal degradation kinetics of Poly (lactic acid) (PLA). Based on VP's cone calorimetry test results, nanoclay would not volatilize within a short time, therefore, it delays the ignition of nanocomposites. The result is consistent with Fig. 19.

6 Conclusions and Future Prospective

Nanoclays play an important role to enhance composite performance through the way of enhancing their properties such as thermal stability, mechanical strength, and barrier properties. The content, shape, size, and the affinity towards matrix material are some of the characteristic factors contribute most to modify the properties of the composites. With their enhanced performance, nanoclay filled polymer matrix based nanocomposites have drawn much attention in the materials industry. A common recognition in this area is that exfoliated clays perform superior strength, modulus and higher barrier properties compared to pure polymer matrix. As nanotechnology developed very fast, great interest has been drawn in the correlation of material properties with filler size. Nanoclays have attracted a great deal of interest due to the fact that they have demonstrated cutting-edge properties that cannot be achieved with their traditional micro-scale counter parts.

References

Alamri, H., Low, I.M.: Effect of water absorption on the mechanical properties of n-SiC filled recycled cellulose fibre reinforced epoxy eco-nanocomposites. Polym. Testing **31**(6), 810–818 (2012)

Alamri, H., Low, I.M., Alothman, Z.: Mechanical, thermal and microstructural characteristics of cellulose fibre reinforced epoxy/organoclay nanocomposites. Compos. Part B-Eng. **43**(7), 2762–2771 (2012). doi:10.1016/j.compositesb.2012.04.037

Ali, E.S., Ahmad, S.: Bionanocomposite hybrid polyurethane foam reinforced with empty fruit bunch and nanoclay. Compos. B Eng. **43**(7), 2813–2816 (2012)

Ávila, A.F., Donadon, L.V., Duarte, H.V.: Modal analysis on nanoclay epoxy-based fiber-glass laminates. Compos. Struct. **83**(3), 324–333 (2008)

Avrami, M.: Kinetics of phase change. I General theory. J. Chem. Phys. **7**(12), 1103–1112 (1939)

Avrami, M.: Kinetics of phase change. II transformation-time relations for random distribution of nuclei. J. Chem. Phys. **8**(2), 212–224 (1940)

Avrami, M.: Granulation, phase change, and microstructure kinetics of phase change.III. J. Chem. Phys. **9**(2), 177–184 (1941)

Bajpai, P.K., Singh, I., Madaan, J.: Joining of natural fiber reinforced composites using microwave energy: experimental and finite element study. Mater. Design **35**, 596–602 (2012). doi:10.1016/j.matdes.2011.10.007

Bajpai, P.K., Singh, I., Madaan, J.: Frictional and adhesive wear performance of natural fibre reinforced polypropylene composites. P. I. Mech. Eng. J.-J. Eng. **227**(J4), 385–392 (2013). doi:10.1177/1350650112461868

Bajpai, P.K., Singh, I., Madaan, J.: Development and characterization of PLA-based green composites: a review. J. Thermoplast. Compos. **27**(1), 52–81 (2014). doi:10.1177/0892705712439571

Bartholmai, M., Schartel, B.: Layered silicate polymer nanocomposites: new approach or illusion for fire retardancy? Investigations of the potentials and the tasks using a model system. Polym. Adv. Technol. **15**(7), 355–364 (2004)

Benhamou, K., Kaddami, H., Magnin, A., Dufresne, A., Ahmad, A.: Bio-based polyurethane reinforced with cellulose nanofibers: a comprehensive investigation on the effect of interface. Carbohyd. Polym. **122**, 202–211 (2015). doi:10.1016/j.carbpol.2014.12.081

Bensadoun, F., Kchit, N., Billotte, C., Bickerton, S., Trochu, F., Ruiz, E.: A study of nanoclay reinforcement of biocomposites made by liquid composite molding. Int. J. Polym. Sci. (2011)

Bilotti, E., Fischer, H., Peijs, T.: Polymer nanocomposites based on needle-like sepiolite clays: effect of functionalized polymers on the dispersion of nanofiller, crystallinity, and mechanical properties. J. Appl. Polym. Sci. **107**(2), 1116–1123 (2008)

Biswal, M., Mohanty, S., Nayak, S.K.: Influence of organically modified nanoclay on the performance of pineapple leaf fiber-reinforced polypropylene nanocomposites. J. Appl. Polym. Sci. **114**(6), 4091–4103 (2009). doi:10.1002/app.31121

Biswal, M., Mohanty, S., Nayak, S.K.: Mechanical, thermal and dynamic-mechanical behavior of banana fiber reinforced polypropylene nanocomposites. Polym. Compos. **32**(8), 1190–1201 (2011). doi:10.1002/pc.21138

Biswal, M., Mohanty, S., Nayak, S.K.: Banana fiber-reinforced polypropylene nanocomposites: effect of fiber treatment on mechanical, thermal, and dynamic-mechanical properties. J. Thermoplast. Compos. **25**(6), 765–790 (2012). doi:10.1177/0892705711413626

Cantero, G., Arbelaiz, A., Mugika, F., Valea, A., Mondragon, I.: Mechanical behavior of wood/polypropylene composites: effects of fibre treatments and ageing processes. J. Reinf. Plast. Compos. **22**(1), 37–50 (2003)

Carrasco, F., Pérez-Maqueda, L.A., Sánchez-Jiménez, P., Perejón, A., Santana, O., Maspoch, M. L.: Enhanced general analytical equation for the kinetics of the thermal degradation of poly (lactic acid) driven by random scission. Polym. Testing **32**(5), 937–945 (2013)

Carreau, P.J., De Kee, D., Chhabra, R.P.: Rheology of Polymeric Systems: Principles and Applications. Hanser Publishers Munich (1997)

Chen, J.M., Yan, N.: Mechanical properties and dimensional stability of organo-nanoclay modified biofiber polymer composites. Compos. Part B-Eng. **47**, 248–254 (2013). doi:10.1016/j.compositesb.2012.11.015

Chen, Z., Wu, W.P., Chen, Z.F., Cong, X.N., Qiu, J.L.: Microstructural characterization on ZrC doped carbon/carbon composites. Ceram. Int. **38**(1), 761–767 (2012). doi:10.1016/j.ceramint.2011.08.002

Chen, Z., Chen, Z., Yang, Z., Hu, J., Yang, Y., Chang, L., Lee, L.J., Xu, T.: Preparation and characterization of vacuum insulation panels with super-stratified glass fiber core material. Energy **93**(Part 1), 945–954 (2015a). doi:10.1016/j.energy.2015.08.105

Chen, Z., Wu, C., Yang, Y., Shi, J., Hu, J., Yang, Z., Chen, Z.: Synthesis and drug delivery of mesoporous silica nanoparticles for cancer therapy. Eur. J. BioMed. Res. **1**(3), 30–36 (2015b)

Di Maio, E., Iannace, S., Sorrentino, L., Nicolais, L.: Isothermal crystallization in PCL/clay nanocomposites investigated with thermal and rheometric methods. Polymer **45**(26), 8893–8900 (2004). doi:10.1016/j.polymer.2004.10.037

Doyle, C.: Estimating isothermal life from thermogravimetric data. J. Appl. Polym. Sci. **6**(24), 639–642 (1962)

Duran, J., Ramos-Tejada, M., Arroyo, F., Gonzalez-Caballero, F.: Rheological and electrokinetic properties of sodium montmorillonite suspensions: I. Rheological properties and interparticle energy of interaction. J. Colloid Interface Sci. **229**(1), 107–117 (2000)

Faruk, O., Matuana, L.M.: Nanoclay reinforced HDPE as a matrix for wood-plastic composites. Compos. Sci. Technol. **68**(9), 2073–2077 (2008). doi:10.1016/j.compscitech.2008.03.004

Feng, Y., Wang, B., Wang, F., Zhao, Y., Liu, C., Chen, J., Shen, C.: Thermal degradation mechanism and kinetics of polycarbonate/silica nanocomposites. Polym. Degrad. Stab. **107**, 129–138 (2014)

Flynn, J.: The isoconversional method for determination of energy of activation at constant heating rates: corrections for the Doyle approximation. J. Therm. Anal. Calorim. **27**(1), 95–102 (1983)

Flynn, J.H., Wall, L.A.: General treatment of the thermogravimetry of polymers. J. Res. Nat. Bur. Stand. **70**(6), 487–523 (1966)

Fornes, T., Yoon, P., Hunter, D., Keskkula, H., Paul, D.: Effect of organoclay structure on nylon 6 nanocomposite morphology and properties. Polymer **43**(22), 5915–5933 (2002)

Fornes, T.D., Paul, D.R.: Crystallization behavior of nylon 6 nanocomposites. Polymer **44**(14), 3945–3961 (2003). doi:10.1016/S0032-3861(03)00344-6

Frankowski, D.J., Capracotta, M.D., Martin, J.D., Khan, S.A., Spontak, R.J.: Stability of organically modified montmorillonites and their polystyrene nanocomposites after prolonged thermal treatment. Chem. Mater. **19**(11), 2757–2767 (2007)

Friedman, H.L.: Kinetics of thermal degradation of char-forming plastics from thermogravimetry. Application to a phenolic plastic. In: Journal of Polymer Science Part C: Polymer Symposia, vol. 1, pp 183–195. Wiley Online Library (1964)

Gu, R., Kokta, B.V., Michalkova, D., Dimzoski, B., Fortelny, I., Slouf, M., Krulis, Z.: Characteristics of wood–plastic composites reinforced with organo-nanoclays. J. Reinf. Plast. Compos. (2010)

Guo, L., Chen, F.X., Zhou, Y.S., Liu, X., Xu, W.L.: The influence of interface and thermal conductivity of filler on the nonisothermal crystallization kinetics of polypropylene/natural protein fiber composites. Compos. Part B-Eng. **68**, 300–309 (2015). doi:10.1016/j.compositesb.2014.09.004

Halpin, J.C., Kardos, J.: The Halpin-Tsai equations: a review. Polym. Eng. Sci. **16**(5), 344–352 (1976)

Han, G., Lei, Y., Wu, Q., Kojima, Y., Suzuki, S.: Bamboo-fiber filled high density polyethylene composites: effect of coupling treatment and nanoclay. J. Polym. Environ. **16**(2), 123–130 (2008). doi:10.1007/s10924-008-0094-7

Haq, M., Burgueno, R., Mohanty, A.K., Misra, M.: Hybrid bio-based composites from blends of unsaturated polyester and soybean oil reinforced with nanoclay and natural fibers. Compos. Sci. Technol. **68**(15–16), 3344–3351 (2008). doi:10.1016/j.compscitech.2008.09.007

Haq, M., Burgueño, R., Mohanty, A.K., Misra, M.: Processing techniques for bio-based unsaturated-polyester/clay nanocomposites: tensile properties, efficiency, and limits. Compos. A Appl. Sci. Manuf. **40**(4), 394–403 (2009)

Hetzer, M., De Kee, D.: Wood/polymer/nanoclay composites, environmentally friendly sustainable technology: a review. Chem. Eng. Res. Des. **86**(10), 1083–1093 (2008)

Hetzer, M., Naiki, J., Zhou, H., Poloso, T., De Kee, D.: Thermal dependence of Young's modulus of wood/polymer/clay nanocomposites. J. Compos. Mater. **43**(20), 2285–2301 (2009)

Huang, X., Netravali, A.: Characterization of flax fiber reinforced soy protein resin based green composites modified with nano-clay particles. Compos. Sci. Technol. **67**(10), 2005–2014 (2007)

Islam, M.S., Ahmad, M.B., Hasan, M., Aziz, S.A., Jawaid, M., Haafiz, M.M., Zakaria, S.A.: Natural fiber-reinforced hybrid polymer nanocomposites: effect of fiber mixing and nanoclay on physical, mechanical, and biodegradable properties. BioResources **10**(1), 1394–1407 (2015)

Kord, B.: Effect of nanoparticles loading on properties of polymeric composite based on Hemp Fiber/Polypropylene. J. Thermoplast. Compos. **25**(7), 793–806 (2012). doi:10.1177/0892705711412815

Kord, B., Kiakojouri, S.M.H.: Effect of nanoclay dispersion on physical and mechanical properties of wood flour/polypropylene/glass fiber hybrid composites. Bioresources **6**(2), 1741–1751 (2011)

Kovacevic, Z., Bischof, S., Fan, M.: The influence of Spartium junceum L. fibres modified with montmorrilonite nanoclay on the thermal properties of PLA biocomposites. Compos. B Eng. **78**, 122–130 (2015)

Kumar, V., Sharma, N.K., Kumar, R.: Dielectric, mechanical, and thermal properties of bamboo-polylactic acid bionanocomposites. J. Reinf. Plast. Comp. **32**(1), 42–51 (2013). doi:10.1177/0731684412461290

Lee, J.-H., Jung, D., Hong, C.-E., Rhee, K.Y., Advani, S.G.: Properties of polyethylene-layered silicate nanocomposites prepared by melt intercalation with a PP-g-MA compatibilizer. Compos. Sci. Technol. **65**(13), 1996–2002 (2005)

Lee, Y.H., Kuboki, T., Park, C.B., Sain, M.: The effects of nanoclay on the extrusion foaming of wood fiber/polyethylene nanocomposites. Polym. Eng. Sci. **51**(5), 1014–1022 (2011)

Lee, Y.H., Kuboki, T., Park, C.B., Sain, M., Kontopoulou, M.: The effects of clay dispersion on the mechanical, physical, and flame-retarding properties of wood fiber/polyethylene/clay nanocomposites. J. Appl. Polym. Sci. **118**(1), 452–461 (2010). doi:10.1002/app.32045

Lei, Y., Wu, Q., ClemonS, C.M., Yao, F., Xu, Y.: Influence of nanoclay on properties of HDPE/Wood composites. J. Appl. Polym. Sci. **106**(6), 3958–3966 (2007). doi:10.1002/app.27048

Lenes, M., Gregersen, Ø.W.: Effect of surface chemistry and topography of sulphite fibres on the transcrystallinity of polypropylene. Cellulose **13**(4), 345–355 (2006)

Leng, J., Lau, A.K.T.: Multifunctional Polymer Nanocomposites. CRC Press, Boca Raton (2011)

Mallick, P.K.: Fiber-Reinforced Composites: Materials, Manufacturing, and Design, 3rd edn. CRC Press, Boca Raton, FL (2008)

Marcovich, N.E., Reboredo, M.M., Kenny, J., Aranguren, M.I.: Rheology of particle suspensions in viscoelastic media. Wood flour-polypropylene melt. Rheolog. Acta **43**(3), 293–303 (2004)

Mohanty, A.K., Misra, M., Drzal, L.T.: Natural Fibers, Biopolymers, and Biocomposites. Taylor & Francis, Boca Raton, FL (2005)

Morote-Martínez, V., Torregrosa-Coque, R., Martín-Martínez, J.M.: Addition of unmodified nanoclay to improve the performance of unsaturated polyester resin coating on natural stone. Int. J. Adhes. Adhes. **31**(3), 154–163 (2011)

Morrison, F.A.: Understanding Rheology. Oxford University Press (2001)

Mravčáková, M., Boukerma, K., Omastová, M., Chehimi, M.M.: Montmorillonite/polypyrrole nanocomposites. The effect of organic modification of clay on the chemical and electrical properties. Mater. Sci. Eng. C **26**(2), 306–313 (2006)

Najafi, A., Kord, B., Abdi, A., Ranaee, S.: The impact of the nature of nanoclay on physical and mechanical properties of polypropylene/reed flour nanocomposites. J. Thermoplast. Compos. **25**(6), 717–727 (2012). doi:10.1177/0892705711412813

Nazare, S., Kandola, B., Horrocks, A.: Flame-retardant unsaturated polyester resin incorporating nanoclays. Polym. Adv. Technol. **17**(4), 294–303 (2006)

Nguong, C., Lee, S., Sujan, D.: A review on natural fibre reinforced polymer composites. World Acad. Sci. Eng. Technol. **2013**, 1123–1130 (2013)

Nourbakhsh, A., Ashori, A.: Influence of nanoclay and coupling agent on the physical and mechanical properties of polypropylene/bagasse nanocomposite. J. Appl. Polym. Sci. **112**(3), 1386–1390 (2009). doi:10.1002/app.29499

Osman, M.A., Mittal, V., Morbidelli, M., Suter, U.W.: Epoxy-layered silicate nanocomposites and their gas permeation properties. Macromolecules **37**(19), 7250–7257 (2004)

Ozawa, T.: A new method of analyzing thermogravimetric data. B Chem. Soc. Jpn. **38**(11), 1881–1886 (1965)

Park, S.-J., Li, K., Hong, S.-K.: Preparation and Characterization of layered silicate-modified ultrahigh-molecular-weight polyethylene nanocomposites technology. **8**, 10 (2005)

Patel, M., Bastioli, C., Marini, L., Würdinger, E.: Life-cycle assessment of bio-based polymers and natural fiber composites. Biopolymers (2005)

Prasad, A.V.R., Rao, K.B., Rao, K.M., Ramanaiah, K., Gudapati, S.P.K.: Influence of nanoclay on the mechanical performance of wild cane grass fiber-reinforced polyester nanocomposites. Int. J. Polym. Anal. Ch. **20**(6), 541–556 (2015). doi:10.1080/1023666X.2015.1053335

Pratheep Kumar, A., Pal Singh, R.: Novel hybrid of clay, cellulose, and thermoplastics. I. Preparation and characterization of composites of ethylene–propylene copolymer. J. Appl. Polym. Sci. **104**(4), 2672–2682 (2007)

Qiu, R., Ren, X., Fifield, L.S., Simmons, K.L., Li, K.: Hemp-fiber-reinforced unsaturated polyester composites: optimization of processing and improvement of interfacial adhesion. J. Appl. Polym. Sci. **121**(2), 862–868 (2011)

Qiu, R., Ren, X., Li, K.: Effect of fiber modification with a novel compatibilizer on the mechanical properties and water absorption of hemp-fiber-reinforced unsaturated polyester composites. Polym. Eng. Sci. **52**(6), 1342–1347 (2012)

Rahman, M.R., Hamdan, S., Hashim, D.M.A., Islam, M.S., Takagi, H.: Bamboo fiber polypropylene composites: effect of fiber treatment and nano clay on mechanical and thermal properties. J. Vinyl Add. Tech. (2014). doi:10.1002/vnl.21407

Rajini, N., Winowlin Jappes, J., Rajakarunakaran, C., Siva, I.: Tensile and flexural properties of MMT-clay/unsaturated polyester using robust design concept. In: Nano Hybrids, pp. 87–101. Trans Tech Publ (2012)

Rajini, N., Jappes, J.W., Jeyaraj, P., Rajakarunakaran, S., Bennet, C.: Effect of montmorillonite nanoclay on temperature dependence mechanical properties of naturally woven coconut sheath/polyester composite. J. Reinf. Plast. Compos. **32**(11), 811–822 (2013a)

Rajini, N., Jappes, J.W., Rajakarunakaran, S., Jeyaraj, P.: Dynamic mechanical analysis and free vibration behavior in chemical modifications of coconut sheath/nano-clay reinforced hybrid polyester composite. J. Compos. Mater. **47**(24), 3105–3121 (2013b)

Ray, S.S., Okamoto, M.: Polymer/layered silicate nanocomposites: a review from preparation to processing. Prog. Polym. Sci. **28**(11), 1539–1641 (2003)

Ray, S.S., Yamada, K., Okamoto, M., Ueda, K.: Control of biodegradability of polylactide via nanocomposite technology. Macromol. Mater. Eng. **288**(3), 203–208 (2003)

Ren, X., Qiu, R., Li, K.: Modifications of kenaf fibers with N-methylol acrylamide for production of kenaf-unsaturated polyester composites. J. Appl. Polym. Sci. **125**(4), 2846–2853 (2012a)

Ren, X., Qiu, R., Fifield, L.S., Simmons, K.L., Li, K.: Effects of surface treatments on mechanical properties and water resistance of kenaf fiber-reinforced unsaturated polyester composites. J. Adhes. Sci. Technol. **26**(18–19), 2277–2289 (2012b)

Ren, X., Li, K.: Investigation of vegetable-oil-based coupling agents for kenaf-fiber-reinforced unsaturated polyester composites. J. Appl. Polym. Sci. **128**(2), 1101–1109 (2013)

Ren, X., Li, C., Li, K.: Investigation of acrylamide-modified melamine-formaldehyde resins as a compatibilizer for kenaf-unsaturated polyester composites. Polym. Eng. Sci. **53**(8), 1605–1613 (2013)

Rout, J., Misra, M., Tripathy, S., Nayak, S., Mohanty, A.: The influence of fiber surface modification on the mechanical properties of coir-polyester composites. Polym. Compos. **22** (4), 468 (2001)

Saheb, D.N., Jog, J.P.: Natural fiber polymer composites: a review. Adv. Polym. Tech. **18**(4), 351–363 (1999). doi:10.1002/(Sici)1098-2329(199924)18:4<351:Aid-Adv6>3.3.Co;2-O

Sajna, V.P., Mohanty, S., Nayak, S.K.: Hybrid green nanocomposites of poly(lactic acid) reinforced with banana fibre and nanoclay. J. Reinf. Plast. Comp. **33**(18), 1717–1732 (2014). doi:10.1177/0731684414542992

Sattler, K.D.: Handbook of Nanophysics. Functional nanomaterials. Taylor & Francis, Boca Raton (2011)

Shen, L., Lin, Y., Du, Q., Zhong, W., Yang, Y.: Preparation and rheology of polyamide-6/ attapulgite nanocomposites and studies on their percolated structure. Polymer **46**(15), 5758–5766 (2005)

Shroff, R., Mavridis, H.: Long-chain-branching index for essentially linear polyethylenes. Macromolecules **32**(25), 8454–8464 (1999)

Smart, G., Kandola, B.K., Horrocks, A.R., Nazare, S., Marney, D.: Polypropylene fibers containing dispersed clays having improved fire performance. Part II: Characterization of fibers and fabrics from PP–nanoclay blends. Polym. Adv. Technol. **19**(6), 658–670 (2008)

Song, M., Wong, C., Jin, J., Ansarifar, A., Zhang, Z., Richardson, M.: Preparation and characterization of poly (styrene-co-butadiene) and polybutadiene rubber/clay nanocomposites. Polym. Int. **54**(3), 560–568 (2005)

Tabari, H.Z., Nourbakhsh, A., Ashori, A.: Effects of nanoclay and coupling agent on the physico-mechanical, morphological, and thermal properties of wood flour/polypropylene composites. Polym. Eng. Sci. **51**(2), 272–277 (2011). doi:10.1002/pen.21823

Taj, S., Munawar, M.A., Khan, S.: Natural fiber-reinforced polymer composites. Proc.-Pak. Acad. Sci. **44**(2), 129 (2007)

Tjong, S.: Structural and mechanical properties of polymer nanocomposites. Mater. Sci. Eng.: R: Rep. **53**(3), 73–197 (2006)

Uma Devi, L., Joseph, K., Manikandan Nair, K., Thomas, S.: Ageing studies of pineapple leaf fiber–reinforced polyester composites. J. Appl. Polym. Sci. **94**(2), 503–510 (2004)

VP, S., Mohanty, S., Nayak, S.K.: A study on thermal degradation kinetics and flammability properties of poly (lactic acid)/banana fiber/nanoclay hybrid bionanocomposites. Polym. Compos. (2015)

Wang, C.H., Shieh, Y.T., Nutt, S.: The effects of soft-segment molecular weight and organic modifier on properties of organic-modified MMT-PU nanocomposites. J. Appl. Polym. Sci. **114**(2), 1025–1032 (2009)

Wang, Z., Massam, J., Pinnavaia, T.: Epoxy-clay nanocomposites. Polym.-Clay Nanocompos. **48** (2000)

Withers, G.J., Yu, Y., Khabashesku, V.N., Cercone, L., Hadjiev, V.G., Souza, J.M., Davis, D.C.: Improved mechanical properties of an epoxy glass-fiber composite reinforced with surface organomodified nanoclays. Compos. Part B-Eng. **72**, 175–182 (2015). doi:10.1016/j.compositesb.2014.12.008

Wu, Z., Zhou, C., Qi, R.: The preparation of phenolic resin/montmorillonite nanocomposites by suspension condensation polymerization and their morphology. Polym. Compos. **23**(4), 634–646 (2002)

Xu, Y., Van Hoa, S.: Mechanical properties of carbon fiber reinforced epoxy/clay nanocomposites. Compos. Sci. Technol. **68**(3–4), 854–861 (2008). doi:10.1016/j.compscitech.2007.08.013

Yeh, S., Ortiz, D., Al-Mulla, A., Gupta, R.: Mechanical and thermal properties of wood/layered silicate/plastic composites. In: Proceedings of the 8th International Conference on Wood fiber-Plastic Composites (2005)

Ying-Chen, Z., Hong-Yan, W., Yi-Ping, Q.: Morphology and properties of hybrid composites based on polypropylene/polylactic acid blend and bamboo fiber. Bioresour. Technol. **101**(20), 7944–7950 (2010)

Yuan, Q., Awate, S., Misra, R.D.K.: Nonisothermal crystallization behavior of polypropylene-clay nanocomposites. Eur. Polym. J. **42**(9), 1994–2003 (2006). doi:10.1016/j.eurpolymj.2006.03.012

Zafeiropoulos, N., Baillie, C., Matthews, F.: A study of transcrystallinity and its effect on the interface in flax fibre reinforced composite materials. Compos. A Appl. Sci. Manuf. **32**(3), 525–543 (2001)

Zhang, X., Xie, T., Yang, G.: Isothermal crystallization and melting behaviors of nylon 11/nylon 66 alloys by in situ polymerization. Polymer **47**(6), 2116–2126 (2006)

Zhong, Y., Poloso, T., Hetzer, M., De Kee, D.: Enhancement of wood/polyethylene composites via compatibilization and incorporation of organoclay particles. Polym. Eng. Sci. **47**(6), 797–803 (2007)

Processing Techniques of Nanoclay Based Natural Fibre Reinforced Polymer Composites

Deepak Verma

Abstract Natural fiber composite materials, also known as the bio-composite materials, are the most versatile materials which are used not only in the infrastructures purpose but also used in automotive applications. This usage generally depends upon their superior properties such as low weight and high strength. The other major problem related to environmental point of view can also solved by reinforcing the agro wastes (natural fibers) into the matrix system. In this chapter we are going to discuss about the nanoclay reinforcement in the polymer matrix system and its effect on various mechanical properties, such as tensile strength, tensile modulus, flexural strength, flexural modulus, impact strength etc. Various conventional processing techniques for the development of the nanoclay reinforced composites and nanocomposites fabrication techniques are also discussed in the current chapter. The morphological studies (SEM) of the nanoclay composites are also shown in this chapter by considering the previous studies or literature.

Keywords Polymer composites · Processing techniques · Mechanical properties · Scanning Electron Microscopy (SEM) · Nano clay

1 Introduction

Natural fibers are abundantly available in the world in the form of agricultural residues/waste or in the form of the plant fibers. These fibers are used to make the renewable materials and are mainly utilized by the plastic firms/industries. These industries are using different types of natural fibers from agricultural fields and reinforced these fibers to make the composite materials. The manufacturing of these composite not only helps to improve our environment from pollutants (agro wastes)

D. Verma (✉)
Department of Mechanical Engineering, Graphic Era Hill University,
Dehradun 248002, Uttarakhand, India
e-mail: dverma.mech@gmail.com

M. Jawaid et al. (eds.), *Nanoclay Reinforced Polymer Composites*,
Engineering Materials, DOI 10.1007/978-981-10-0950-1_9

but also improve the economy of the country. In ancient period peoples are using agricultural waste such as rice husk and wheat straw for making the bricks for their houses. Now at present the natural fiber reinforcement in composites become most popular because of the easy availability of these fibers which also solves the environmental problems (Satyanarayana et al. 2009). Natural fibres reinforced composites have various advantages such as low weight, high strength and degradability etc. A good varieties of natural fibers are available in the world which acan be used as a reinforcer or filler for the development of the composites. As the name itself says the natural fibers so there is no health hazards problem occurs from these composites also these are more cheaper as compared to the synthetic fibers one (Jawaid and Abdul Khalil 2011). Natural fibers can also be used to make the acoustic insulator materials. Nano clays are also used with natural fibers to make composites. Nano clays are the nano particles obtained from the layered mineral silicates. Nano clays are generally divided into some classes such as Montmorillonite, bentonite, kaolinite, hectorite and hectosyte, these all depends upon the chemical composition and morphology of the nanoparticle. It is generally observed that the improvement in the properties such as physical and mechanical properties of the nanoclay filled polymer composites depends upon the dispersion of the nanoclay into the matrix system. So to attain the maximum advantage from the nano clay, exfoliated processing and morphologies have to be used. On the other side it is very difficult to get a combination of the exfoliated and interaction morphologies (Pourabas and Raeesi 2005; Ludueña et al. 2007). The said combination is also depends upon the processing technique used and the agreement of the nano clay with the matrix system. In other study it has been noticed that by increasing the clay loading in the matrix system also demands to increase the solvent solution for the proper exfoliation which ultimately requires a huge amount of energy for the removal of the solvent and results in the thermal degradation of the polymers.

In this chapter various conventional techniques for the manufacturing of the natural fiber/nanoclay reinforced composites are explained. Various techniques for the development of the nanocomposites are also briefly discussed in the current chapter. Chapter also illustrated the various mechanical properties and morphological studies of the composites based on the past research and literature.

2 Conventional Composite Fabrication Techniques for Nanocomposites

The various conventional techniques are used for the fabrication/manufacturing of the nanoclay reinforced composites, some of them are as follows:

2.1 Two Roll Mill Technique

The two roll mill machine is generally a mixing machine. This machine is having two rolls which are operating at dissimilar speeds. The front roll is operating at slower speed and the back roll is operating at faster speed. By considering the flow properties of the matter or materials and the rollers surface conditions, the material frequently bonded to the slower first roll or to the back roll. The main disadvantage of this process is that it is more rigorous process and also the production rate of this process is very low. If we consider the case of the fiber reinforced composites then the first step of the milling process is to remove the moisture content of the fibers by drying it into an oven (the moisture content rate should be approximately below 0.1 %). The other option for removing the moisture from the fiber is by the chemical treatments. Chemical treatment removes the cellulose contents from the fibers and makes them more suitable with the matrix content. In second step the treated fibers are mixed with the matrix formulation during the mixing operation in two roll mill. The product we have got from the two roll mill process is a homogeneous matrix reinforced with the fibers.

2.2 VARTM (Vacuum Assisted Resin Transfer Molding) Process

VARTM is an advanced manufacturing process for composite materials (polymer based), and generally used in commercial and defense applications. The main advantage of the VARTM process over the RTM process is that it lessens the volatile emission and permit the use of lower injection pressure of resin. The VARTM process can be completed in three steps. These are (a) Rigid tool plate surface (consists of fiber preform), neighboring a formable vacuum bag; (b) Resin impregnation which is injected by a single or multiple inlet and transferred into the preform by gravity and capillary effects; and (c) Impregnated preform curing (Chittajallu 2004) (Fig. 1).

2.3 Twin Screw Extruder

Twin screw extrusion or extruder is generally used for mixing, compounding, or making polymer based composite materials. This process manufactured of highly homogeneous material and finely structured composite materials. The twin screw extruder machine consists of two intermeshing, co-rotating screws which are mounted on splined shafts in a closed barrel. The various process functions can be

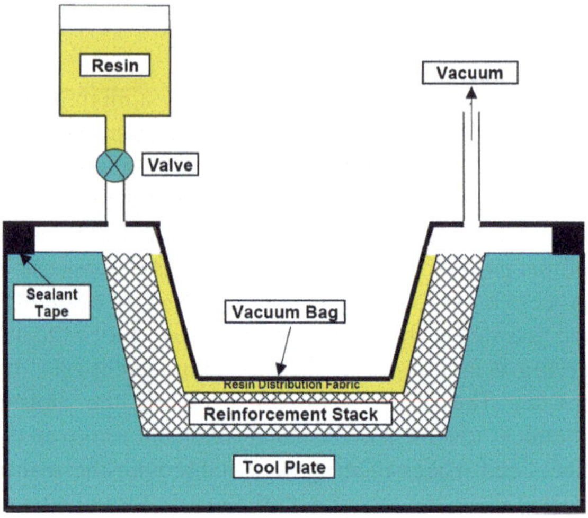

Fig. 1 VARTM process (Chittajallu 2004)

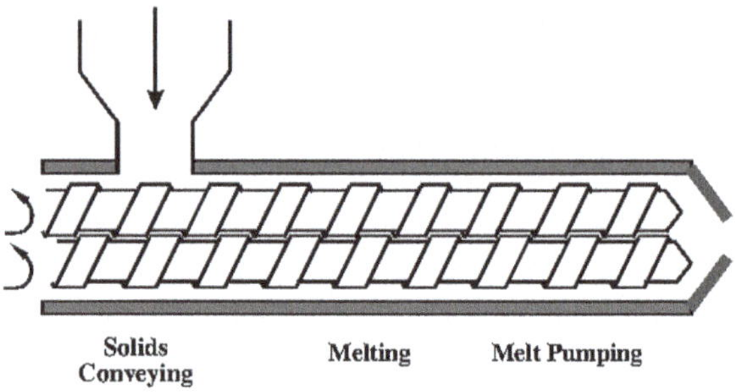

Fig. 2 Twin screw extruder

settled as according to the requirement on the said technique. The twin screw extruder generally used and flexible to perform compressing, heating, shearing, shaping etc. The main advantage of intermeshed co-rotating twin screw extruders is their noticeable mixing ability which gives surprisingly features to extruded products and added a notable worth to processing units. The raw materials which can be used in twin screw extrusion process can be solids, liquids and possibly gases. The extruded products can be plastics or rubber compounds, modified polymers (chemically or physically), cellulose pulps, etc. (clextral technologies.com) (Fig. 2).

2.4 Compression Molding

The compression molding process for composite fabrication generally aims to design the routes for the surplus resin. The first step in compression molding

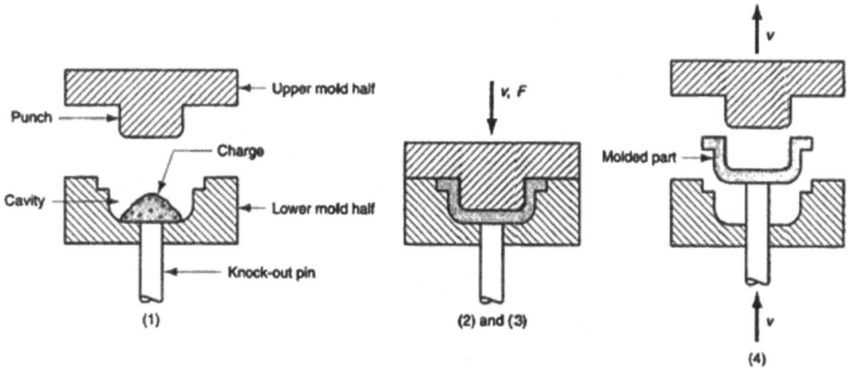

Fig. 3 Compression molding process *Source* Sinotech Corporation (http://www.sinotech.com/compressionAndTransferMolding.html)

process is the same as the hand lay-up process. Compression molding process generally used male and female molds for the manufacturing of the composites. A releasing agent is required and applied on the surfaces of the molds. After that the required amount of fibers (may be in slashed form) are placed on the surfaces of these molds. After this process the resin (catalyzed) is applied on these fibers in excess form leading to the joining of the molds. When the two molds are joined together then an appropriate pressure is applied (mechanically or pneumatically) on to the molds. The desired amount of the resin is distributed throughout the fibers and comes out to the exit channels (Verma et al.) (Fig. 3).

2.5 Injection Molding Process

Injection molding process generally used to make injection molded composite parts and is a purely plastic injection process. This process generally uses the thermoplastic resin such as polypropylene, polyethylene and polymethylmethacrylate (PMMA) etc. This process can also be used to make the fiber reinforced composites (specifically short fibers). The fibers can be incorporated into the resin to get reinforced plastic components. The fibers are mixed with resin and injected together. Injection moulding can be performed by using materials such as metals, glasses, elastomers etc. In this process the material (usually plastic thermoset or thermoplastics) is filled into a heated barrel then mixed, and forced into a mold cavity where after some times it cools and hardens to the configuration of the cavity (Fig. 4).

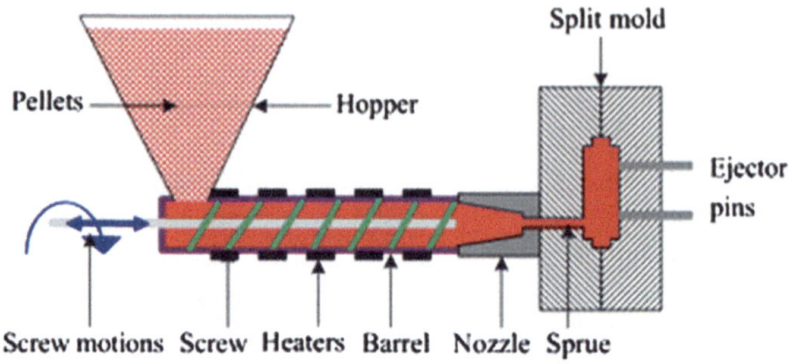

Fig. 4 Injection molding *Source* Vijay Plastic Corporation (http://www.veejayplastic.com/injection-molding.html)

2.6 *Filament Winding*

Filament winding process includes a band of continuous resin impregnated fibers which is wrapped around a revolving mandrel and then cured at room temperature or in an oven for the production of the final product (Fig. 5).

The revolving mandrel can be cylindrical or round or any shape. The applications of filament winding are cylindrical and spherical pressure vessels, pipe lines, oxygen and other gas cylinders, large underground storage tanks etc. This process is not only limited to symmetric structures (axis symmetric) but also the complex structures and prismatic shapes can be made easily. The mechanical strength of the filament wound composites generally depends upon the composition of the material and the process parameters such as winding angle, fibre tension, and curing cycle. Filament winding process includes a number of fibres pulled out from series of reels dipped into liquid resin and catalyst and other chemical compounds. The Fibre tension is controlled by the guides placed between reel and resin bath. The excess resins from the fibers are removed by the wiping device which is placed at the end of the resin tank or bath. Squeeze rollers are generally acts as a wiping device. In this the placement of the top roller is behave as a regulator which not only controls

Fig. 5 Filament winding

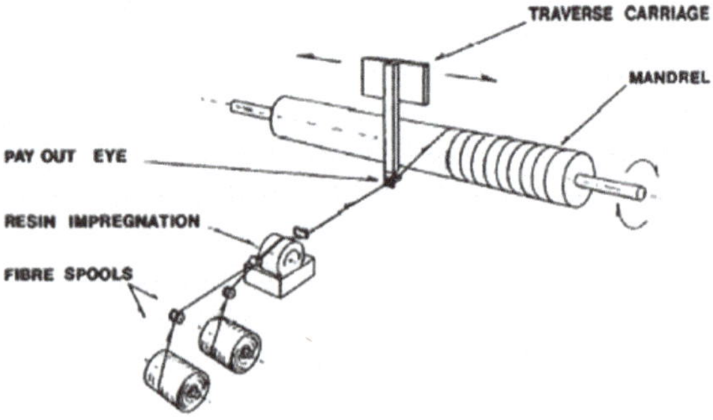

the resin flow but also the tensions in the fibre. The desired winding angles can be made by controlling the speed of the carriage and the winding speed of the mandrel. At last the rotating filament wound mandrel is cured to maintain the uniformity of resin around the circumference and the product is removed from the mandrel by mechanical extractor.

3 Nano Composite Fabrication Techniques

There are many processes available for the manufacturing of nano composites some of them are explained as follows:

3.1 Sonication

Sonication is a technique which is used for the manufacturing of the nanocomposites. Sonication technique uses sound waves for the stirring of the particles in solution. Such method can be used to dissolve the solid into the solution very fast and separate the dissolved gas from the liquid. The ultrasound waves are generally used in the sonication process having frequencies above 20 kHz and increase of the frequency led to the increase in the agitation of particles in the solution. Ultimately this frequency increases the cycle of pressure which again leads to the formation of the microscopic vacuum bubbles and the collapse into the solution. Then this vibration breaks the cluster of particles leading to the mixing into the solution. The sonicator system generally consists of three main components; these are (a) Generator (b) Converter and (c) Probe.

The function of the **generator** is to convert the AC line into high frequency electrical energy. Similar ways the **converter** is connected to the generator and converts the high voltage electrical energy into mechanical vibration. And the **probe** is connected to the converter which amplified the vibration along the length of the probe.

3.2 Intercalation Technique

This method involves the inclusion of the polymer matrix or polymer chains into the layered structures in crystallographically regular fashion with very less distance (about nono meters) without concerning to the ratio of the polymer and clay with respect to layered structures. Intercalation nanocomposites consists of a few layer of polymers (at molecular level) which are interlayered between the layered structures. The properties of these composites are just similar to the ceramic composites.

3.3 Flocculated Nanocomposites

The flocculated nanocomposites are considered to be similar to intercalated nanocomposites. This generally considered to be the edge to edge intercalation of the silicatelayers. That is Hydroxylated edge–edge interaction of the clay layers occurs in this technique.

3.4 Exfoliated Nanocomposites

This technique is generally deals with the amount of clay loading. In this each clay layers are isolated in a continuous polymer matrix by an average distances. It is also noticed that the amount of clay content in exfoliated nanocomposites is much lower than that of the intercalated nanocomposites (Ray and Okamoto 2003).

3.5 Melt Intercalation

In this technique no solvent is needed and the layered silicate is reinforced in the polymer matrix (in molten state) (Dennis et al. 2001). The polymer generally used in for making nanocomposites by this method are is thermoplastic. In first step the said polymer is mixed with by using conventional techniques with organoclay filler at high temperature (Kornmann et al. 2001). The polymer chains are formed (by intercalation or exfoliation) leading to the manufacture of the nanocomposites. This is most accepted method for making thermoplastic reinforced nanocomposites. This method is used for polymers which are not suitable insitu polymerization (Ray and Okamoto 2003).

4 Nanoclay Reinforced Polymer Composites: A Past Research

Chowdury et al. (2006) performed the flexural tests of VARIM (vacuum assisted resin infusion method) samples (with or without thermal post curing) for the evaluation of bulk stiffness and strength of carbon-epoxy nanocomposites. These samples were subjected to with or without thermal post curing. Figures 6 and 7 shows the stress strain behavior of the flexural test. Authors found the consequence of post curing (thermal) after comparing both of the figures. From Figs. 6 and 7 it can be easily noticed that the composite strength is increased up to 2 wt% of nanoclay loading.

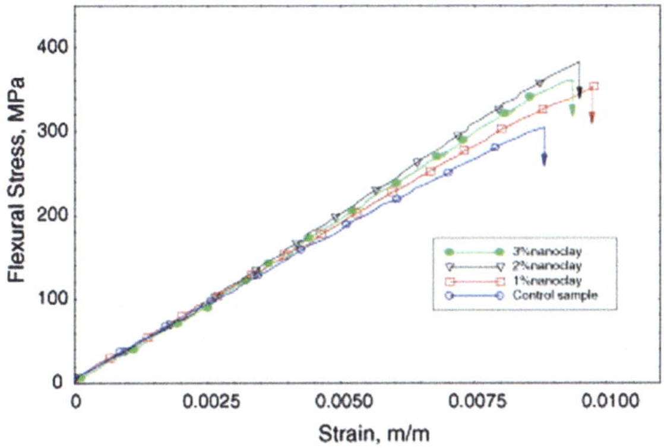

Fig. 6 Flexural stress–strain plot of VARIM samples without post curing. Reproduced with permission from Elseveir Ltd. (Chowdury et al. 2006)

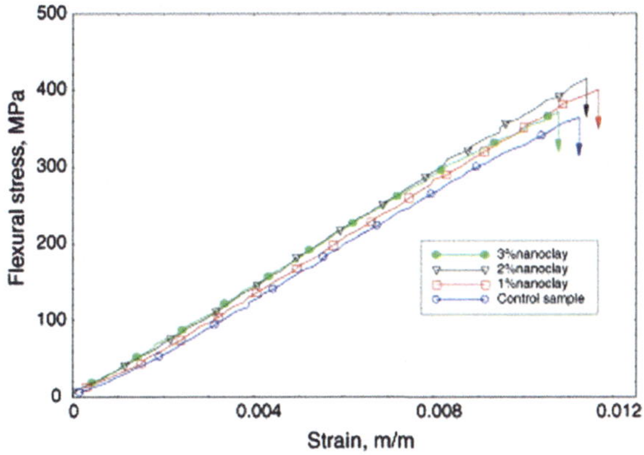

Fig. 7 Flexural stress–strain plot of post cured VARIM samples. Reproduced with permission from Elseveir Ltd. (Chowdury et al. 2006)

Nguyen and Baird (2007) evaluated the tensile properties of different nanocomposites made by using different processing methods are showed in Table 1 and Fig. 8. The polymer used in this nanocomposites is pure polypropylene having **Young's modulus of 1.374–0.133 GPa**. The reinforcement of approximately 4 wt% Cloisite 20A (prepared via method #1) results an increase in Young's modulus value of nanocomposites as compared to the pure PP. By adapting the same process at various contents of 6.7, 10, and 14 wt% of C20A, the values of Young's moduli of nanocomposites were found to be 1.753–0.045, 1.725–0.130, and 1.787–0.037 GPa, respectively.

Huang and Netravali (2007) evaluated and showed (Fig. 9) the stress–strain behaviors of the modified Soy Protein Concentrate (**SPC**) resin filled with different clay contents and made by 30 % glycerol at pH of 7.9. From study it is noticed that the addition of the nanoclay in SPC resin system notably improves the stiffness value of the SPC resin. The availability of the nano-clay platelets into the system makes it more difficult for the molecules of soy protein to deform which ultimately results in higher stiffness.

Table 1 Tensile properties of various nanocomposites prepared using different processing methods

Materials	Young's modulus (GPa)	S.D.	% Increase	% Increase	S.D.	% Elongation	S.D.	HDT (°C)	S.D.
PP—pure	1.374	0.133	–	–	0.47	115.3	19.79	84	2.3
Method #1: 4 wt%	1.611	0.059	17	18	0.32	32	28.71	92.5	3.1
Method #1: 6.7 wt%	1.753	0.045	28	29	0.35	26.89	15.41	NA	NA
Method #1: 10 wt%	1.726	0.13	26	27	0.77	16.59	2.42	NA	NA
Method #1: 14 wt%	1.787	0.037	30	31	0.34	10.16	0.77	NA	NA
Method #2: 4 wt%	1.716	0.136	25	26	0.39	34.59	6.92	NA	NA
Method #2: 6.7 wt%	1.81	0.122	32	33	0.33	24.09	13.92	NA	NA
RTP TSE: 4 wt%	1.769	0.167	29	30	0.56	48.85	33.77	NA	NA
RTP TSE: 6.5 wt%	1.861	0.068	35	36	0.42	17.67	5.57	NA	NA
RTP TSE: 10 wt%	1.85	0.185	35	36	0.48	4.11	2.45	100	2.1
Method #3: 4 wt%	1.848	0.107	34	35	0.32	80.4	0.1	95.5	3.2
Method #3: 6.6 wt%	2.118	0.077	54	55	0.89	14.82	5.9	104	1.9

Reproduced with permission from Elseveir Ltd. (Nguyen and Baird 2007)

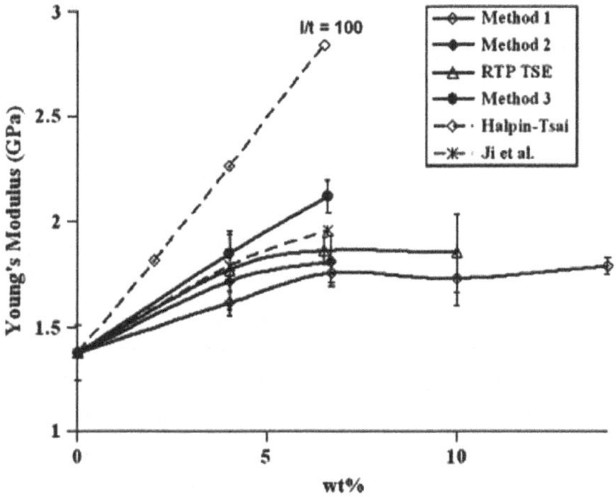

Fig. 8 Young's modulus of different nanocomposites versus clay weight percent as a function of different processing techniques. Reproduced with permission from Elseveir Ltd. (Nguyen and Baird 2007)

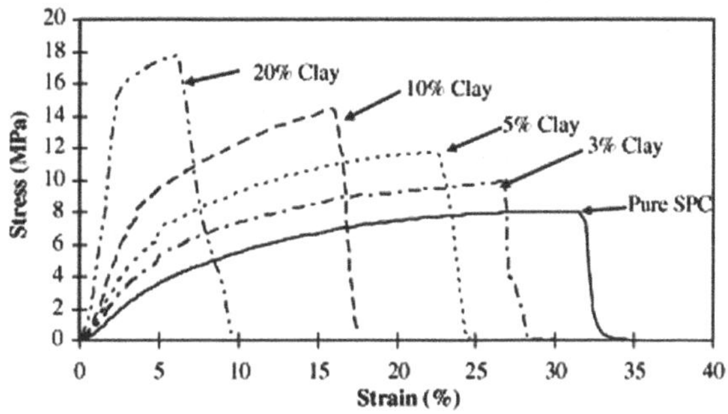

Fig. 9 Effect of nano-clay loading on the tensile properties of the modified SPC resins. Reproduced with permission from Elseveir Ltd. (Huang and Netravali 2007)

Haq et al. (2008) evaluated and showed the mechanical properties, specifically tensile and notched impact strength in Fig. 10. The toughness of the biocomposite is increased by the addition of the EMS in the matrix system. The resin system without nanoclay addition showed an average improvement in ultimate strain of about 30 and 65 % for biocomposites B (10 % EMS) and C (20 % EMS), respectively. In the similar way the impact strengths of the biocomposites B and C are increased by 10 and 17 %.

Marras et al. (2008) evaluated the mechanical properties of the nanocomposites and presented in Table 2. It is seen that the addition of the inorganic particles notably improves the tensile modulus of the polymer. Due to the small size of particles, a large surface area is provided by the silicate layers which ultimately restrict the potency of the polymer chains which results an increased material's stiffness. **PCL** retains a higher ductility just after the inclusion of the inorganic particles due to the perfect interfacial bonding between the organically modified filler and the matrix. The nanocomposites made by reinforcing 15 wt% clay showed an elongation at the break of around 500 %.

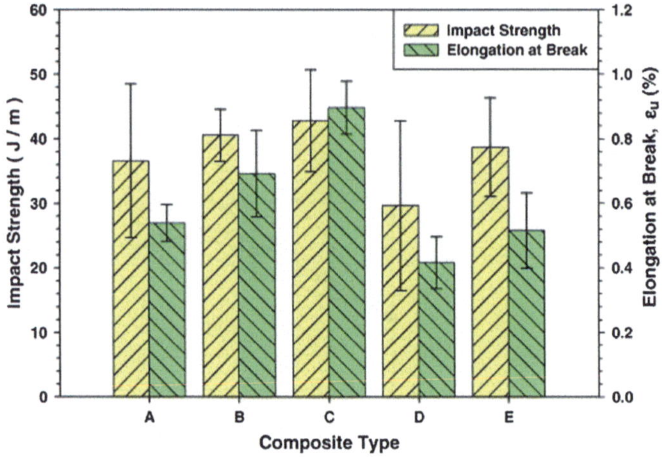

Fig. 10 Impact strengths from notched Izod tests and tensile strains at failure. Reproduced with permission from Elseveir Ltd. (Haq et al. 2008)

Table 2 Effect of organoclay content on the tensile properties of the nanocomposite films

Sample	Yield strength (MPa)	Strength at break (MPa)	Elongation at break (%)	Young's modulus (MPa)
Pure PCL	14.5 ± 1	26.5 ± 2.5	702 ± 28	305 ± 17
1 wt% Cl25A	16.6 ± 0.7	23.8 ± 1.7	632 ± 37	321 ± 15
3 wt% Cl25A	16.9 ± 1.2	22.7 ± 1.2	588 ± 24	342 ± 21
5 wt% Cl25A	18 ± 1.2	21 ± 2	561 ± 30	394 ± 28
9 wt% Cl25A	16.3 ± 1.5	20.3 ± 2	532 ± 28	467 ± 18
15 wt% Cl25A	14.9 ± 1.7	18.1 ± 2.3	494 ± 32	563 ± 40

Reproduced with permission from Elseveir Ltd. (Marras et al. 2008)

Haq et al. (2009) investigated the tensile modulus results for unmodified and bio-based resin systems with different clay loadings are shown in Fig. 11. Also the tensile Modulus and Poisson's ratio of neat resins (without clay) are shown in Fig. 12. From study the reduction in the tensile modulus by inclusion of bio resin Epoxidized methyl soyate (EMS) was observed. This decrease is found to be 28 % after blending 10 % EMS and 42 % for the 20 % EMS blend with reference to neat Unsaturated Polyesters (UPE). On the other side the inclusion of 1.5 wt% nanoclay showed an increasing trend in tensile modulus of 15–20 % as compared to unfilled UPE. It is also noticed that increasing the EMS content ultimately increases the poisson's ratio of the nanocomposites (Fig. 3).

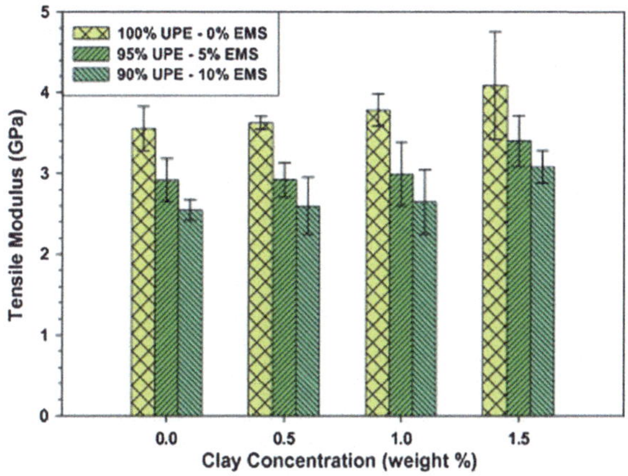

Fig. 11 Experimental tensile modulus of bio-blend resin systems with varying clay contents. Reproduced with permission from Elseveir Ltd. (Haq et al. 2009)

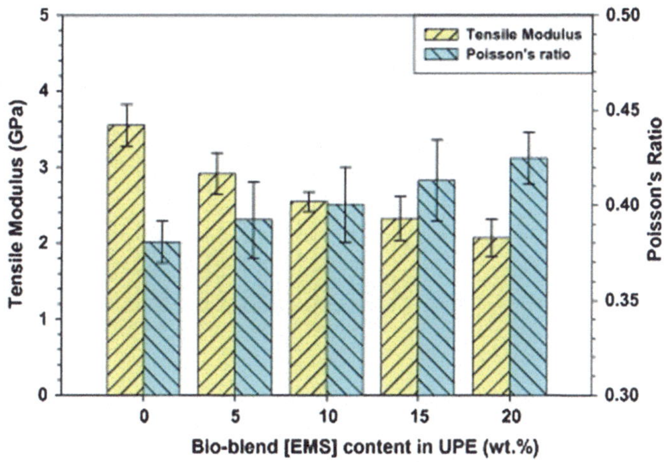

Fig. 12 Experimental tensile modulus and Poisson's ratio of neat resin (no clay) systems. Reproduced with permission from Elseveir Ltd. (Haq et al. 2009)

Ferreira et al. (2011) showed (Fig. 13) the tensile stress–strain curves of four material compositions with three different clay treatments (B2, B3 and B4). For the assessment of the impact energy, authors performed two different series of tests. The first test uses the specimen obtained from the injection molding process and other one sample is the 2 mm depth notch machine drilled, in order to examine not only the effect of the nanoclay, but also the effect of the curvature radius at the notch root. The test results are shown in Fig. 14 and Table 3.

Fig. 13 Comparison of stress–strain curves for nanoclayed and unfilled materials. Reproduced with permission from Elseveir Ltd. (Ferreira et al. 2011)

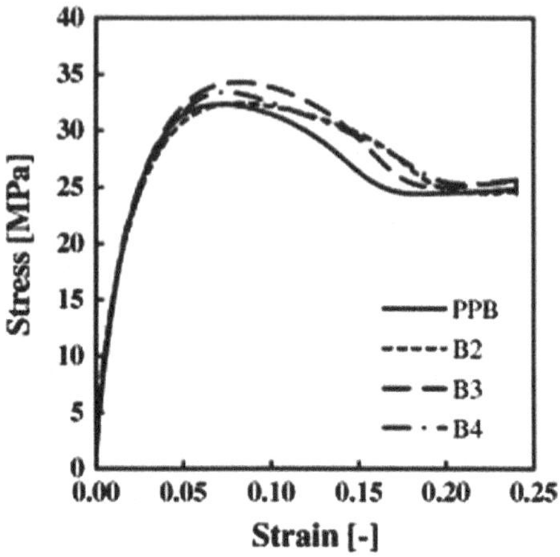

Fig. 14 Absorbed impact energy: influence of the material and notch manufacture. Reproduced with permission from Elseveir Ltd. (Ferreira et al. 2011)

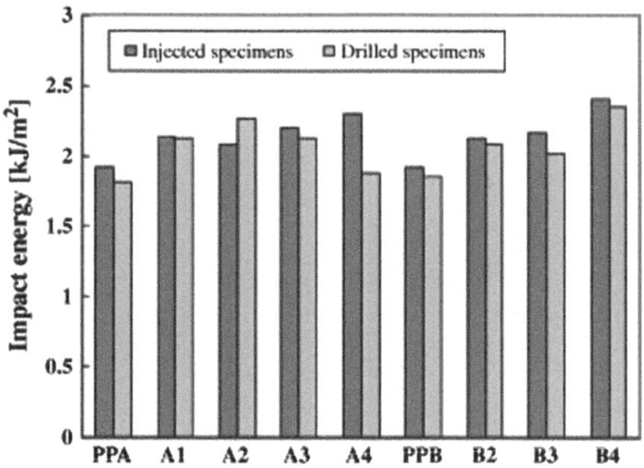

Table 3 Average mechanical properties

S. No.	Reference codes	Average tensile strength (MPa)	Average apparent modulus of elasticity (GPa)	Average toughness Gc (kJ/m^2)	Impact energy (injected specimens) (kJ/m^2)	Impact energy (machined notches) (kJ/m^2)
1	PPA	33.62	1.365	11.53	1.92	1.81
2	A1	31.38	1.085	10.15	2.14	2.13
3	A2	31.95	1.23	8.78	2.08	2.27
4	A3	33.66	1.061	9.38	2.2	2.13
5	A4	32.2	1.221	9.23	2.3	1.87
6	PPB	32.36	1.265	11.25	1.92	1.85
7	B2	32.56	1.279	8.67	2.13	2.08
8	B3	34.51	1.302	9.1	2.17	2.02
9	B4	32.84	1.201	8.76	2.41	2.35

Reproduced with permission from Elseveir Ltd. (Ferreira et al. 2011)

Chan et al. (2011) reported an increase in Young's Modulus of nanoclay/epoxy composites which is shown in Fig. 15 and Table 4. The increase in Young's modulus is depends upon the amount of nanoclay reinforcement. It has been noticed that the stiffness of the samples were improved by different nano clay content such as 3, 4 and 5 wt% and found to be 24, 31 and 34 % respectively. In the similar way Table 5 showed that all the nanoclay/epoxy composites attains maximum ultimate tensile strength in contrast to the new sample. Also the strength of the composites were increased by 12, 16 and 25 % respectively by adding 3, 4 and 5 wt% of nanoclay in the matrix.

Mohan and Kanny (2011) showed the water absorption behavior of the clay reinforced and unreinforced composites series in Fig. 16. It has been observed that the water absorption rate is continuously increases with time for all composite samples. It is also found that the water absorption of unreinforced composite sample

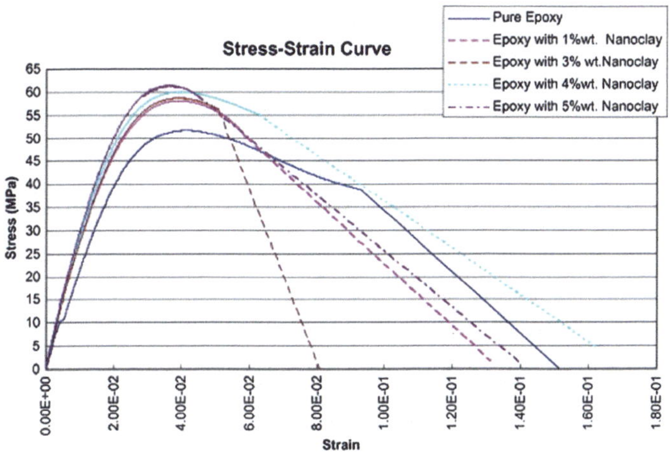

Fig. 15 Stress–Strain curve extracted from tensile test of nanocomposites with different percentage of nanoclay particles. Reproduced with permission from Elseveir Ltd. (Chan et al. 2011)

Table 4 Measured Young's modulus of the samples

Sample	Young modulus (MPa)	Percentage improvement (%)
Pristine epoxy	2120.3	0
1 wt% of nanoclay inside epoxy	2474.06	16.68
3 wt% of nanoclay inside epoxy	2625.09	23.81
4 wt% of nanoclay inside epoxy	2771.74	30.72
5 wt% of nanoclay inside epoxy	2841.28	34
7 wt% of nanoclay inside epoxy	3334	57.24
9 wt% of nanoclay inside epoxy	2431.98	14.7

Reproduced with permission from Elseveir Ltd. (Chan et al. 2011)

Table 5 Measured ultimate strength of all samples

Sample	Mean of ultimate strength (MPa)	Percentage improvement (%)
Pristine epoxy	52.4	0
1 wt% of nanoclay inside epoxy	58.02	10.72
3 wt% of nanoclay inside epoxy	58.49	11.23
4 wt% of nanoclay inside epoxy	60.72	15.88
5 wt% of nanoclay inside epoxy	65.22	24.47
7 wt% of nanoclay inside epoxy	54.43	3.87
9 wt% of nanoclay inside epoxy	32.82	−43.43

Reproduced with permission from Elseveir Ltd. (Chan et al. 2011)

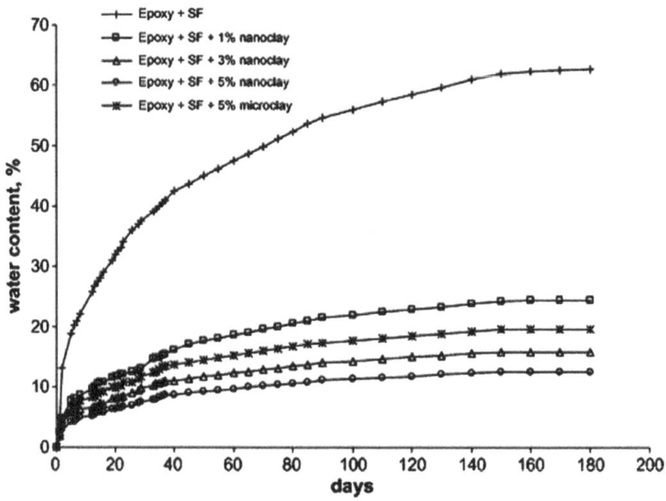

Fig. 16 Water uptake of unfilled and clay filled composite series. Reproduced with permission from Elseveir Ltd. Mohan and Kanny (2011))

was more as compared to the clay reinforced composites samples. It is also reported that maximum decrease in water absorption was noticed for 5 wt% nanoclay reinforced composites. Figure 17 shows the tensile stress–strain plots of the clay reinforced and unreinforced composites. It is observed that the modulus of the composites samples were continuously increases by increasing the nanoclay loading, nevertheless, strength and strain values decreases above 3 wt% of clay addition. The 5 wt% of organoclay addition showed an improvement in modulus values approximately about 47 %.

Haq et al. (2011) evaluated and showed the tensile modulus of neat (UPE) and bio-based polymer systems with different clay loadings in Fig. 18. In Fig. 18 the dashed horizontal line represents the baseline UPE. The direct tensile tests showed a decreasing trend of tensile modulus by reinforcement of EML and also a recuperation

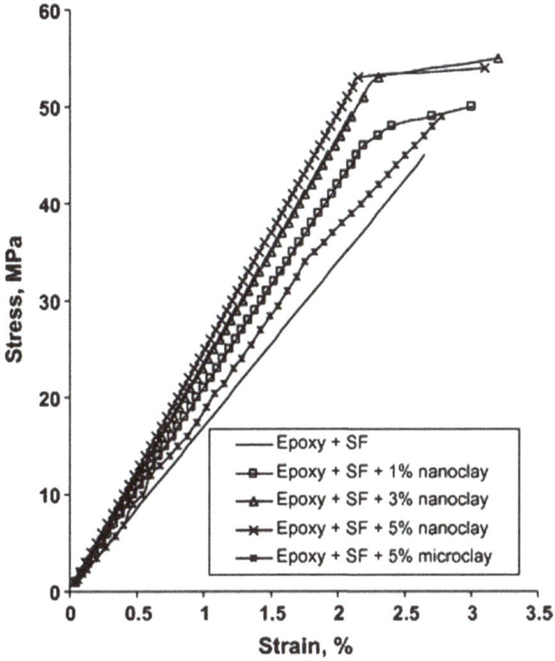

Fig. 17 Tensile stress strain curves of unfilled and clay filled composite series before placing in water and before degassing. Reproduced with permission from Elseveir Ltd. Mohan and Kanny (2011)

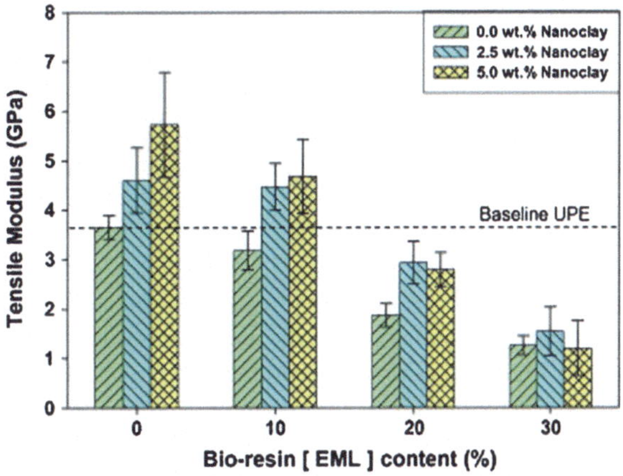

Fig. 18 Experimental tensile modulus of bio-based polymer systems with varying nanoclay content. Reproduced with permission from Elseveir Ltd. (Haq et al. 2011)

of lost stiffness by addition of nanoclay. It has been noticed that the loss of stiffness up to 10 wt% of EML content is recovered and enhanced by nonoclay addition. A notable improvement in tensile modulus was noticed with nanoclay inclusion for resin systems with 20 % EML in relation to the similar unfilled resins with 20 % EML. Figure 19 shows the ultimate tensile strengths (UTS) results of the composites. It has been observed that a reduction in UTS occurred by the inclusion of both EML and nanoclay. This reduction in ultimate tensile strength was due to the addition of

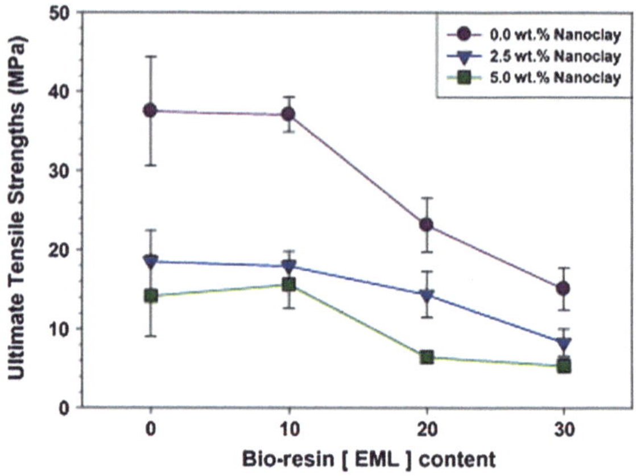

Fig. 19 Experimental ultimate tensile stresses for various bio-based polymer/clay nanocomposites. Reproduced with permission from Elseveir Ltd. (Haq et al. 2011)

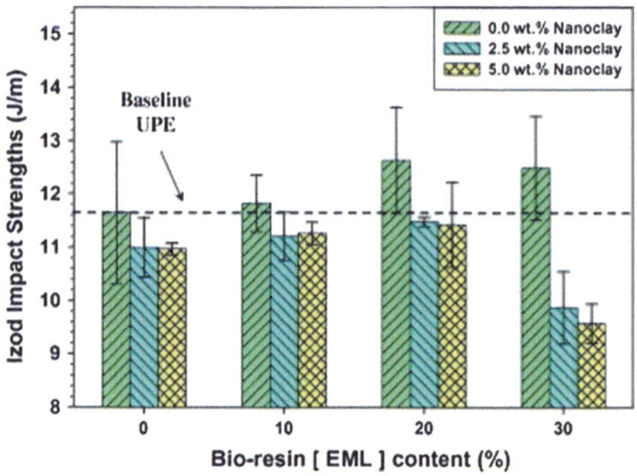

Fig. 20 Variation of Izod impact strength with varying bio-resin (EML) and nanoclay concentration. Reproduced with permission from Elseveir Ltd. (Haq et al. 2011)

the lowest content of the EML (10 wt%). On the other side it is noticed that the reduction in UTS also occurs because of the addition of nanoclay. This is due to the fact that the nanoclay makes the nanocomposites more brittle because of its stiffening effect. Figure 20 shows the Impact strength test results of all the developed composites. Authors have also concluded that the inclusion of bio resin ultimately increases the impact strength property and also the toughness of the polymer matrix system. On the other side the nanoclay reinforcement ultimately reduces the impact strength of the developed composites. The 10 and 20 % EML based nanocomposites, the impact strength were 2–4 % lesser than that the base UPE.

Iman and Maji (2012) showed the mechanical properties (such as flexural and tensile properties) of glutaraldehyde (GA) and clay reinforced composites in Table 6. From table it has been noticed that by increasing the glutaraldehyde content in the matrix system ultimately increases both the flexural and tensile properties of the composites. This increase is due to the inclusion of glutaraldehyde which ultimately crosslinks the polar group of starch with jute. From Table 6 It is

Table 6 Flexural and tensile properties of (a) S/J, (b) S/J/G30, (c) S/J/G40, (d) S/J/G50, (e) S/J/G60, (f) S/J/G70, (g) S/J/G50/M1, (h) S/J/G50/M3 and (i) S/J/G50/M5

Sample	Flexural properties		Tensile properties	
	Strength (MPa)	Modulus (MPa)	Strength (MPa)	Modulus (MPa)
S/J	27.67 (±1.22)	1141.9 (±1.57)	12.59 (±1.26)	755.38 (±16.74)
S/J/G30	28.31 (±1.34)	1336.2 (±1.16)	18.62 (±1.84)	1117.10 (±13.46)
S/J/G40	37.8 (±2.31)	1641.6 (±2.11)	19.88 (±0.96)	1192.53 (±17.32)
S/J/G50	39.5 (±2.14)	2234.0 (±1.96)	22.54 (±1.03)	1352.18 (±15.34)
S/J/G60	45.8 (±1.12)	2575.6 (±1.32)	22.65 (±1.31)	1358.81 (±10.79)
S/J/G70	51.4 (±1.32)	2620.1 (±1.13)	24.64 (±2.07)	1478.53 (±14.08)
S/J/G50/M1	69.8 (±1.31)	3777.0 (±2.09)	28.42 (±0.74)	1705.47 (±11.67)
S/J/G50/M3	82.0 (±1.76)	4516.3 (±1.54)	32.40 (±1.37)	1944.78 (±17.62)
S/J/G50/M5	90.7 (±2.54)	6434.1 (±2.09)	40.53 (±2.31)	2344.67 (±12.78)

Reproduced with permission from Elseveir Ltd. Iman and Maji (2012)

also observed there is no noticeable improvement seen on the mechanical properties such as flexural modulus, tensile strength and tensile modulus past 50 % of GA.

Arrakhiz et al. (2013) reported the variation in the tensile properties of the composites in the form of Young's modulus and tensile strength. This is shown in Fig. 21. From Fig. 21 it is noticed that by increasing the clay content (E1) in the matrix system ultimately increases the Young's modulus of the composites. On the other side it is seen that Young's modulus of the composite reinforced with 30 wt% fibers, is higher than that of unfilled PP (Polypropylene). Also the composites with 30 wt% of clay and 30 wt% of pine cone fibers have higher values of Young's moduli (about 1857.7 and 1272.5 MPa), as compared to the neat PP (1011.2 MPa). Figure 22 showed the behavior and properties of the hybrid composites.

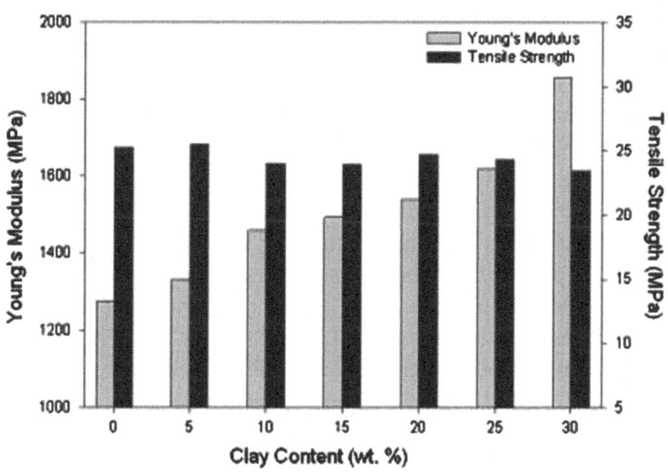

Fig. 21 Young's modulus and tensile strength of PP hybrid composites P.co fibers/clay particles. Reproduced with permission from Elseveir Ltd. Arrakhiz et al. (2013)

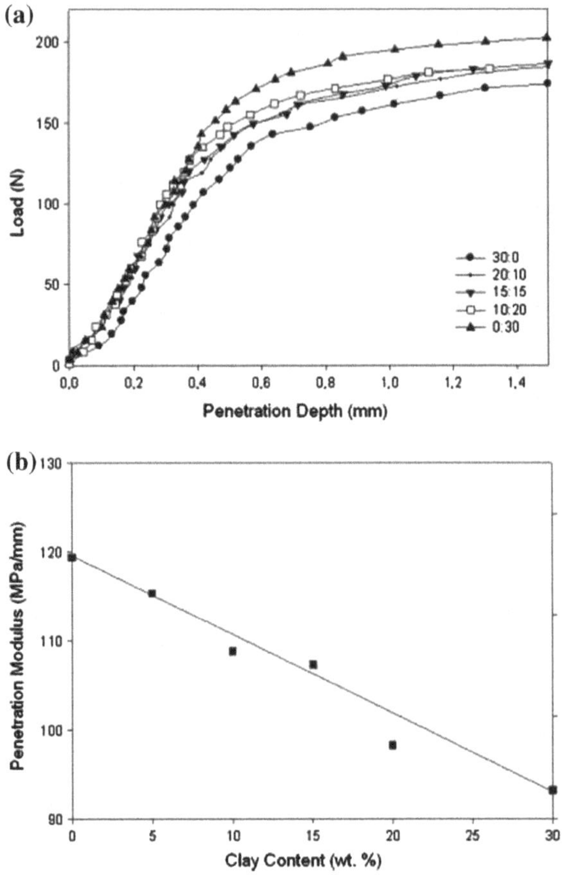

Fig. 22 The evolution of **a** load-penetration curves and **b** penetration modulus of the hybrid composite. Reproduced with permission from Elseveir Ltd. Arrakhiz et al. (2013)

Majeed et al. (2013) suggested that the addition of the nanoclay into the HDPE/rice husk system ultimately increases the tensile strength, tensile modulus, storage modulus and loss modulus of the composites. It is also observed that the composites having 2 phc (parts per hundred contents) of the nanoclay showed the good morphology as compared to the other formulations.

Babaei et al. (2014) developed and evaluated the physical and mechanical properties of foamed HDPE/wheat straw flour/nanoclay hybrid composite. Figures 23 and 24 showed the tensile properties (strength and modulus) of the composites containing different contents/loading of **NC and AZD**. From figures it has been noticed that the addition of AZD into matrix without NC generally shows the decreasing nature of the properties of the HPDE/WSF composites specifically tensile strength and modulus. The Izod impact resistance of composites fabricated by various amounts of NC and CFA is shown in Fig. 25. Impact resistance is used to measure the energy which is used to develop the cracks in the composites. The results of impact resistance (without NC) showed that the impact resistance of composite containing 2 and 4 phr (parts per hundred) of AZD showed the

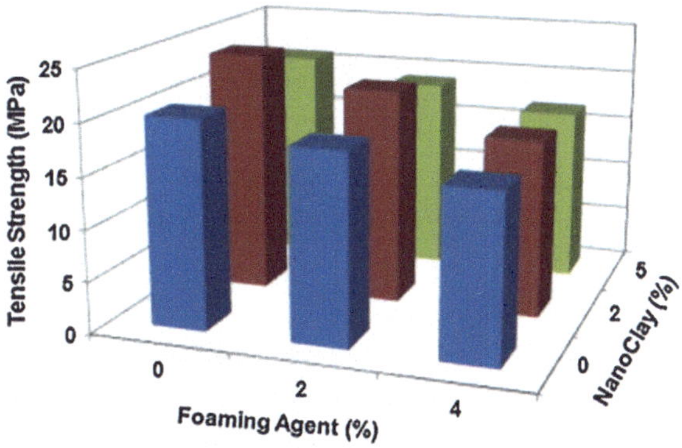

Fig. 23 Tensile strength of HDPE/WSF composites at different AZD and NC loading. Reproduced with permission from Elseveir Ltd. (Babaei et al. 2014)

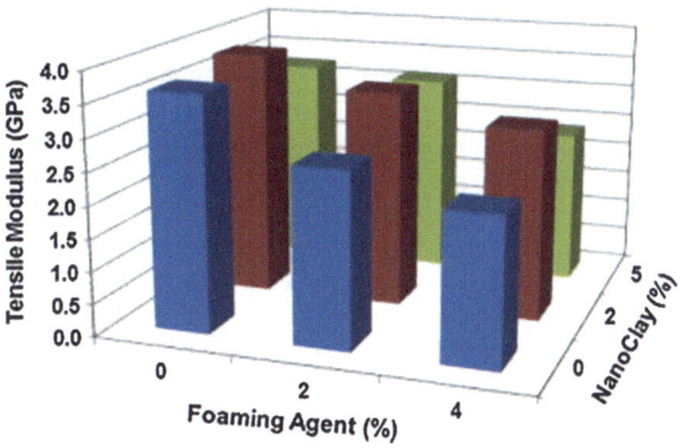

Fig. 24 Tensile modulus of HDPE/WSF composites at different AZD and NC loading. Reproduced with permission from Elseveir Ltd. (Babaei et al. 2014)

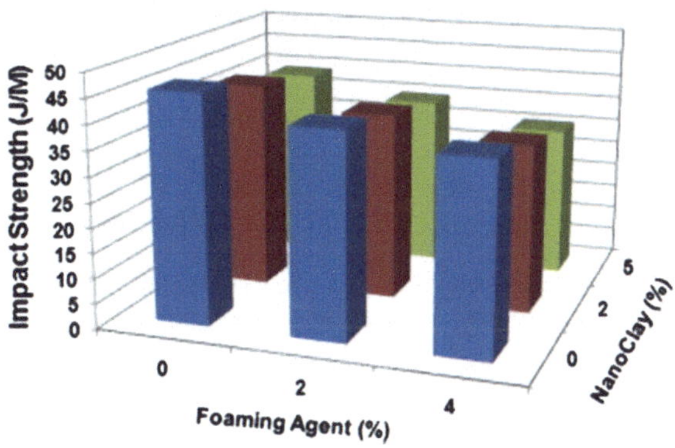

Fig. 25 Impact strength of HDPE/WSF composites at different AZD and NC loading. Reproduced with permission from Elseveir Ltd. (Babaei et al. 2014)

reductions of about 17.1 and 10.8 % which is lower as compared to the non-foamed composites, respectively.

Kanny and Mohan (2014) (Fig. 26) performed the dynamic mechanical analysis (DMA) of nanoclay reinforced and unreinforced composites. The values of these composites are shown in Table 7. The storage modulus and glass transition

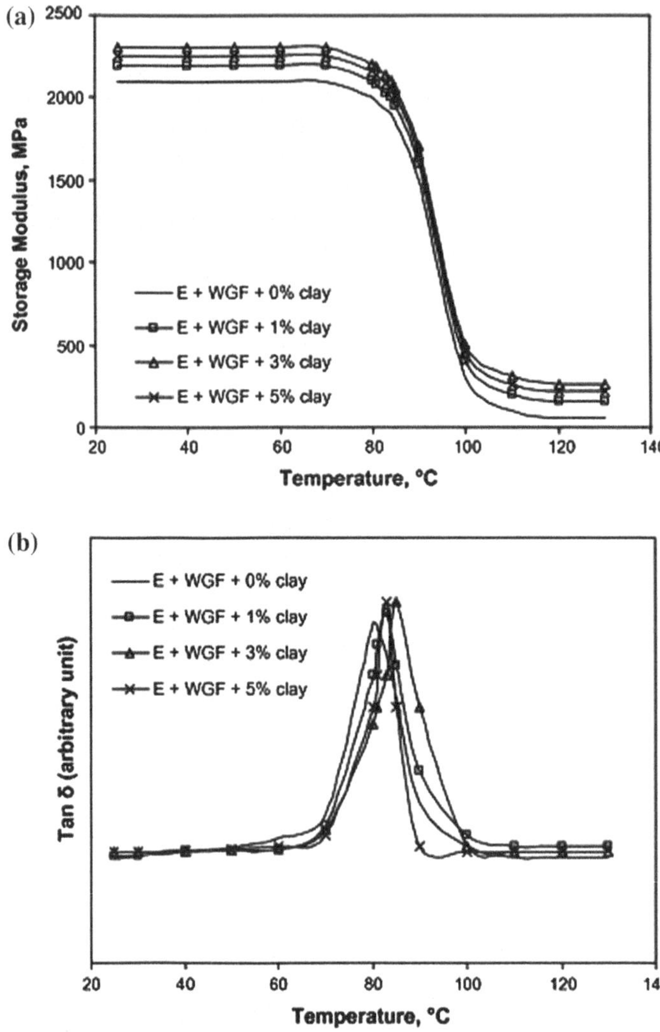

Fig. 26 Effect of temperature on **a** storage modulus and **b** Tan δ of laminate series. Reproduced with permission from Elseveir Ltd. (Kanny and Mohan 2014)

temperatures (Tg) were evaluated. The experimental results confirmed that the storage modulus generally depends upon the temperature and decreases by increasing the temperature. The glass transition temperature of polymer showed the maximum value of Tan δ. the storage modulus and the glass transition temperature (Tg) of the polymer composite is affected by the inclusion of the clay. The both of these two values depends upon the clay content. It has been observed that the values of storage modulus and Tg increases up to 3 wt% of nano clay content after which a decrease in modulus and Tg occurred.

Table 7 DMA properties of laminate series

S.No.	Material	Storage modulus at 25 °C (MPa)	Tg (°C)
1	E + WGF + 0 % clay	2100	81
2	E + WGF + 1 % clay	2201	83
3	E + WGF + 2 % clay	2257	84
4	E + WGF + 3 % clay	2308	86
5	E + WGF + 4 % clay	2271	84
6	E + WGF + 5 % clay	2260	83

Reproduced with permission from Elseveir Ltd. (Kanny and Mohan 2014)

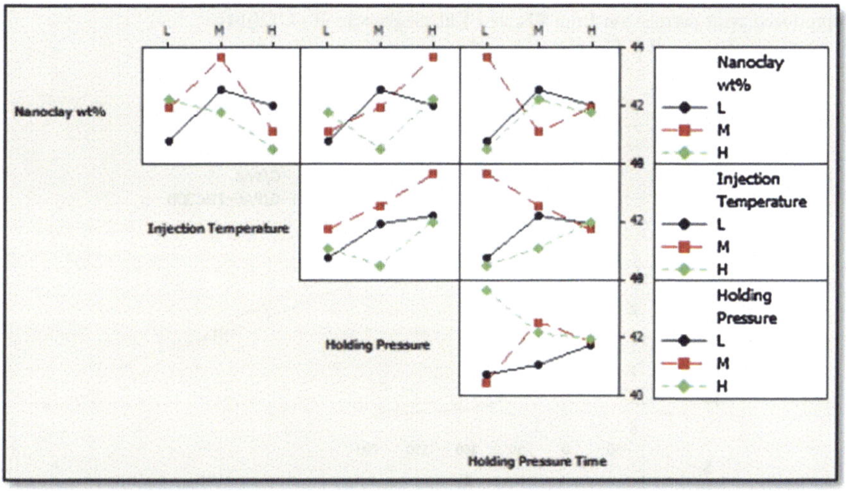

Fig. 27 Interactions plot of processing parameter for tensile strength. Reproduced with permission from Elseveir Ltd. (Shishven et al. 2014)

Shishavan et al. (2014) developed and investigated the consequence of nanoclay and processing variables on the mechanical properties such as tensile strength and hardness Acrylonitrile Butadiene Styrene–organoclay nanocomposites made by injection molding. Figure 27 represents the processing parameters used to evaluate the tensile strength of the material. In order to achieve the maximum tensile strength, the injection pressure should be medium. Also the other parameters such as high holding pressure and less holding pressure time play an important role to get the higher values of tensile strength (Fig. 27). The stress strain curve of pure and nanoclay-filled specimens under tensile test is shown in Fig. 28. From Fig. 28 it has been clearly noticed that 2 and 4 wt% nanoclay composites showed reduction in elongation at break as compared to the pure ABS composite sample.

Monti et al. (2015) evaluated the Dynamic mechanical analysis of the neat and glass fiber reinforced systems and showed in the form of storage modulus (E0) versus temperature Fig. 29. The systems (neat and Cloisite 30B (C30B-doped))

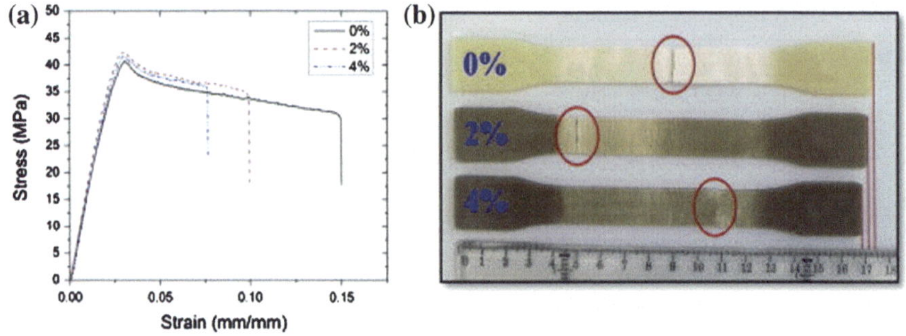

Fig. 28 **a** Stress–strain curves and **b** elongation at break of different specimens under tensile test. Reproduced with permission from Elseveir Ltd. Shishavan et al. (2014)

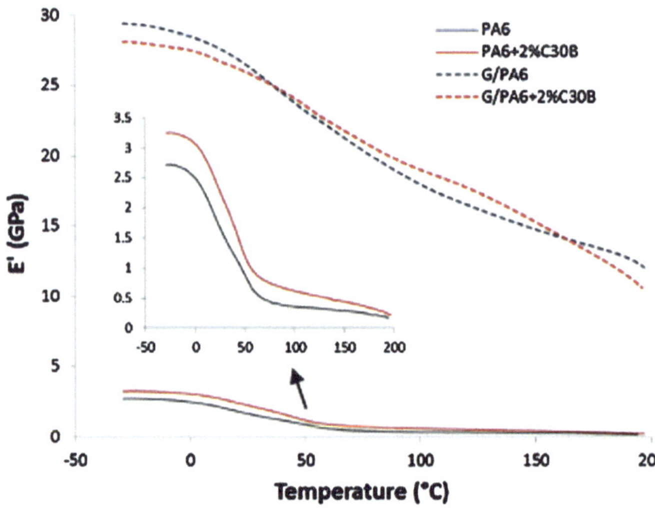

Fig. 29 Storage modulus curves obtained by DMA for G/PA6 and G/PA6 + 2 %C30B, compared against their respective unreinforced systems. Reproduced with permission from Elseveir Ltd. (Monti et al. 2015)

showed a sharp decline in the modulus with correspondence to the glass transition temperature of Polyamide 6 (PA6). At room temperature the two modulus values of the systems are found to be 2.7 and 3.3 GPa and permit authors to evaluate the rigidity induced by the montmorillonite. With regard to the glass fiber laminates, both systems (neat and C30B-doped) exhibited similar values and curve profiles. The results showed that modulus values are higher than that of the unreinforced systems.

5 Morphological Studies of the Nanoclay Composites

Morphological analysis of the composites can be carried out by using various techniques such as Scanning Electron Microscopy, Transmission Electron Microscopy, Atomic Force Microscopy etc. SEM (Scanning Electron Microscopy) is one of the popular technique and is frequently used by the researchers. Some of the morphological analysis performed by researchers using SEM are as follows:

Huang and Netravali (2007) used the three resins for the development of the composites and shows the SEM image in Fig. 30. From fig it is observed that the resin is still attached on the yarn surface after the failure of composites (yarn reinforced), which shows a good adhesion between the flax fiber and the modified SPC resin. Haq et al. (2008) studied the interfacial characteristic of the pull-out failures of the composites. From Fig. 31a, b it is noticed that increasing the bio-resin contents for 10 % EMS (biocomposite B) and 20 % EMS (biocomposite C) results an increase in the interfacial gap, respectively. Haq et al. (2009) showed the SEM image of unfilled UPE (Fig. 32a). On the other hand Fig. 32b shows the SEM image of clay reinforced (1.5 wt%), 10 % EMS blended composite. it is also noticed that the EMS

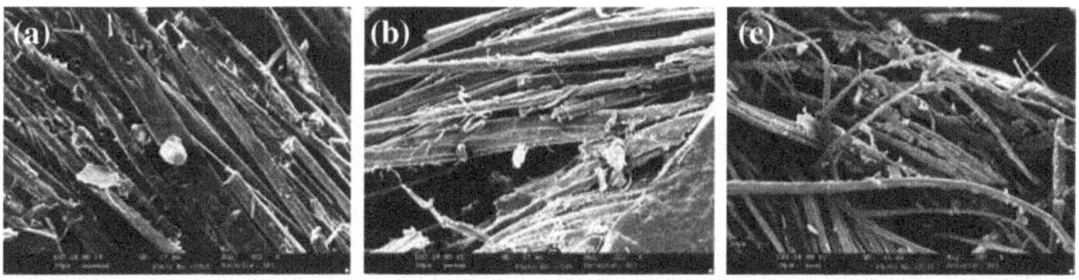

Fig. 30 SEM photomicrographs of failure surfaces of flax yarn reinforced composites with different resins: **a** SPC resin; **b** C-SPC resin and **c** GAC-SPC resin. Reproduced with permission from Elseveir Ltd. (Huang and Netravali 2007)

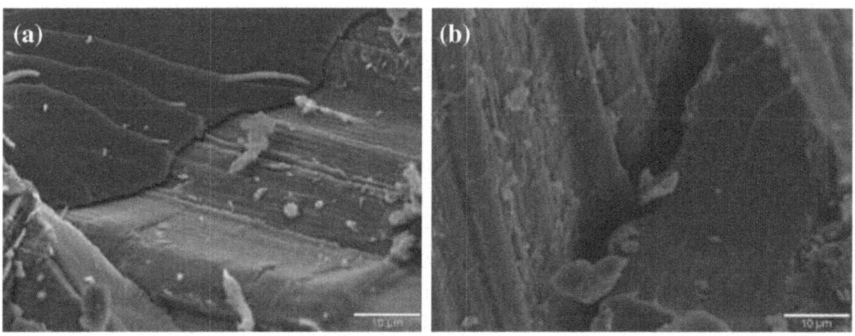

Fig. 31 SEM micrographs of tensile fracture surfaces showing interfacial gaps between fiber and matrix. **a** Biocomposite B: 10 % EMS and no nanoclay in UPE, **b** biocomposite C: 20 % EMS and no nanoclay in UPE. Reproduced with permission from Elseveir Ltd. (Haq et al. 2008)

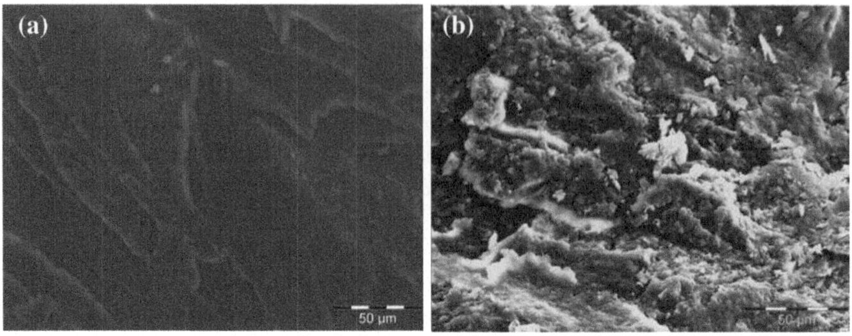

Fig. 32 SEM micrographs of tensile failure surfaces: **a** Neat UPE without inclusions, scale bar = 50 μm, **b** 10 % bio-blend [EMS] in UPE with 1.5 wt% clay inclusions, scale bar = 50 μm. Reproduced with permission from Elseveir Ltd. (Haq et al. 2009)

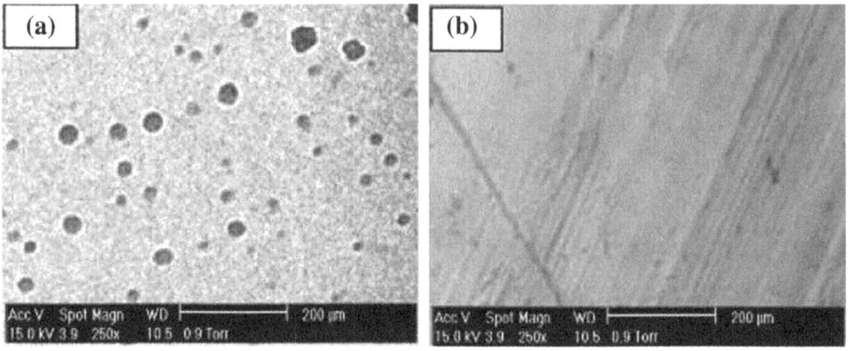

Fig. 33 SEM of 5 wt% clay filled composite: **a** before degassing and **b** after degassing. Reproduced with permission from Elseveir Ltd. Mohan and Kanny (2011)

blends and clay content generally affects the roughness of the fracture surfaces. The EMS blends and clay loading increases the toughness of the composite.

Mohan and Kanny (2011) performed morphological analysis such as SEM (Scanning Electron Microscopy) of 5 wt% nanoclay reinforced composite. Figure 33a shows the SEM image of the developed composite, in which we can clearly see the presence of the voids. This is because after blending the nanoclay into the resin then it is required that the hardener should be mixed instantly into the matrix system. This leads to the entrapment of the air in the solution which forms the voids in the system. Also the formation of the voids also led to absorb water in the specimen. So it is advised that the matrix should be degas by the degassifier before the addition of the hardener, which ultimately removes the voids from the solution. This is shown in Fig. 33b (without voids). Arrakhiz et al. (2013) shows a SEM image of the clay reinforced pine cone polymer composite in Fig. 34. The pine cone used in this study was chemically treated i.e. Alkali treated and also the

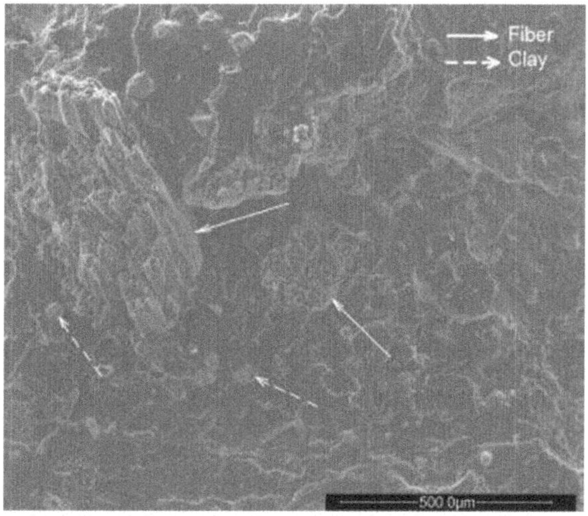

Fig. 34 SEM image of clay filled hybrid system from the PP/alkali pine cone/clay (15:15). Reproduced with permission from Elseveir Ltd. (Arrakhiz et al. 2013)

polymer used in the development of the composite was grafted with 8 wt% of SEBSg-MA coupling agent. The fillers in the matrix system of the composites can be distinctly seen in the figure.

6 Application of Nanoclay Composites

The application of nanoclay composites generally found not only in automotive sectors but also in industrial sectors. Following are some of the common applications of nanoclay composites

a. Food packaging.
b. Drive belt covers.
c. Car seat backs.
d. Body parts of trucks and cars.
e. Step assist for Chevrolet Astro van.
f. Body side molding for Chevrolet Impala.
g. Cargo bed of General Motor Hummmer.

7 Conclusion/Summary

This chapter concludes that the nano clay can be used as filler in polymer composites. Various types of the processes used for the manufacturing of nano clay composites have also been discussed in this chapter. The current chapter also showed and evidenced from the past research that chemical coupling agents improves the adhesion properties between the nano clay particles and the matrix material which is most important for achieving the good mechanical properties and applications. The major applications of the nano reinforced composites were also listed in the current chapter. The main applications were found in the field of automobiles industry and reported good results.

References

Arrakhiz, F.Z., Benmoussa, K., Bouhfid, R., Qaiss, A.: Pine cone fiber/clay hybrid composite: mechanical and thermal properties. Mater. Des. **50**, 376–381 (2013)

Babaei, I., Madanipour, M., Farsi, M., Farajpoor, A.: Physical and mechanical properties of foamed HDPE/wheat straw flour/nanoclay hybrid composite. Comp. B **56**, 163–170 (2014)

Chan, M., Lau, K., Wong, T., Ho, M., Hui, D.: Mechanism of reinforcement in a nanoclay/polymer composite. Comp. B **42**, 1708–1712 (2011)

Chittajallu, K.M.: Computational Modeling of the Vacuum Assisted Resin Transfer Molding (VARTM) Process. Thesis, M.S., Clemson University, USA (2004)

Chowdhury, F.H., Hosur, M.V., Jeelani, S.: Studies on the flexural and thermomechanical properties of woven carbon/nanoclayepoxy laminates. Mater. Sci. Eng. A **421**, 298–306 (2006)

Dennis, H.R., Hunter, D., Chang, D., Kim, S., Paul, D.R.: Effect of melt processing condition on the extent of exfoliation in organoclay-based nanocomposites. Polymer **42**, 9513–9522 (2001)

Ferreira, J.A.M., Reis, P.N.B., Costa, J.D.M., Richardson, B.C.H., Richardson, M.O.W.: A study of the mechanical properties on polypropylene enhanced by surface treated nanoclays. Comp. B **42**, 1366–1372 (2011)

Haq, M., Burgueño, R., Mohanty, A.K., Misra, M.: Hybrid biobased composites from blends of unsaturated polyester and soybean oil reinforced with nanoclay and natural fibers. Comp. Sci. Tech. **68**, 3344–3351 (2008)

Haq, M., Burgueño, R., Mohanty, A.K., Misra, M.: Biobased unsaturated polyester/layered silicate nanocomposites: characterization and thermophysical properties. Comp. A **40**, 540–547 (2009)

Haq, M., Burgueño, R., Mohanty, A.K., Misra, M.: Biobased polymer nanocomposites from UPE/EML blends and nanoclay: development, experimental characterization and limits to synergistic performance. Comp. A **42**, 41–49 (2011)

Huang, X., Netravali, A.: Characterization of flax fiber reinforced soy protein resin based green composites modified with nanoclay particles. Comp. Sci. Tech. **67**, 2005–2014 (2007)

Iman, M., Maji, T.K.: Effect of crosslinker and nanoclay on starch and jute fabric based green nanocomposites. Carbohy. Poly. **89**, 290–297 (2012)

Jawaid, M., Abdul Khalil, H.P.S.: Cellulosic/synthetic fibre reinforced polymer hybrid composites: a review. Carbohy. Polym. **86**, 1–18 (2011)

Kanny, K., Mohan, T.P.: Resin infusion analysis of nanoclay filled glass fiber laminates. Comp. B **58**, 328–334 (2014)

Kornmann, X., Linderberg, H., Bergund, L.A.: Synthesis of epoxy-clay nanocomposites: influence of the nature of the curing agent on structure. Polymer **42**, 4493–4499 (2001)

Ludueña, L.N., Alvarez, V.A., Vazques, A.: Processing and microstructure of PCL/clay nanocomposites. Mater. Sci. Eng. A **460–461**, 121–129 (2007)

Marras, S.I., Kladi, K.P., Tsivintzelis, I., Zuburtikudis, I., Panayiotou, C.: Biodegradable polymer nanocomposites: the role of nanoclays on the thermomechanical characteristics and the electrospun fibrous structure. Acta Biomater. **4**, 756–765 (2008)

Majeed, K., Jawaid, M., Hassan, A., Abu Bakar, A., Abdul Khalil, H.P.S., Salema, A.A., Inuwa, I.: Potential materials for food packaging from nanoclay/natural fibres filled hybrid composites. Mater Des. **46**, 391–410 (2013)

Mohan, T.P., Kanny, K.: Water barrier properties of nanoclay filled sisal fibre reinforced epoxy composites. Comp. A **42**, 385–393 (2011)

Monti, M., Tsampas, S.A., Fernberg, S.P., Blomqvist, P., Cuttica, F., Fina, A., Camino, G.: Fire reaction of nanoclaydoped PA6 composites reinforced with continuous glass fibers and produced by commingling technique. Poly. Deg. Stab. **121**, 1–10 (2015)

Nguyen, Q.T., Baird, D.G.: An improved technique for exfoliating and dispersing nanoclay particles into polymer matrices using supercritical carbon dioxide. Polymer **48**, 6923–6933 (2007)

Pourabas, B., Raeesi, V.: Preparation of ABS/montmorillonite nanocomposites using a solvent/non solvent method. Polymer **46**, 5533–5540 (2005)

Ray, S.S., Okamoto, M.: Polymer/layered silicate nanocomposite: a review from preparation to processing. Prog. Poly Sci. **28**, 1539–1641 (2003)

Shishavan, S.M., Azdast, T., Ahmadi, S.R.: Investigation of the effect of nanoclay and processing parameters on the tensile strength and hardness of injection molded Acrylonitrile Butadiene Styrene–organoclay nanocomposites. Mater. Des. **58**, 527–534 (2014)

Satyanarayana, K.G., Arizaga, G.G.C., Wypych, F.: Biodegradable composites based on lignocellulosic fibers—an overview. Prog Poly. Sci. **34**, 982–1021 (2009)

Mechanical Performance of Nanoclay-Cellulose Fibre Particulate Composites Fabricated by Modified Two Stage Wet/Hand Layup Method

Nadendla Srinivasababu

Abstract Polymer composites are multiphase materials, where the reinforcement in the form the short/long fibre, filler, particulate enveloped by a matrix. In case of particulate composites, different shapes and sizes of it are dispersed into matrix in random manner. This class of materials is introduced to cheapen the price of polymer and achieve specific behavior may be improved mechanical performance, electrical resistance etc. Firstly cellulose fibre, nanoclay is analyzed for its particle size and size of particles is also found from Scanning Electron Micrograph image to get the clear, rough idea about particulates respectively. An experimental effort is made to manufacture cellulose-nanoclay, nanoclay reinforced vinyl ester composites lamina by Modified Two Stage Hand/Wet Layup Method (MTSHLM) and the specimens for various tests viz. tensile, flexural and bending are machined according to ASTM standards. This innovative method of making composites are experimentally ensured the presence of entire particulates i.e. cellulose, nanoclay. Further the dimension of the specimen made in stage 1, in thickness direction, along its length is measured using micrometer to understand the capability of lamina manufactured by MTSHLM. After characterizing of composites under mechanical load revealed that cellulose acts as a supplement in cellulose-nanoclay composites. An increase in nanoclay content in nanoclay composites is also result in reasonably good improvement of mechanical properties, particularly under bending load. The highest tensile strength, modulus of 32.65, 383.55 MPa is achieved in 5 % cellulose fibre—6.67 % nanoclay composition. In three-point-bend test the composition contains 5 % Cellulose fibre—6.67 % nanoclay lead to 104.13 MPa, 3.73 GPa flexural strength, modulus respectively. But the composition of 6.67 % nanoclay vinyl ester composites has exhibited Charpy impact strength of 2.17 kJ/m^2. Tested composite specimens are analyzed under scanning electron microscope to know dispersion of particulate, interfacial bond between reinforcement and matrix.

N. Srinivasababu (✉)
Department of Mechanical Engineering, Vignan's Lara Institute of Technology & Science, Vadlamudi 522 213, Andhra Pradesh, India
e-mail: cnjlms22@yahoo.co.in

M. Jawaid et al. (eds.), *Nanoclay Reinforced Polymer Composites*, Engineering Materials, DOI 10.1007/978-981-10-0950-1_10

Keywords Modified Two Stage Hand/Wet Layup Method (MTSHLM) · Mechanical properties · Scanning election microscope · Cellulose fibre · Nanoclay · Particle size

1 Introduction

One of the earthy materials occurring in the nature is clay and possesses plasticity when it is wet. At this stage clay can be molded to any form and is retained after drying (Ries 1906). When the molded items/articles are heated, the material becomes very hard e.g. brick(s) used in construction, drinking water stored in pots.

The main uses of the clay have been recognized since the earliest periods of civilization. Many references are available in the ancient Assyrian and Egyptians records to the employment of clay for manufacture of bricks and for fulling or whitening cloth (Searle 1912).

A careful observations is made and review conducted to understand the manufacturing method of composites, introduction of reinforcements and role i.e. cellulose, nanoclay on different matrices like bio, thermoplastic, thermoset. Further mechanical performance of composites is understood from the past works systematically. Microcrystalline cellulose powder of different concentrations was partially dissolved in lithium chloride/N, N-dimethylacetamide to produce cast nanocomposite film. Then by using Zwick 20 kN Universal Testing Machine tensile tests was conducted on samples at a cross head displacement rate of 1 mm/min to determine their strength, modulus (Gindl and Keckes 2005). Bionanocomposites films were prepared with the combination of poly lactic acid as matrix and microfibrillated cellulose as reinforcement through solvent evaporation at room temperature (Tingaut et al. 2010). Fully green composites were laminated and pressed at 190 °C from polylactic acid, bamboo fibre, microfibrillated cellulose. The specimens of 40 × 3 × 1.5 mm were tested at a cross head speed of 2 mm/min to determine bending properties (Okubo et al. 2005).

Composite films made form micro to nano fibrillated cellulose and polyurethane were subjected to tensile load for knowing strength, modulus (Ozgur and Seydibeyoglu 2008). Atactic polyvinyl alcohol solution contains different concentrations of microfibrillated cellulose is sonicated (1, 5 min) and stirred for 24 h. Then the films (100 μm) were cast on to glass plate with controlled leveling and dried at room temperature for 7 days to determine tensile properties using miniature materials tester Minimat at a cross head speed of 5 mm/min (Jue et al. 2008). With the help of Instron 3365 universal materials tester tensile test was conducted on specimens of 38 × 4.5 × 0.35 mm made from films to find young's modulus, strength and strain at fracture (Nakagaito and Yano 2008).

High Density Poly Ethylene (HDPE, grade HD6605), pine flour and organic clay were blended in thermo kinetic high-shear mixer i.e. K-mixer at 5000 rpm and is discharged at 190 °C. Then the granulated blends were injection molded through

Vista Sentry VSX at 199 °C and a mold temperature of 100 °C to determine tensile, flexural and impact performance (Lei et al. 2007). Melt, direct dry blending processes were adopted to introduce nanoclay into HDPE to make wood-plastic, wood-flour composites by compounding. Tensile, flexural tests were carried out using an Instron 5585 H at a test head speed of 50, 1.3 mm/min respectively (Faruk and Matuana Laurent 2008).

Using Barbender mixing machine chopped wood, coir fibre (2 mm length), poly propylene granules, montmorillonite nanoclay were heated in Hot Press Machine at 170 °C to make hybrid composites for tensile test (Islam et al. 2015). Medium Density Fibreboard (MDF) was manufactured using aspen fibre, Urea Formaldehyde resin, monmorillonite type nanoclay at different contents to test their mechanical performance (Ashori and Nourbakhsh 2009).

Two different batches of composites films were made using Poly-L-Lactide, Montmorillonite clay (Cloisite 25 A) by extrusion. Then tensile test was conducted at a speed of 10 mm/min on Instron 4481 mechanical tester (Lewitus et al. 2006). Poly Vinyl Chloride, hectorite, bentonite clay based composites were made and tensile tested (Awad Walid et al. 2009). Modified soy protein concentrate (SPC) resin was prepared by blending SPC with nano-clay particles, then cross linked with glutaraldehyde. This matrix was combined with flax yarn, fabric to prepare composites and find their tensile, flexural properties (Huang and Netravali 2007).

The effect of montmorillonite nanoclay reinforcement in epoxy resin was investigated through flexural test (Zainuddin et al. 2010). Bisphenol A diglycidyl ether (D.E.R. 332), Nanocor Organophilic clay mixture was ultrasonically stirred using JAC ultrasonic 1505 bath for 1, 3 h to find the dispersion effect by Dynamic Mechanical Test (Wang and Qin 2007). Ultrasound sonication effect in nanoclay clusters of nanoclay/araldite epoxy composites was investigated (Lam et al. 2005). Nanoclay was firstly introduced into epoxy resin and is manually mixed with glass rod before it was mechanically stirred at 500 rpm for 1 h. The mixture was heated in a hot water bath at 50 °C to reduce viscosity of resin and then the hardener is introduced. Using MTS machine alliance RT/50 tensile test was conducted at a cross head speed of 2 mm/min on dog bone shaped composites (Ho et al. 2006). With the aid of mechanical shearing nanocomposites were prepared by in situ polymerization method. The composites were tested for tensile strain, strength and modulus (Qi et al. 2006).

From the above and literature review of about 2000 papers from various journals, it is understood that Two Stage Hand/Wet Layup was firstly introduced, which ensures the complete presence of dora hemp particulate fibre into polyester resin. Then the composites were mechanically tested (Srinivasababu 2015).

In the current work an attempt is made to reinforce nanoclay alone, cellulose-nanoclay into vinyl ester matrix and the composites mechanical performance is also evaluated along with the analysis of particulates i.e. cellulose, nanoclay. The terms cellulose fibre and cellulose are used invariably in this chapter.

2 Materials, Manufacturing and Testing

The materials used in this work are vinyl ester resin, cellulose, nanoclay, procured from Ecmas Resins Pvt. Ltd., Hyderabad, Telangana and Sigma-Aldrich Corporation Bangalore, India respectively.

2.1 Matrix—Vinyl Ester Resin

Ecmalon 9911 is bisphenol epoxy based vinyl ester resin is used as matrix. This kind of resins is suitable for the fabrication of chemical equipment because they offer good resistance to degradation in contact with variety of chemicals which include strong acid and alkalies. Paper and pulp industry use Bisphenol epoxy vinyl ester resins. This resin is suited for wet layup, pultrusion and other types of molding. The properties of the liquid resin are tested according to IS6746-1994 by the manufacturer, are given in Table 1.

Shelf life of resin is 3 months from the date of manufacture when it is stored at temperature below 25 °C. The resin shelf life drops significantly at higher temperatures. The specimens were made from the cast sheets cured 24, 4 h at room temperature and 90 °C respectively. Mechanical properties of the tested specimens were provided by Ecmas Resins, given in Table 2.

Table 1 Properties of Ecmalon 9913 vinyl ester resin (provided by manufacturer)

Property	Description/value
Appearance	Clear, yellow liquid
Viscosity @25 °C (cps)	400 (Brookfield viscometer)
Specific gravity (25/25 °C)	1.05
Acid value (mg KOH/g)	10
Volatiles @150 °C (%)	42
Gel time @25 °C	25 min

Table 2 Mechanical properties of Ecmalon 9913 vinyl ester resin cast specimens (provided by manufacturer)

Property	Unit	Typical value	Test method
Tensile strength	Mpa	70	ISO 527
Tensile modulus	Mpa	3200	ISO 527
Elongation at break	%	4.0	ISO 527
Flexural strength	Mpa	115	ISO 178
Flexural modulus	Mpa	3000	ISO 178
Heat deflection temp	°C	102	ISO 75
Hardness	Barcol	35	IS6746-'94

Table 3 Specification of cellulose medium fibre (provided by manufacturer)

Test	Specification
Appearance (color)	White to off white
Appearance (form)	Powder or solid
Moisture content	≤10 %

Table 4 Specification of nanoclay (provided by manufacturer)

Test	Specification
Appearance (color)	White to off-white
Appearance (form)	Powder
Loss on drying	3.0 %
Size	≤20 μm
Density (bulk density)	200–500 kg/m^3

2.2 *Reinforcements—Cellulose Fibre, Nanoclay*

One of the reinforcement used in this work is cellulose fibers (CF). Sigma Aldrich has specified the fiber and is given in Table 3. The applications of cellulose fibres are for partition chromatography. Cellulose (medium fibers) is also used in metabolic pathway and carbohydrate applications. It has been used to study biofuel and biorefinery applications. The relative density of fibre is 0.6 gm/cm^3. Surface modified nanoclay (Montmorillonite clay) contains 0.5 wt% aminipropyltriethoxysilane, 15–35 wt% octadecylamine and the properties provided by supplier is given in Table 4. This nanoclay (NC) was recommended by manufacturer for melt compounding with polypropylene (PP), polyethylene (PE), and ethylene vinyl acetate (EVA) and provided following properties when it was reinforced.

- Improve modulus
- Increase barrier
- Enhance chemical resistance
- Improve flame retardancy
- Increase heat deflection temperature (HDT)

2.3 *Manufacturing*

In this experimentation two batches of specimens are fabricated by using Two Stage Wet/Hand Layup manufacturing method published elsewhere (Srinivasababu 2015) but with some modification i.e. MTSHLM and are described subsequently.

Initially vinyl ester resin is poured in 1000 ml capacity Borosil beaker and the nanoclay is introduced into it slowly by stirring with a glass rod of 6 in. length. Once the resin-nanoclay is mixed manually, it is further stirred in a mechanical

Table 5 Composition of cellulose fibre, nanoclay

Composition label in graph	Composition label	Cellulose fibre content (%)	Nanoclay content (%)
A	A1	5	3.33
B	B1	5	6.67
C	C1	5	10
A	A2	–	3.33
B	B2	–	6.67
C	C2	–	10

mixer up to 5 min. Now the mixture is poured through a Borosil beaker with 10 mm hole into the mold till it is completely occupied on entire surface and allowed to cure at room temperature up to 30 min. Once the resin-cellulose mixture in partially soft solid form, resin is poured into the mold again till complete thickness is attained as per the dimensions specified by ASTM standards and is allowed to cure at room temperature for 24 h. During preparation of lamina promoter, accelerator, catalyst 2 % of each is mixed in the resin one after another in stage-1 lamina preparation. But the percentage of the above chemicals is 2.2 in stage-2 lamina preparations to enhance the bond between the two layers.

Secondly, the next batch of specimens is prepared with the following additional steps. Cellulose fibre, nanoclay are introduced into vinyl ester matrix one after another by ensuring thorough mixing of the reinforcements in resin. The compositions of CF, NC used in this experimentation is given in Table 5.

The prepared lamina is heated in NSW-143 Oven Universal up to 5 h and allowed to cool at room temperature and is shown in Fig. 1. Using mechanically operated cut machine tensile, flexural and impact test specimens are obtained from lamina as per the specifications of ASTM D638-10, ASTM D 790-07 and ASTM D6110-08 for tensile, flexural and Charpy impact test respectively. With the help of belt grinding machine the specimens are ground to ensure flatness and finishing of edges.

2.4 *Testing*

Cellulose, nanoclay particle size is analyzed using Particle Size Analyser at Kelvin Labs, Hyderabad, Telangana, India. Further reinforcement samples are characterized for its morphology and size using Scanning Electron Microscope.

Tensile, flexural tests are conducted using PC 2000 Electronic Tensometer at a crosshead speed of 2, 8 mm/min respectively. A notch is cut to a depth of 2.5 mm using Notch Cutter and Charpy impact test is conducted on Computerized Izod/Charpy impact tester. In all the mechanical tests five specimens are tested in each case and the average value is taken.

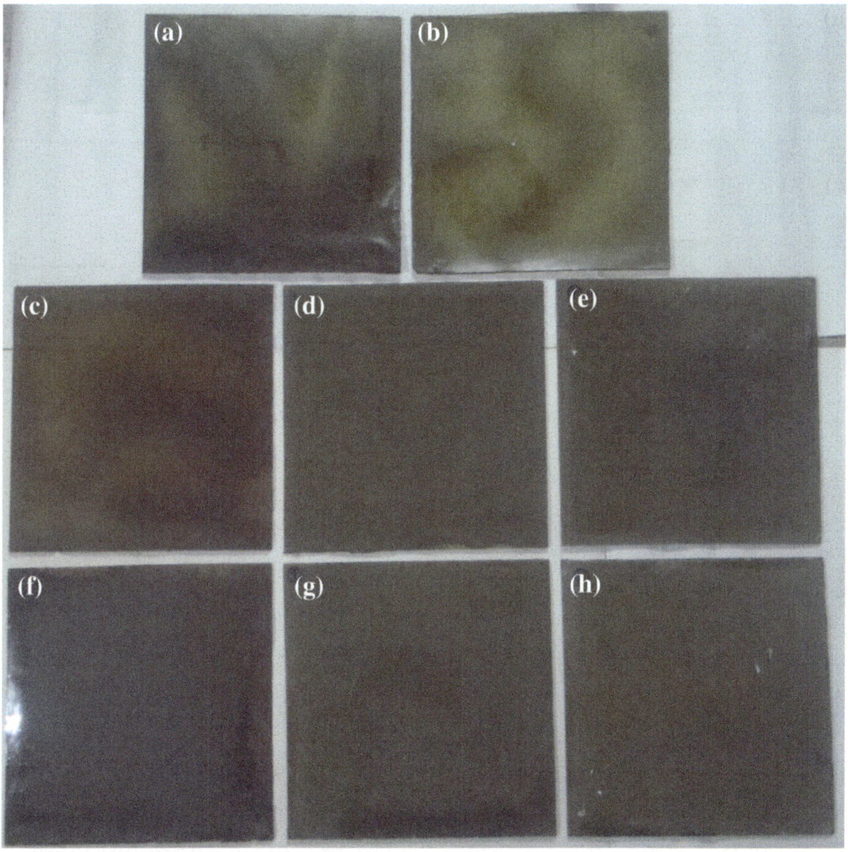

Fig. 1 **a**, **b** Vinyl ester Plain, **c** 3.33 wt% NC, **d** 6.67 wt% NC, **e** 10 wt% NC, **f** 5 wt% CF-3.33 wt% NC, **g** 5 wt% CF-6.67 wt% NC, **h** 5 wt% CF-10 wt% NC Composite lamina manufactured by Two Stage Hand/Wet Layup Manufacturing Method

3 Results and Discussion

The reinforcement viz. cellulose fibre, nanoclay are analyzed for size, morphology and two batches of composites are tested for understanding their manufacturing ability, mechanical performance.

3.1 Cellulose Fibre, Nanoclay—Size, Morphology

The nanoclay, cellulose fibre is tested for its size, run 1, 2, 3 and average is reported in Figs. 2a–d and 3a–d respectively. In case of nanoclay the peak diameter varied

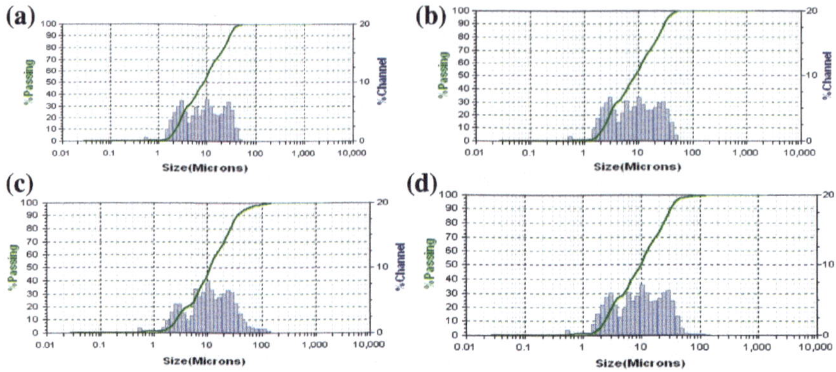

Fig. 2 Particle size analysis of Nanoclay **a** Run 1, **b** Run 2, **c** Run 3, **d** Average

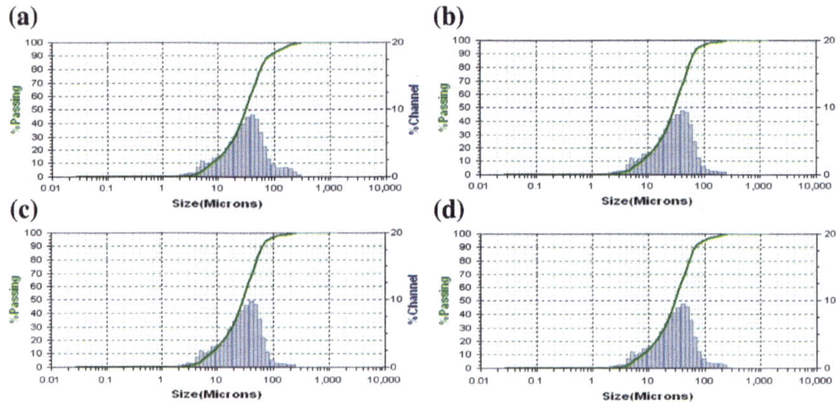

Fig. 3 Particle size analysis of Cellulose Fibre **a** Run 1, **b** Run 2, **c** Run 3, **d** Average

from 2.531 to 24.55 nm whereas 50 percentile particles are of 9.45 μm. Formation of clusters with variable size is clearly visible from Scanning Electron Micrograph (SEM), Fig. 4. Now the size of more/less big size clusters is measured and its value is 154–229 μm, the shape of the particle is nearly spherical. Formation of soft heap solid mass is recognized from Fig. 5 i.e. cellulose fibre SEM. In this case fibre group size varies as 119–230 μm. Due to micron size of the cellulose fibre it seems to be spherical in its shape.

3.2 Mechanical Properties

In this work plain, CF-NC, NC vinyl ester composite specimens prepared from lamina are tested as per ASTM standards and the average value of five specimens is

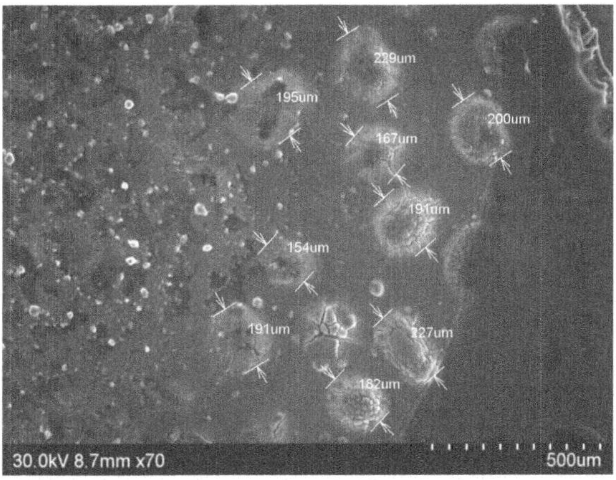

Fig. 4 Scanning Electron Microscope image of Nanoclay with particle size

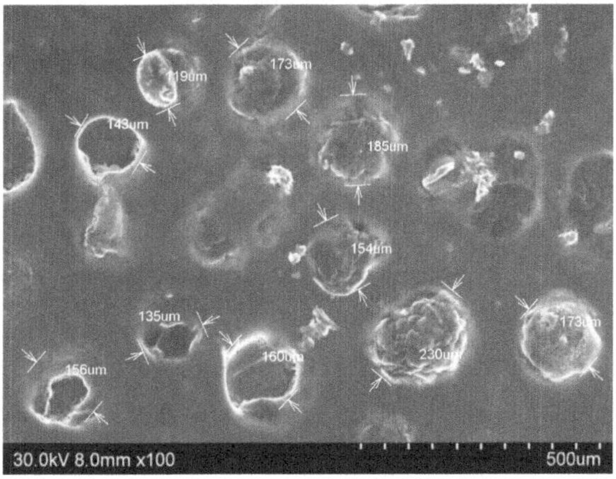

Fig. 5 Scanning Electron Microscope image of contains particle size, geometry of Cellulose Fibre

calculated in each test. The tensile strength of the CF-NC, NC vinyl ester composites is decreased with increase in NC content due to the lack of sufficient matrix. The reinforcement effect of CF is clearly recognizable from Fig. 6, where the tensile strength of CF-NC composites at A1, B1, C1 compositions is greater than A2, B2, C2 respectively. The highest value of 32.65 MPa is achieved at composition A1. Elongation is clearly decreased with increase in NC content at all compositions and moreover the composition reinforced with CF have more elongation when compared to others composites possess no CF. Tensile modulus of the composites against composition is shown in Fig. 7. Highest tensile modulus (383.55 MPa) in tension is obtained for the composites at A1 composition.

In all the composites tensile strength, modulus is varied with respect to composition. An expected reasonable amount of properties enhancement is not observed due to disparity in thickness of lamina prepared in stage-1 of the process along its length, width. With an eye observation it is clear that the lamina prepared

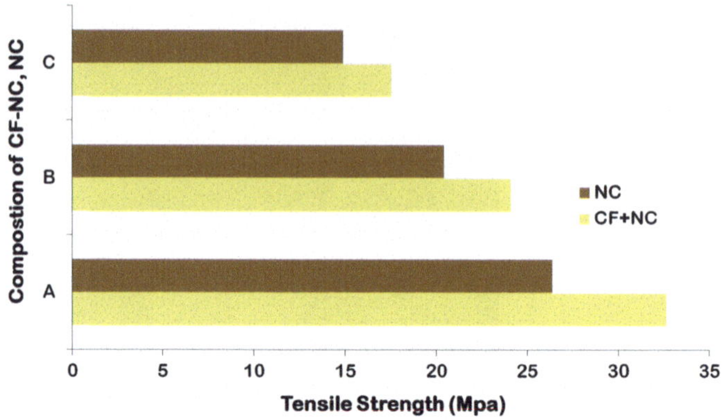

Fig. 6 Tensile strength of CF-NC, NC reinforced vinyl ester composites at different compositions

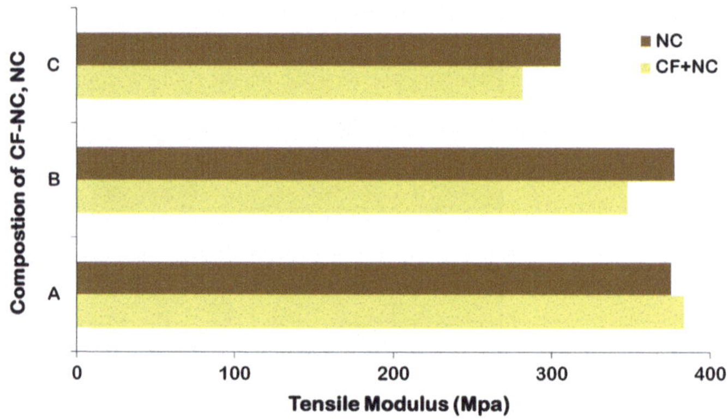

Fig. 7 Tensile modulus of CF-NC, NC reinforced vinyl ester composites at different compositions

at highest content of NC unable to spread over the mold surface due to high viscosity possessed by the resin-reinforcement mixture. Hence thickness of the specimens manufactured in stage-1 is measured at five different locations along its length and is given in Table 6 for CF-NC, NC composites.

Miniaturized tensile test was conducted (Ozgur and Seydibeyoglu 2008) on different compositions of polyurethane and cellulose fibre composites, where polyurethane and cellulose are hydrophilic. Poly urethane-cellulose nano fibre1 has exhibited maximum tensile strength.

Flexural strength, modulus of the CF-NC, NC composites is increased with NC content except at its initial, highest level. At initial level the reinforcement acts as filler, high level lack of matrix in stage-1. Figure 8 and 9 shows the flexural strength, modulus at different composition of CF/NC, NC contents. The highest flexural strength of 104.13 MPa is attained for the composition B1. This tells that

Table 6 Thickness of tensile test specimens manufactured in stage-1

Specimen number	Thickness (mm)				
	Position 1	Position 2	Position 3	Position 4	Position 5
Composition A1: 5 % CF + 3.33 % NC					
1	3.169	3.192	3.566	3.461	3.180
2	2.559	3.028	3.408	3.387	3.305
3	3.040	3.437	3.791	3.157	2.772
4	2.763	2.304	2.516	2.953	3.253
5	2.947	3.326	3.115	3.087	3.331
Composition B1: 5 % CF + 6.67 % NC					
6	4.588	3.370	3.479	3.208	2.255
7	4.910	4.198	3.831	3.700	3.550
8	4.196	3.774	3.120	2.715	3.448
9	2.949	2.550	2.551	3.817	3.97
10	2.570	3.113	3.493	3.873	2.733
Composition C1: 5 % CF + 10 % NC					
11	2.205	3.281	3.154	1.901	2.802
12	3.542	2.351	2.385	2.466	2.288
13	2.898	1.929	1.878	1.928	2.813
14	3.005	3.152	2.388	2.432	2.402
15	3.381	2.020	2.706	2.143	2.675
Composition A2: 3.33 % NC					
16	2.755	2.909	3.528	3.420	2.591
17	3.497	2.798	2.685	2.476	2.279
18	4.016	3.638	3.088	3.320	2.998
19	2.989	3.068	2.607	3.927	3.682
20	2.254	2.488	2.971	3.311	3.385
Composition B2: 6.67 % NC					
21	3.240	2.698	2.608	2.560	2.579
22	3.286	2.899	2.762	2.762	2.358
23	3.030	2.929	2.885	2.626	2.942
24	2.926	2.935	2.891	2.630	2.721
25	2.604	2.461	2.693	2.545	2.405
Composition C2: 10 % NC					
26	2.233	1.593	3.247	3.281	3.292
27	3.435	2.207	3.130	3.112	2.898
28	2.530	3.126	2.196	2.397	2.894
29	3.507	2.998	1.506	2.117	1.798
30	2.569	3.259	3.879	3.878	3.044

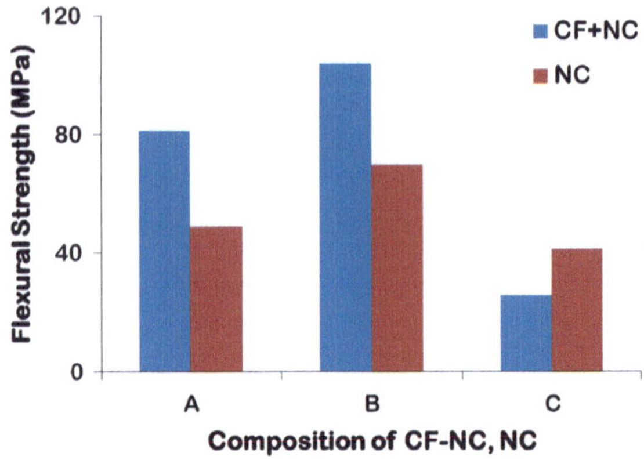

Fig. 8 Flexural strength of CF-NC, NC reinforced vinyl ester composites at different compositions

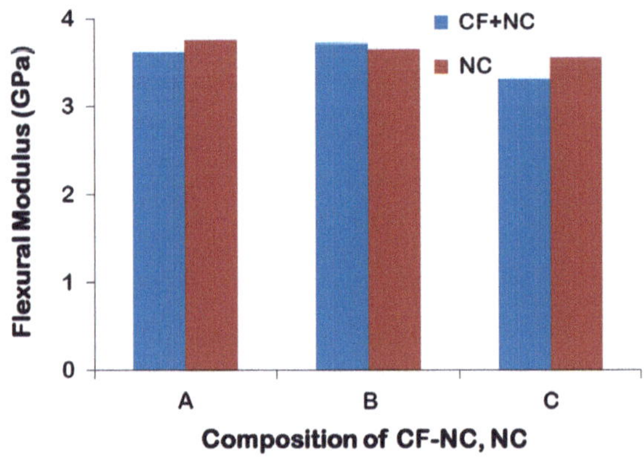

Fig. 9 Flexural modulus of CF-NC, NC reinforced vinyl ester composites at different compositions

increase in NC has enhanced the flexural strength of composites and bears load. All composites are failed due to bending only, is identified after the test. NC at highest compositions i.e. C1, C2 thickness of first stage process itself occupied maximum mold and thereby second stage process of manufacturing has very minimum thickness results in decrease in flexural strength, modulus of the CF/NC, NC composites. Interestingly the flexural strength at C2 is 37.62 % more when compares with composites have composition C1. The highest flexural modulus 3.73, 3.76 is achieved by the compositions B1, A1 composites respectively. Thickness of flexural test specimens is given in Table 7 which is taken at five different sites.

Impact strength of the CF-NC, NC reinforced vinyl ester composites are increased with its content(s) except at C1, C2 compositions due to insufficient presence of the matrix. The highest Chrapy impact strength of 1.9, 2.17 kJ/m^2 is achieved for the composites with the compositions B1, B2 and is visible from Fig. 10. All the composites have exhibited 'C' type failure as per ASTM D6110-08. In these test specimens the thickness is measured at five different positions and is tabulated in Table 8.

Table 7 Thickness of flexural test specimens manufactured in stage-1

Specimen number	Thickness (mm)				
	Position 1	Position 2	Position 3	Position 4	Position 5
Composition A1: 5 % CF + 3.33 % NC					
1	2.598	2.448	2.385	2.313	2.643
2	2.673	2.598	2.519	2.513	2.428
3	3.046	2.962	2.496	2.493	2.448
4	3.307	2.791	3.683	2.709	2.818
5	3.106	3.216	3.259	4.14	2.875
Composition B1: 5 % CF + 6.67 % NC					
1	4.553	4.557	4.394	2.995	2.415
2	2.72	2.38	2.806	2.672	2.681
3	3.324	3.543	3.711	4.136	4.13
4	4.278	4.266	4.285	3.754	3.405
5	4.56	4.344	3.881	3.737	2.977
Composition C1: 5 % CF + 10 % NC					
1	3.14	4.465	2.241	2.604	3.965
2	4.167	3.735	2.974	3.644	4.536
3	3.535	2.923	3.389	2.672	3.903
4	3.193	2.962	2.822	2.964	3.328
5	2.053	3.342	3.183	2.073	1.929
Composition A2: 3.33 % NC					
1	3.546	2.936	2.596	2.762	3.579
2	3.45	3.639	3.89	3.416	3.886
3	4.193	3.883	3.978	4.052	4.409
4	3.579	3.285	3.141	2.632	2.118
5	4.896	4.618	4.381	4.09	3.643
Composition B2: 6.67 % NC					
1	3.112	2.796	2.492	2.875	2.834
2	3.006	2.939	2.819	2.409	3.114
3	3.298	3.396	3.313	3.487	3.81
4	3.072	3.163	2.84	3.018	3.03
5	3.63	3.435	3.282	3.155	3.207
Composition C2: 10 % NC					
1	2.905	3.339	2.245	1.992	2.445
2	2.564	2.686	2.178	3.151	2.791
3	3.935	4.267	3.403	3.144	3.031
4	3.106	3.564	3.381	1.958	2.713
5	3.43	2.61	3.41	3.569	3.65

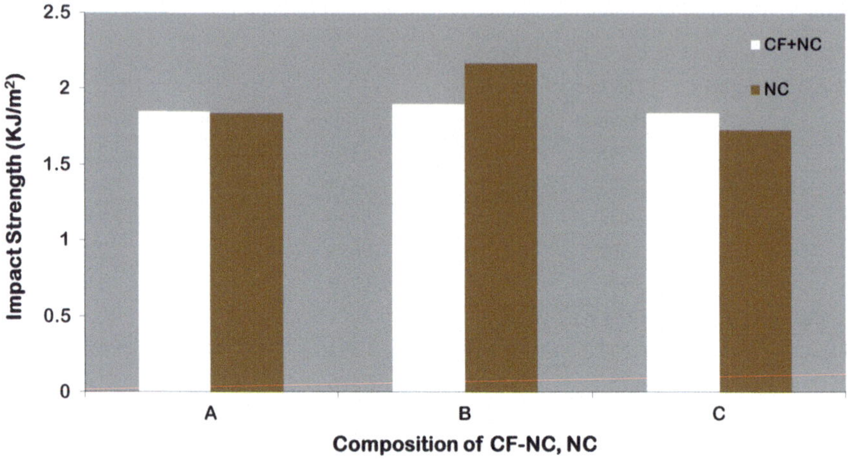

Fig. 10 Charpy impact strength of CF-NC, NC reinforced vinyl ester composites at different compositions

Table 8 Thickness of Charpy impact test specimens manufactured in stage-1

Specimen number	Thickness (mm)				
	Position 1	Position 2	Position 3	Position 4	Position 5
Composition A1: 5 % CF + 3.33 % NC					
1	2.488	2.652	2.648	2.478	2.38
2	3.059	3.101	3.778	2.241	2.229
3	2.465	2.747	2.438	2.368	2.662
4	2.955	2.681	2.786	2.725	2.596
5	2.63	3.077	3.046	2.926	3.657
Composition B1: 5 % CF + 6.67 % NC					
6	2.867	2.141	2.37	2.155	2.001
7	1.966	2.654	2.565	2.484	2.447
8	2.776	2.204	2.469	2.571	1.964
9	2.728	3.073	3.102	3.812	4.637
10	2.766	2.578	2.611	2.451	2.08
Composition C1: 5 % CF + 10 % NC					
11	2.772	4.518	3.79	4.02	3.798
12	2.517	2.667	2.68	2.246	1.579
13	3.176	3.01	2.789	2.861	2.861
14	3.307	2.294	3.695	2.861	1.929
15	2.207	2.551	2.819	1.888	2.278

(continued)

Table 8 (continued)

Specimen number	Thickness (mm)				
	Position 1	Position 2	Position 3	Position 4	Position 5
Composition A2: 3.33 NC					
16	3.652	4.208	4.23	4.12	4.308
17	4.314	3.331	2.23	2.707	3.114
18	3.73	2.989	3.399	3.75	3.776
19	3.136	3.046	3.01	3.932	3.28
20	4.166	4.12	3.96	3.794	3.859
Composition B2: 6.67 % NC					
21	4.309	4.296	4.209	3.814	4.036
22	4.1	4.019	3.608	3.766	3.893
23	3.991	3.627	3.621	3.549	2.881
24	3.974	3.91	3.784	3.361	2.644
25	3.374	3.369	3.313	4.378	4.427
Composition C2: 10 % NC					
26	3.612	3.973	3.864	2.372	2.012
27	2.914	2.643	2.262	3.072	2.349
28	2.926	3.678	3.065	2.547	2.797
29	3.768	3.96	2.418	3.202	2.433
30	2.101	2.071	2.601	3.949	4.716

3.3 Morphology Study

After testing the composites at different load viz. tensile, flexural and Charpy impact, the fractured specimen surface is examined under Scanning Electron Microscope and the following observations are made. In performing morphology study overall view, four different locations are observed in order to understand the dispersion, distribution of reinforcements and bond between them and vinyl ester matrix.

The presence of white lines (Fig. 11a) over entire surface of the sample given belief that the nanoclay presence along with cellulose. Figure 11a–d tells that the agglomerated particulates are comes peel out from the matrix, whose size varies. A rougher fracture surface is identified from the surface of the specimen, visible in Fig. 11b–e.

Two layers (stage-1: composite; stage-2: plain matrix) in the composite lamina are clearly distinguishable from Fig. 12 a, b. A rougher and smooth fractured

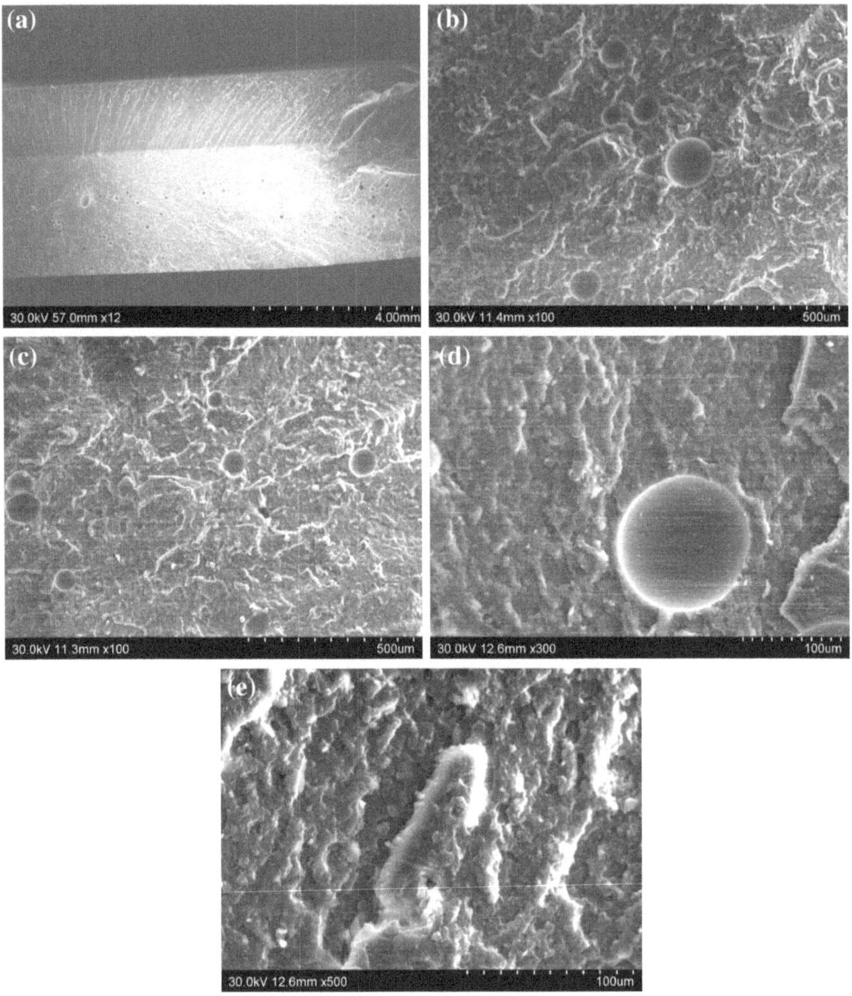

Fig. 11 SEM image of tensile tested 5 % CF-3.33 % NC reinforced vinyl ester composites **a** overall view, **b** location 1, **c** location 2, **c** location 3, **d** location 4, **e** location 5

surface and some amount agglomerates are observed in Fig. 12b, c respectively. Narrow cleavage is visible from Fig. 12d, e.

Specimens are failed at its outer most layer due initiation of crack in three-point-bend test and identified from the SEM image of Fig. 13a, which contains smooth, rough fractured surfaces along with the voids that indicates the presence of particulates before test. Rougher surface, diversified dimples with

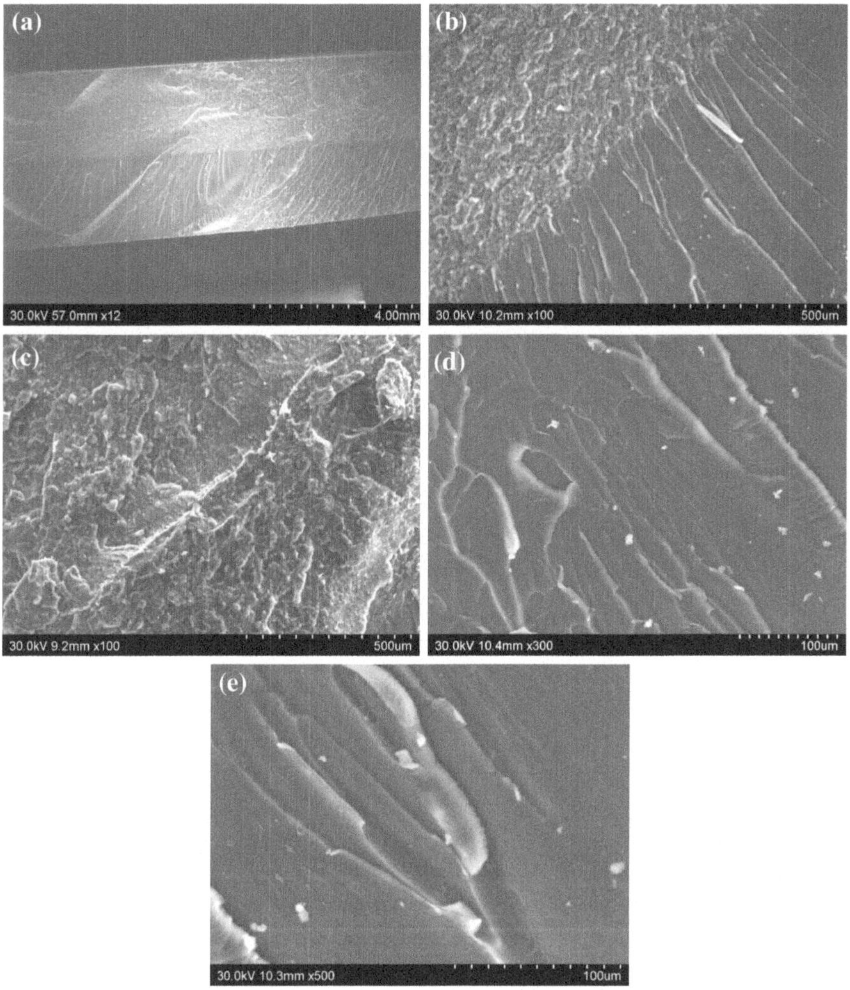

Fig. 12 SEM image of tensile tested 3.33 % NC reinforced vinyl ester composites **a** overall view, **b** location 1, **c** location 2, **c** location 3, **d** location 4, **e** location 5

circular projected agglomerates are visible from Fig. 13b–e. Nearly spherical shaped voids are identified from the fractured surfaces of nanoclay composites, Fig. 14a–e.

SEM image of fractured surface of the impact tested CF-NC, NC reinforced vinyl ester composites are shown in Figs. 15a–e and 16a–e respectively. A rougher uneven surface, even surface is observed in Figs. 15 and 16 respectively.

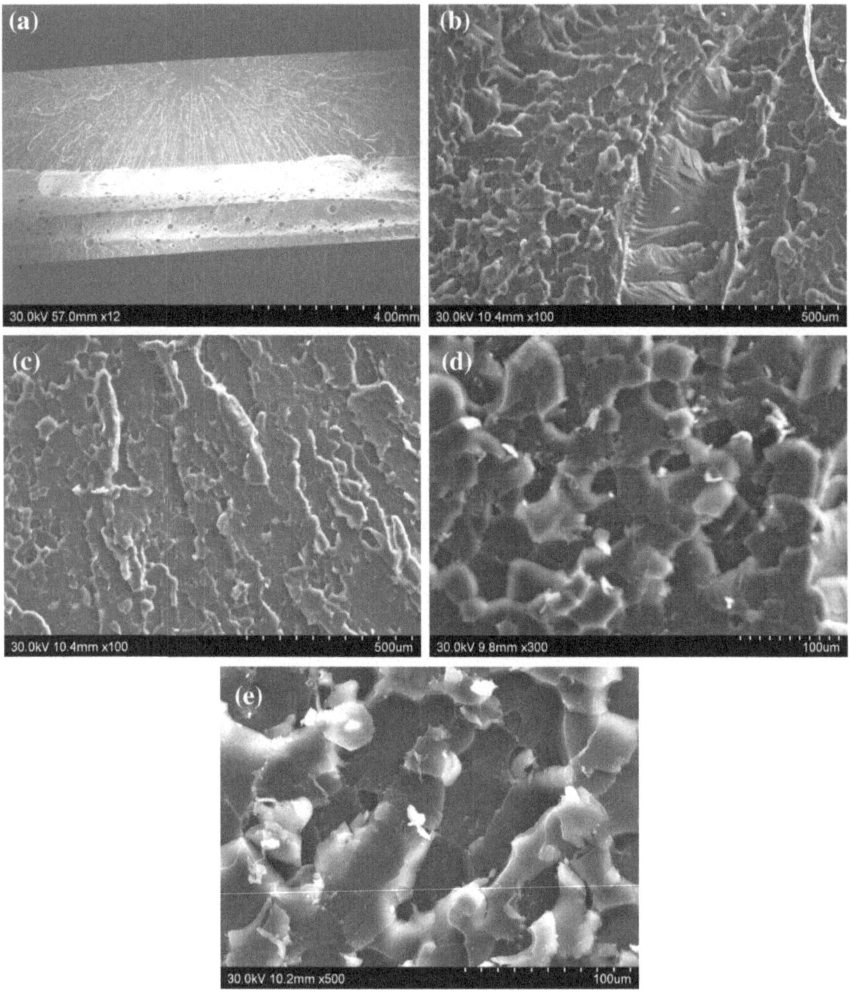

Fig. 13 SEM image of flexural tested 5 % CF-6.67 % NC reinforced vinyl ester composites **a** overall view, **b** location 1, **c** location 2, **c** location 3, **d** location 4, **e** location 5

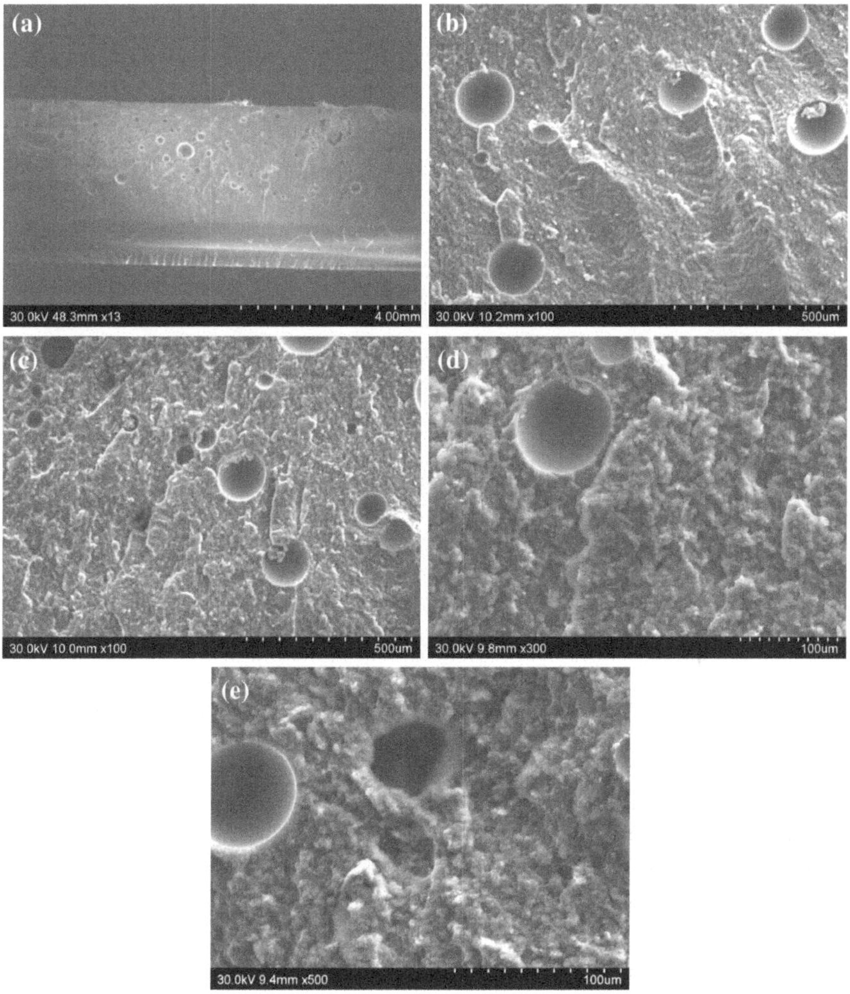

Fig. 14 SEM image of flexural tested 6.67 % NC reinforced vinyl ester composites **a** overall view, **b** location 1, **c** location 2, **c** location 3, **d** location 4, **e** location 5

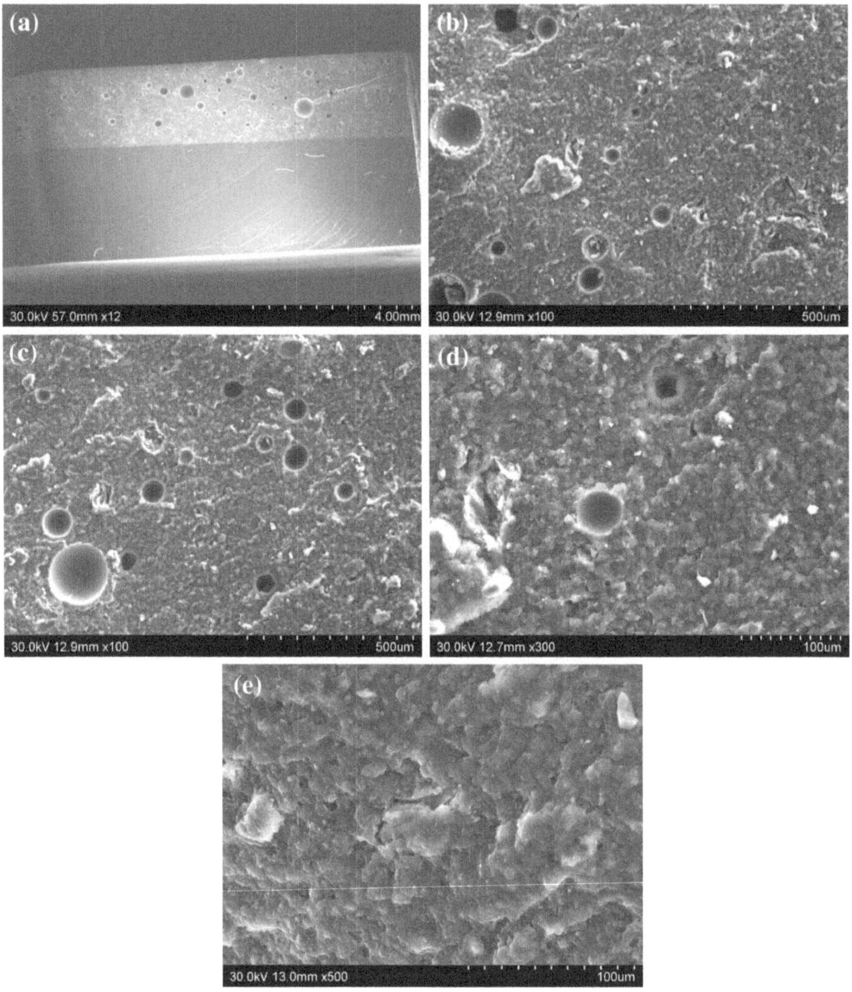

Fig. 15 SEM image of impact tested 5 % CF-6.67 % NC reinforced vinyl ester composites **a** overall view, **b** location 1, **c** location 2, **c** location 3, **d** location 4, **e** location 5

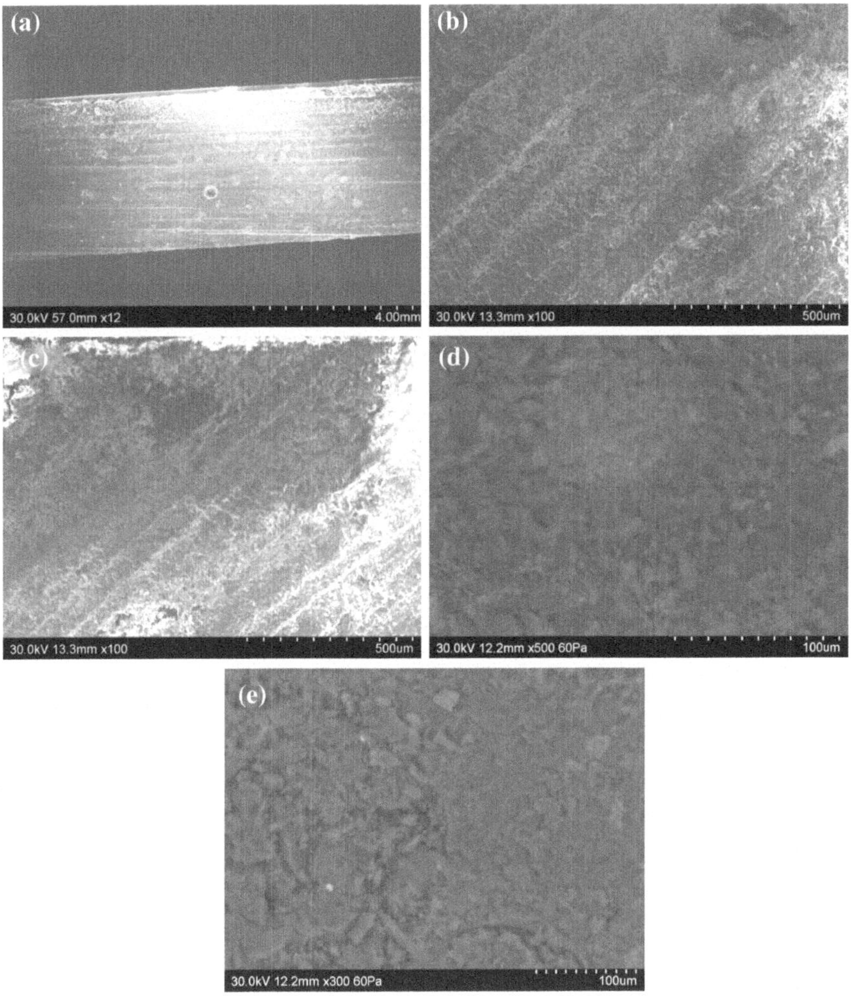

Fig. 16 SEM image of impact tested 6.67 % NC reinforced vinyl ester composites **a** overall view, **b** location 1, **c** location 2, **c** location 3, **d** location 4, **e** location 5

4 Conclusions

CF-NC, NC particulate reinforced vinyl ester composites have successfully prepared by two stage hand/wet layup technique. This process of manufacturing has ensured complete presence of reinforcement in the matrix. Out of all the composites experimentally studied in this work, composites with CF-NC have exhibited superior performance when compared with NC alone reinforced composites. Mechanical stirring of the resin/reinforcement(s) mixture result in better distribution/mixing of reinforcements in the matrix.

Though the present attempt and investigation has given some satisfactory results, there is a serious emergent need to thoroughly investigate the MTSHLM process to enhance mechanical performance of the composites along with uniform distribution of reinforcements.

References

Ashori, A., Nourbakhsh, A.: Effects of nanoclay as a reinforcement filler on the physical and mechanical properties of wood-based composite. J. Compos. Mater. **43**, 1869–1875 (2009)

Awad Walid, H., Beyer, G., Benderly, D., Ijdo, W.L., Songtipya, P., del Mar, M., Jimenez-Gasco, M.E., Wilkie Charles, A.: Material properties of nanoclay PVC composites. Polymer **50**, 1857–1867 (2009)

Faruk, O., Matuana Laurent, M.: Nanoclay reinforced HDPE as a matrix for wood-plastic composites. Compos. Sci. Technol. **68**, 2073–2077 (2008)

Gindl, W., Keckes, J.: All-cellulose nanocomposite. Polymer **46**, 10221–10225 (2005)

Ho, M.-W., Lam, C.-K., Lau, K.-t., Ng Dickon, H.L., Hui, D.: Mechanical properties of epoxy-based composites using nanoclays. Compos. Struct. **75**, 415–421 (2006)

Huang, X., Netravali, A.: Characterization of flax fiber reinforced soy protein resin based green composites modified with nano-clay particles. Compos. Sci. Tehcnol. **67**, 2005–2014 (2007)

Islam, M., Ahmad Mansor, B., Hasan, M., Aziz, S.A., Jawaid, M., Mohamad Haafiz, M.K., Zakaria Siti, A.H.: Natural fiber-reinforced hybrid polymer nanocomposites: effect of fiber mixing and nanoclay on physical, mechanical and biodegradable properties. Bioresources **10**, 1394–1407 (2015)

Jue, L., Wang, T., Drzal Lawrence, T.: Preperation and properties of microfibrillated cellulose polyvinyl alcohol composite materials. Compos. A Appl. Sci. Manuf. **39**, 738–746 (2008)

Lam, C.-k., Lau, K.-t., Cheung, H.-y., Ling, H.-y.: Effect of ultrasound sonication in nanoclay clusters of nanoclay/epoxy composites. Mater. Lett. **59**, 1369–1372 (2005)

Lei, Y., Qinglin, W., Clemons Craig, M., Yao, F., Yanjun, X.: Influence of nanoclay on properties of HDPE/Wood composites. J. Appl. Polym. Sci. **106**, 3958–3966 (2007)

Lewitus, D., McCarthy, S., Ophir, A., Kenig, S.: The effect of nanoclays on the properties of PLLA-modified polymers Part 1: Mechanical and thermal properties. J. Polym. Environ. **14**, 171–177 (2006)

Nakagaito, A.N., Yano, H.: Toughness enhancement of cellulose nanocomposites by alkali treatment of the reinforcing cellulose nnaofibers. Cellulose **15**, 323–331 (2008)

Okubo, K., Fujii, T., Yamashita, N.: Improvement of interfacial adhesion in bamboo polymer composite enhanced with micro-fibrillated cellulose. JSME Int. J. Ser. A **48**, 199–204 (2005)

Ozgur, M., Seydibeyoglu, K.O.: Novel nanocomposites based on polyurethane and micro fibrillated cellulose. Compos. Sci. Technol. **68**, 908–914 (2008)

Qi, B., Zhang, Q.X., Bannister, M., Mai, Y.W.: Investigation of the mechanical properties of DGEBA-based epoxy resin with nanoclay additives. Compos. Struct. **75**, 514–519 (2006)

Ries, H.: Clays: Their Occurrence, Properties and Uses, With Especial Reference to Those of the United States, p. 1. Wiley, New York (1906)

Searle, A.B.: The Natural History of Clay, p. 1. Cambridge University Press (1912)

Srinivasababu, N.: Role of two stage wet layup manufacturing method to fabricate and test chemically treated dora hemp particulate FRP composites. Appl. Mech. Mater. **787**, 548–552 (2015)

Tingaut, P., Zimmermann, T., Lopez-Suevos, F.: Synthesis and characterization of bionanocomposites with tunable properties from poly(lactic acid) and acetylated microfibrillated cellulose. Biomacromolecules **11**, 454–464 (2010)

Wang, J., Qin, S.: Study on the thermal and mechanical properties of epoxy-nanoclay composites: the effect of ultrasonic stirring time. Mater. Lett. **61**, 4222–4224 (2007)

Zainuddin, S., Hosur, M.V., Zhou, Y., Narteh Alfred, T., Kumar, A., Jeelani, S.: Experimental and numerical investigations on flexural and thermal propetrties of nanoclay-epoxy nanocomposites. Mater. Sci. Eng. A **527**, 7920–7926 (2010)

Effect of Plasticizer on Fracture Toughness of Polylactic Acid Reinforced with Kenaf Fibre and Montmorillonite Hybrid Biocomposites

M.N. Nur Aimi and H. Anuar

Abstract Monocomposites and hybrid biocomposites based-polylactic acid (PLA) are developed for specific application. In this study, monocomposites are referred to PLA reinforced with kenaf fibre (KF) and PLA filled with montmorillonite (MMT) organoclay. Hybrid biocomposite is denoted to PLA/KF/MMT. Both monocomposites and hybrid biocomposites are extruded using twin screw extruder and then injection moulded. The amount of kenaf fibre and MMT are fixed at 20 and 3 wt%, respectively. In the case of plasticized composites, amount of 3 wt% polyethylene glycol (PEG) was added. The effect of plasticizer on fracture toughness of monocomposites and hybrid biocomposites was studied. The result showed that the fracture toughness of plasticized PLA/KF/MMT yield 24.2 % higher than non-plasticized hybrid composite. Observation under scanning electron microscope (SEM) revealed that the fibre bridging, fibre pull-out and fibre breakage contributed to high fracture toughness of plasticized hybrid biocomposites. X-ray diffraction (XRD) also evident high percentage of crystallinity for plasticized hybrid biocomposites.

Keywords Hybrid biocomposites · Polylactic acid · Kenaf fibre · Montmorillonite · Fracture toughness

1 Introduction

Paradigm towards the need of quality and environmental friendly materials was cultivated due to the global concern with environmental issues. Therefore, progressive technological research and innovations were developed for the sustainable society. Polymer composites with the incorporation of natural fibre is among the

M.N. Nur Aimi · H. Anuar (✉)
Department of Manufacturing and Materials Engineering,
Kulliyyah of Engineering, International Islamic University Malaysia,
P.O. Box 10, 50728 Kuala Lumpur, Malaysia
e-mail: hazleen@iium.edu.my

M. Jawaid et al. (eds.), *Nanoclay Reinforced Polymer Composites*,
Engineering Materials, DOI 10.1007/978-981-10-0950-1_11

materials that owe such concept of environmental and safety in ecological which lead to the socio-economic development as an alternative to the petroleum-based (Saba et al. 2014). This is due to the advantages of natural fibres that had been time to time improved, such as reduced tool wear, low cost, and low density per unit volume and acceptable specific strength, along with their sustainable, renewable and degradable features which make them suitable as filler in polymer composites (Borba et al. 2014). It is undeniable that natural fibres in polymer composites had some limitation for industrial applications. Polymer reinforced with natural fibres composite is however can be surface modified through either physical or chemical modification, addition of coupling agent; such as maleated grafted anhydride, silane and addition of plasticizer (such as PEG) (Qiu et al. 2013; El Fattah et al. 2014). Such modifications on natural fibre are necessary for well interaction of fibre and polymer matrix.

There has been tremendous research on hybridization of composites which involve the incorporation of several types of reinforcement materials in single matrix (Rios et al. 2015). It was found to be worth and more favourable in term of mechanical properties specifically for the one that based on natural fibre as it is balance between inherent advantages and drawbacks (Ravikumar and Prasad 2014). The combination of hybrid polymer composites with nanofiller are getting popular in term of researched compared to conventional composites. The addition of nanoparticles like montmorillonite organoclay in the conventional and hybrid polymer composites improved the mechanical and physical properties (Panagiotis et al. 2010; Piscitelli et al. 2012). This is due to the fact that nano scale fillers are usually free of defects with large surface area hence results in high matrix-reinforcement interfacial area. Moreover, the uniformity and homogenous dispersion of nanoparticles enhanced and improved the mechanical and thermal properties of the composites produced (Hari and Pukanzsky 2011; Majeed et al. 2013; Zhu et al. 2013).

Kenaf, *Hibiscus cannabinus, L. family Malvaceae*, is an herbaceous annual plant that had wide application in composite industry specifically on its kenaf bast fibre. This is due to its large amount availability, renewability, better mechanical property and lower cost. Hydrophilicity of natural fibers is very important properties need to be considered as it will determine better adhesion specifically when it reinforces with hydrophilic matrix resins which is polylactic acid (PLA). Research on mono composite KF with other polymer matrix besides PLA is very common (Sarifuddin et al. 2013; Saba et al. 2015). However, hybrid composite of KF and PLA with other material is still new and need further progressive research.

To narrow down, the mechanical properties of biocomposite either it is mono composite or hybrid composite is very crucial aspect for many industrial design applications. Whether the materials are classified as brittle or ductile fractures, it is determined by the fracture toughness. Fracture toughness can be defined as the material containing crack to resist fracture that relates to the mechanical separation of a solid material due to the application of stress (Salleh et al. 2012). Many researchers had taken into account the fracture toughness aspect in their polymer biocomposites researches. The improvement of the fracture toughness of numerous nanofiller based on epoxy matrix and reported that debonding and cavitation as the main driving forces in activating energy dissipation mechanisms such as matrix shear deformation, crack deflection, crack twisting and crack bridging of the composite (Domun et al. 2015). The fracture toughness mechanisms were also supported by scanning electron microscopy (SEM) and X-ray diffraction (XRD) analysis.

Hybridization of nanofiller and natural fibre complement each element in the composite subsequently enhanced and improved crack resistance of polymer composites. For stiff and rigid polymer matrix, the presence of plasticizer may further improvised the properties of composites. The present work study the effect of polyethylene glycol (PEG) on the fracture toughness of PLA/KF/MMT hybrid biocomposites.

2 Experimental

2.1 Materials

Kenaf fibre (KF) used in this study was obtained from Kenaf Fibre Industries Sdn. Bhd., Malaysia. KF length ranges from 400–800 mm and diameter from 180–300 μm. KF was dried at 105 °C for 24 h. Polylactic acid (PLA 3051D; density = 1.42 g cm^{-3}) was provided by Natureworks (China). Montmorillonite (MMT, Nanomers® 1.31PS, density = 0.89 g cm^{-3}) and polyethylene glycol (PEG) were purchased from Sigma Aldrich (Malaysia).

2.2 Hybrid Biocomposite Preparation

The nomenclature and composition of PLA and PLA hybrid biocomposites with KF and MMT are listed in Table 1. All plasticized and non-plasticized hybrid biocomposites were prepared by using twin screw extruder (Rheomix 600, Germany). After extrusion, the composites were injection moulded using Battenfeld 600 at 180 °C with pressure 950 bar.

2.3 Hybrid Biocomposite Characterization

Mode I fracture test was carried out on single edge notched hybrid composites according to ASTM D790. The test was performed at a speed of 1 mm/min. Figure 1 illustrates the specimen used for testing. The test was conducted using LLOYD LR 100 K universal testing machine under ambient temperature. An

Table 1 Nomenclature used for various composition of PLA and hybrid biocomposites

Samples	PLA (wt%)	KF (wt%)	MMT (wt%)	PEG (wt%)
PLA	100	0	0	0
PLA/KF	80	20	0	0
PLA/MMT	97	0	3	0
PLA/KF/MMT	77	20	3	0
ρPLA	97	0	0	3
ρPLA/KF	77	20	0	3
ρPLA/MMT	94	0	3	3
ρPLA/KF/MMT	74	20	3	3

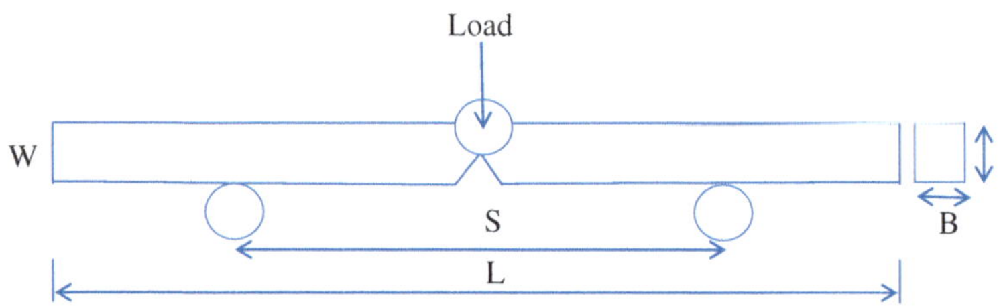

Fig. 1 Illustration of specimen dimension of the single-edge-notched bend (SENB) fracture toughness testing

Table 2 Dimension of PLA and hybrid biocomposites

Composites	Length, L (mm)	Width, W (mm)	Thickness, B (mm)	Span length, S (mm)
PLA	125.74	12.68	3.2	100
PLA/KF	125.65	12.67	3.19	100
PLA/MMT	125. 73	12.66	3.17	100
PLA/KF/MMT	125.74	12.67	3.2	100
ρPLA	125.70	12.65	3.19	100
ρPLA/KF	125.75	12.66	3.19	100
ρPLA/MMT	125.75	12.65	3.17	100
ρPLA/KF/MMT	125.83	12.62	3.16	100

average of five samples were evaluated for calculation of data. Average specimen length (L), span length (S), width (W) and thickness (B) of the testing samples are tabulated in Table 2.

Mode I plain-strain fracture toughness (K_c) of single-edge-notched bend (SENB) specimens can be calculated with the following relation.

$$K_c = \sigma_c Y \sqrt{\alpha} \quad (1)$$

where σ_c is the critical applied stress, Y is a geometric factor and α is a length of the crack.

$$Y = 1.93 - 3.7\left(\frac{\alpha}{W}\right) + 14.53\left(\frac{\alpha}{W}\right)^2 - 25.11\left(\frac{\alpha}{W}\right)^3 + 25.80\left(\frac{\alpha}{W}\right)^4 \quad (2)$$

Critical applied stress σ_c is calculated from:

$$\sigma_c = \frac{3FS}{2BW}. \quad (3)$$

2.4 *Field Emission Scanning Electron Microscopy (FESEM)*

Fracture surface morphology of the hybrid composite was studied using FESEM (JEOL JSM 6500, USA) operated at 10 kV voltage. All samples were ion sputter coated with gold to enhance conductivity. The samples were mounted with carbon tape on aluminium stubs prior to FESEM observation.

2.5 *X-Ray Diffraction (XRD) Analysis*

The crystallinity of hybrid composites was determined using X-ray diffractometer (XRD-6000, Shimadzu) applying CuKα (λ = 1.54) radiation and a graphite monochromator with a current of 30 mA and a voltage 40 kV. The diffraction intensity was in the range of 10°–80° of 2θ (Bragg's angle), and the scanning speed was 0.01°/s. The percentage of crystallinity was calculated by using the following equation:

$$X\% = \frac{I_{200} - I_{am}}{I_{200}} \times 100 \quad (4)$$

where,

$X\%$ = Percentage of crystallinity
I_{200} = Peak intensity of crystalline part
I_{am} = Peak intensity of amorphous part.

3 Results and Discussion

3.1 Characterization of Fracture Toughness

Fracture toughness is the resistances of materials towards fracture. The fracture toughness is depends on manufacturing process that indicates the existences of crack at the maximum load (Borba et al. 2014). The fracture toughness of hybrid composites is calculated using Eqs. (1) and (2) based on SENB instrumented by low speed impact test (1 m/s).

Fracture toughness, K_c of plasticized and non-plasticized PLA hybrid biocomposites is shown in Fig. 2. Fracture toughness is enhanced with the presence of reinforcement and plasticizer. It is likely influence the stress concentration due to the presence of fibre. However there were also studies reported on the decrement of K_c of PLA which could be related to the influence of degree of crystallinity that was affected by fibre content. It was reported that toughness decreased as crystallite increased due to crack may propagate easily through crystalline phase (Wong et al. 2010; Lamon and R'Mili 2014).

Fibre pull-out also contributed to the increased in toughness (Chen et al. 2009; Yousif et al. 2012). Apart from that, several failure mechanisms are activated in fibrous composites during crack growth. These mechanisms include fibre breakage, matrix damage, fibre-matrix debonding, fibre pull-out etc. Brittle fibre and brittle matrix may also lead to high fracture toughness. The fracture energy of epoxy resin increased up to 223 J/m^2 with the addition of carbon nanotubes (Hsieh et al. 2011). Similarly, fracture toughness of PLA is improved from 229.9 to 276.8 kJ/m^2 in the

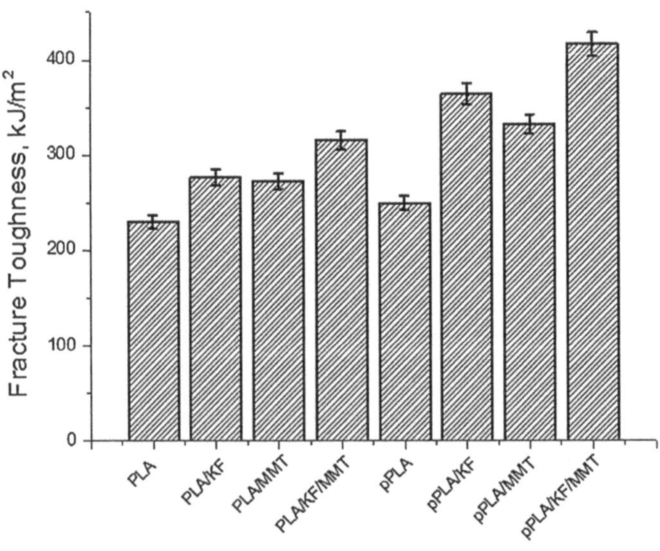

Fig. 2 Fracture toughness of plasticized and non-plasticized PLA hybrid biocomposites

presence of short fibre KF. The mechanism of toughness like ductile metals and fibrous composite also exhibits the so called R-curve behaviour. It explained the energy required for catastrophic or large scale crack extension is larger than initiation or small-scale crack growth (Greco et al. 2013; Ferrari et al. 2014).

From Fig. 2 also, fracture toughness is slightly higher in the case of plasticized PLA/MMT than in PLA/MMT nanocomposite. PLA matrix is brittle thus plastic deformation is undesirable and is highly constrained (Huang et al. 2015). Local stress concentrations may initiate cracks which lead to spontaneous failure. It is therefore the primary aim of many researchers to provide PLA with higher toughness but without significantly sacrificing the other characteristics such as thermo-mechanical properties and modulus, which are desired and require for many applications (Li et al. 2014).

On the other hand, nanoparticles MMT limit the PLA matrix deformation. The interaction of particles with matrix is expected to affect the deformation mechanism in the matrix either the particles is in nano/micro size (Sylvester et al. 2012). MMT could promote formation of a large number of subcritical micro cracks or micro void and retard the collapse into critical crack by coalescence. Improvement in toughness also is contributed by well dispersion and distribution of nanoparticles MMT in PLA matrix. Good adhesion between MMT and PLA can be enhanced by the addition of plasticizer such as PEG which resulted in higher toughness (Anuar et al. 2012; Shi et al. 2015). However, prior to plasticization, rigidity of filler and polymer is must be considered. Filler should possess higher rigidity than polymer to increase stiffness, higher specific surface are, a sufficient filler-matrix bonding to improve strength and to allow a controlled stress transfer from the matrix to the fillers. It is preferably small dimensions of filler to reduce local stress concentration and to generate high toughness and impact resistance (Ayandele et al. 2012).

The smaller the filler dimension in brittle matrix will improve the toughness as potential defects like inclusion and agglomeration in composites is minimized. The minimization of defects serves as the importance strategy to toughen the matrix (Li et al. 2014). It is noted in Fig. 2 that the used of polyethylene glycol resulted in a significant change in toughness. The lower toughness experienced by the non-plasticized PLA composites is expected due to applied stress is not uniformly distributed and matrix bear most of the load applied. Other than that, fibre may act as flaws or weakening points and this reduce the composite strength in non-plasticized PLA and composites (Weitsman 2012). Synergistic combination of KF and MMT in plasticized composite is seen in Fig. 2 where the highest toughness is recorded. The plasticized PLA/KF/MMT hybrid composites showed 24.2 % higher in fracture toughness than PLA/KF/MMT hybrid composite. The plasticized PLA/KF and plasticized PLA/MMT are 23.9 and 17.8 % higher than PLA/KF and PLA/MMT, respectively. The presence of polyethylene glycol improve the interface interaction between matrix and filler in the plasticized composite, and MMT were mixture of exfoliated and intercalated in the polymer matrix (Bonnia et al. 2010).

In mono composite system, it is noticed in Fig. 2 that PLA/KF composite possess higher fracture toughness than PLA and PLA/MMT. This could be

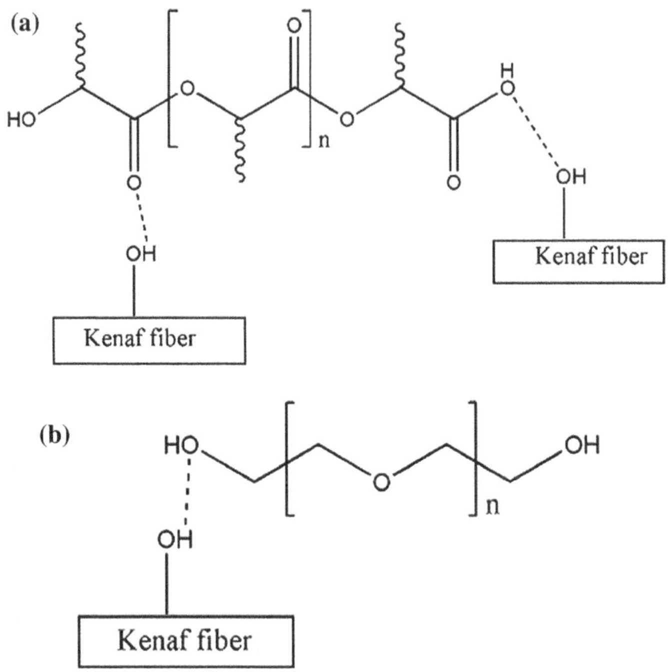

Fig. 3 Possible reaction of **a** PLA-KF and **b** kenaf fibre and PEG

attributed to good interaction between kenaf fibre and PLA matrix due to the formation of hydrogen bonding between hydroxyl groups of kenaf and the terminal hydroxyl, or carbonyl groups as well as the carbonyl groups of the ester group of PLA (Dehbari et al. 2014). The possible reaction between KF and PLA is illustrated in Fig. 3. PLA can undergo thermal hydrolysis or depolymerization during processing (Prasanth et al. 2010). This may lead to increasing number and availability of hydroxyl terminal or carbonyl groups to interact with the hydroxyl group from KF. This improves adhesion of fibre and matrix leading to better matrix-to-fibre stress transfer subsequently enhances composite performance (Prasanth et al. 2010).

In addition, it is also suggested that PEG might also interrupted the interaction between PLA and KF in the plasticized PLA/KF composites. This is evident by comparing the fracture surface morphology of plasticized PLA/KF and PLA/KF composite in Figs. 4b and 5b. PEG is a low molecular weight materials (much lower than PLA) and has hydroxyl groups that can interact with the hydroxyl group on the surface of KF by forming hydrogen bonding. The illustration of interaction is shown in Fig. 3. The adverse effect of PEG on the fibres and matrix adhesion is reflected in the FESEM micrographs. The matrix surface is rough compared to non-plasticize composite, which has smooth surface. The rough surface structure is enough to hold KF thus consumed more energy in order to pull-out of the matrix resin (Liao et al. 2011).

There are however, several works that describe a poor interaction between KF and PLA due to the presence of non-cellulose components on the fibre surface (Lazim et al. 2014). Previous studies also reported that the interfacial adhesion between natural fibres and PLA can be improved through fibre surface modification such as by grafting less polar groups at the fibre surface as well as fibre pre-treatment with alkali (Fernandes et al. 2013; Lazim et al. 2014).

3.2 FESEM Observation

The morphology of non plasticized and plasticized PLA hybrid biocomposites are shown in Figs. 4a–d and 5a–d. Figure 4a shows the FESEM micrograph of fracture surface of PLA. Brittle failure characteristic and wavy structures of PLA are clearly

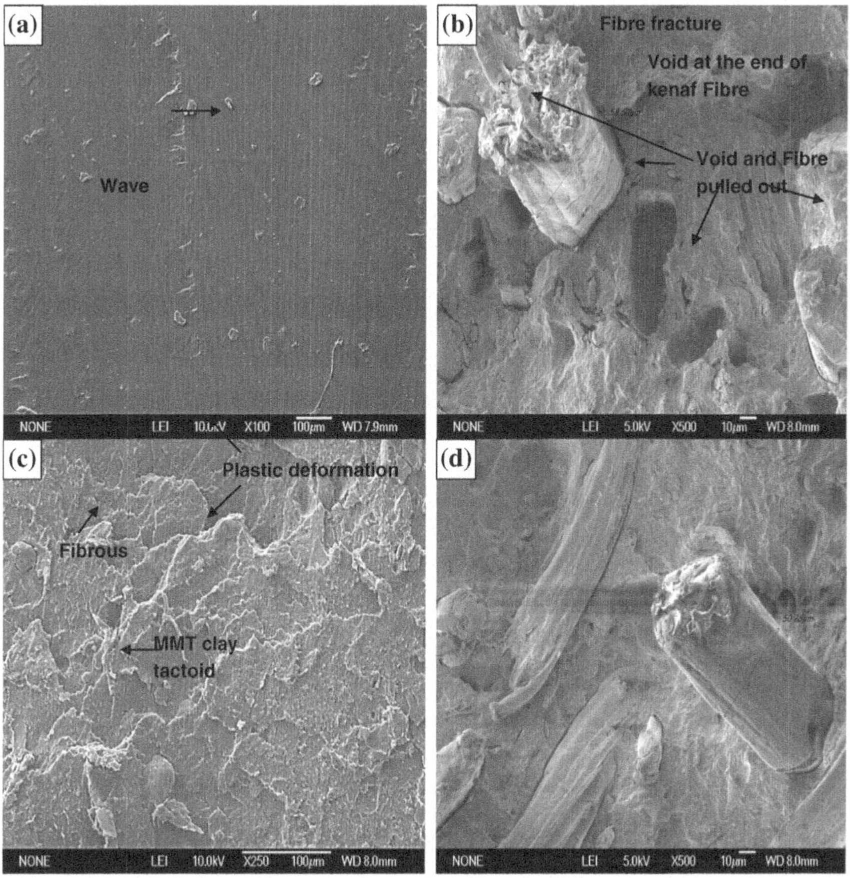

Fig. 4 The morphology of fracture surface of non-plasticized PLA and composites **a** PLA; **b** PLA/KF; **c** PLA/MMT; **d** PLA/KF/MMT

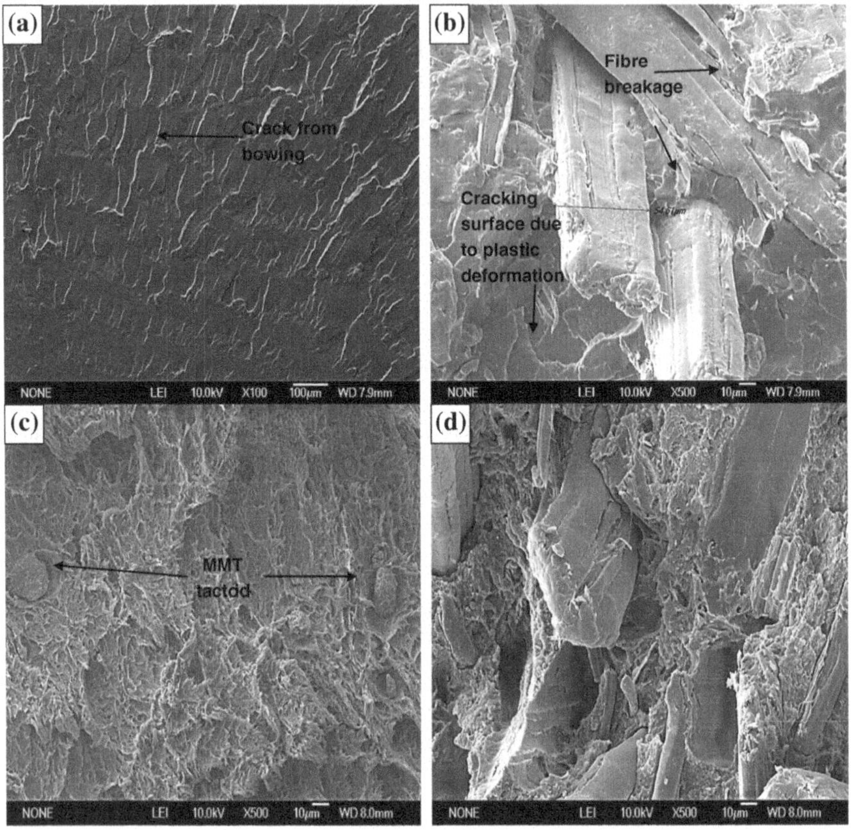

Fig. 5 The morphology of fracture surface of plasticized PLA and composites **a** ρPLA; **b** ρPLA/KF; **c** ρPLA/MMT; **d** ρPLA/KF/MMT

seen in Fig. 4a. The presence of PEG in plasticized PLA in Fig. 5a slightly reduced the brittleness of PLA and hence improved its failure characteristics due to the formation of plastic deformation. The unclear wavy cracking of PLA fracture surface indicates that PLA specimen break relatively fast due to quick development of micro-crack under flexure loading (Zhao et al. 2010). In the previous work, some small platforms on the fracture surface in addition to the waves (Todo and Takayama 2011). The fracture toughness morphology of PLA/MMT and ρPLA/MMT composites are shown in Figs. 4c and 5c. There are fewer drawn fibril on the ρPLA/MMT nanocomposites surface compare to non-plasticized PLA/MMT which is consistent with fracture toughness result presented in Fig. 2. This situation indicates that energy was absorbed more during the fracture process.

From Figs. 4 and 5, it is clear that in general all composite exhibited smooth fibre pulled-out with voids around them. These observations indicate poor adhesion between KF and PLA matrix. Despite lacking of adhesion, ρPLA/KF and ρPLA/KF/MMT still exhibited higher fracture toughness compare to other composites and neat ρPLA and PLA. Even though plasticized composite shows better toughness than non-plasticized composite, PLA/KF and PLA/KF/MMT are still higher compare to PLA nanocomposite (PLA/MMT) and neat PLA. The improvement of toughness due to fibre content would occur in the case of fibre equal or less than the critical fibre length (Indira et al. 2013). The critical fibre length is a function of fibre-matrix adhesion and interfacial shear strength. The extent of this stress transfer are depends on fibre length and the degree of fibre-matrix adhesion. As the fibre-matrix interfacial shear strength or the fibre-matrix interfacial adhesion is low, the critical fibre length is large in value (Liu et al. 2013). Fibres greater than an equal to the critical fibre length are desirable to be used in a composite materials to obtain good composite properties (Indira et al. 2013). The pulled out fibres can be seen in Figs. 4b–d and 5b–d. These fibres pulled-out indicate that the fibre length is lower than critical fibre length (Kumar et al. 2012). Fibres breakage also can be seen in FESEM micrograph in the Fig. 5b. This suggested that, some of the fibre in the composite were stressed to their fracture surface before composites failure resulting in improvement in toughness and strength (Sumalla et al. 2013). Fibre may also stress to their limit resulting high composite strength if the fibre length is equal or greater than critical fibre length. It is revealed in the FESEM micrographs that ranges of fibre from longer to shorter than critical fibre length are observed. The critical fibre length can be estimated from the FESEM micrograph of a composite fracture surface where the average pull-out length of a fibre cannot be longer than half of a critical fibre length (Shah et al. 2012).

The dispersion and distribution of the fibres in the polymer matrix were further analysed as illustrated in Fig. 6. The crack propagation was suppressed due to the fibre bridging as shown in Fig. 6. The interaction between KF and PLA resulted the KF to resist the increasing load acting on it.

3.3 X-Ray Diffraction (XRD) Analysis

XRD analysis of both non-plasticized and plasticized PLA hybrid biocomposites are shown in Fig. 7. X-ray diffractographs of PLA and ρPLA show the same broad peaks in the region $2\Theta = 15°–22.3°$. The broad peaks indicate that both PLA chains are poorly ordered at very low crystalline regions. This condition is supported by the previous researched on mechanical properties of PLA/perlite and PLA/MMT (Tian and Tagaya 2007; Kaiser et al. 2013). However, the XRD diffractograph for PLA become narrowed upon additional of KF at region of $2\Theta = 15°$. The high second peak at $2\Theta = 22.5°$ reflected the ordered structure and crystalline of PLA changes correspond to KF. The changed of peak to the narrower area could be due the shear force as KF distributed in PLA and modified its crystal structure during extrusion process. XRD diffractograph of KF mainly appeared around $2\Theta = 15°$ and $2\Theta = 22.5°$ as seen in Fig. 7. This represents the strong crystalline peak which

Fig. 6 Kenaf fibre bridging appear in PLA/KF/MMT hybrid biocomposite

corresponds to the plane (200) and (012), respectively. The results revealed that the structure of cellulose fibre exhibited cellulose I. This is in the agreement where the diffraction peak of natural fibre are at 2Θ = 14.8°, 16.3° and 22.6° for (110) and (200) planes which assigned to crystal cellulose I and at region of 2Θ = 12.3°, 20.3° and 21.6° for crystal structure II (Poletto et al. 2014). However, no peak is seen at 2Θ = 12.3° and 20.3° in plasticized PLA/KF and plasticized PLA/KF/MMT hybrid composite since the intensity is increased with the addition of KF.

On the other hand, PLA/MMT, PLA/KF/MMT, ρPLA/MMT and ρPLA/KF/MMT composites exhibited small peak at 2Θ = 5° and 2Θ = 29°. This could be due to the shear force during the extrusion process. This also might due to the disappearance of peak of nanoclay which is between 2Θ = 2°–5° after being added to the amorphous PLA. MMT layer was made up of stacking alumina and silicate which generate negative charge layers. This silicate layers are not easy to be

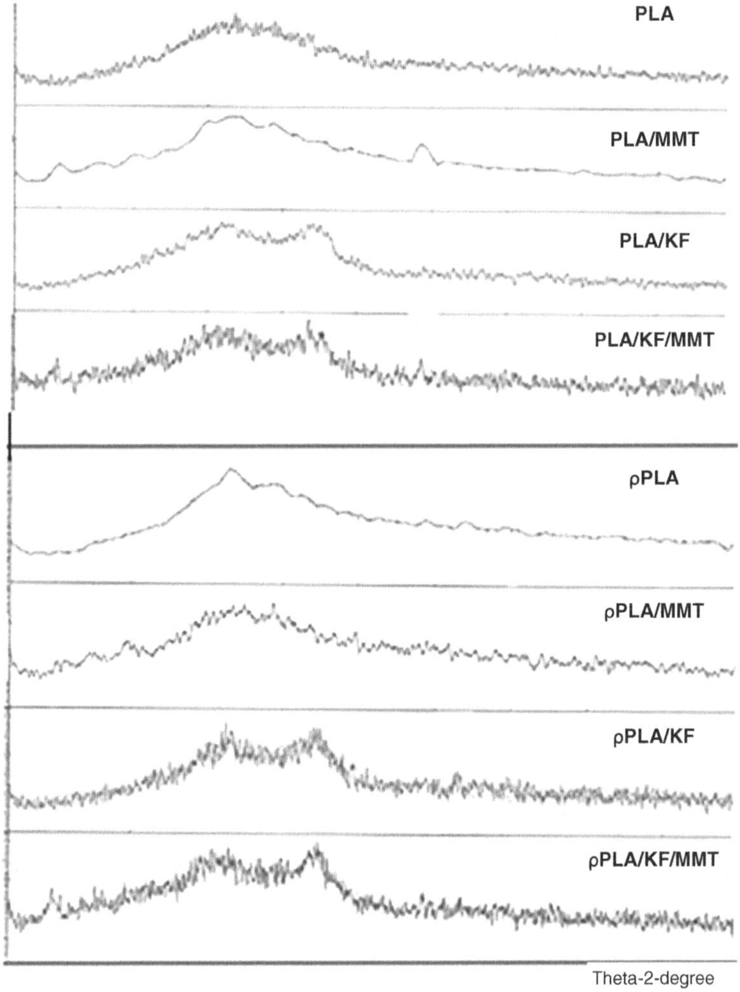

Fig. 7 XRD diffractograph of PLA hybrid biocomposites

separated into interlayer space of MMT. Furthermore, it is expected that the small amount of 3 wt% MMT was significantly covered in the composition of the reinforcement of KF in PLA.

From Fig. 7, shifting of peak is seen as PLA was extruded in the presence of KF and MMT. This indicates that there is physical change in MMT due to the dispersion of some silicate tactoid in polymer matrix to form platelets of silicates (Agubra et al. 2013). This proved that there is no exfoliation but intercalation of MMT in PLA matrix. The comparison on d-spacing by taking into consideration of the peak around $2\Theta = 2°–3°$ for the composites with MMT like PLA/MMT, ρPLA/MMT, PLA/KF/MMT, and ρPLA/KF/MMT) are tabulated in Table 3.

Table 3 The comparison of d-spacing of composite based on MMT organoclay

Sample(s)	d-spacing (Å)	2Θ
MMT	22.1272	3.99
PLA/MMT	30.1295	2.93
PLA/KF/MMT	27.4166	3.22
ρPLA/MMT	17.5543	5.03
ρPLA/KF/MMT	27.8929	3.16

Table 4 Crystallinity percentage of composite

Sample(s)	Crystallinity (%)
PLA	18.1023
PLA/KF	20.0246
PLA/MMT	15.2298
PLA/KF/MMT	19.9291
ρPLA	15.7653
ρPLA/KF	22.4376
ρPLA/MMT	17.1886
ρPLA/KF/MMT	19.8409

The d-spacing of crystallographic plane of MMT in plasticised biocomposites is increased in interlayer spacing. For ρPLA/KF/MMT the spacing increased from 22.1272 Å (2.2127 nm) to 27.8929 Å (2.7892 nm). It is expected that MMT is intercalated as there was a combination of shear force and speed during extrusion which dispersed the large MMT particles into stacks of silicate tactoids (Shiraz et al. 2013).

The crystallinity of PLA hybrid biocomposites are calculated and tabulated in Table 4. It can be seen that the crystallinity of PLA and ρPLA are the lowest in comparison to the rests of the PLA biocomposites. This is in consistent with the fracture toughness described earlier as well as the FESEM micrographs in Figs. 4a and 5a. On the other hand, the broad peak of PLA shown in Fig. 7 indicates that molecular structures of PLA are poorly ordered or highly amorphous. However, when KF was added to PLA, the crystallinity is considerably increased about 9.5 % compared to unfilled PLA. The crystallinity of ρPLA/KF is 29 % higher compared to ρPLA. This could be due to the KF is acting as nucleating agent to crystallise the PLA. In contrast, from Table 4, the presence of MMT in composites with and without plasticizer not significantly affected the crystallinity percentage. This could be related to the small percentage of MMT that avoid it to act as nucleating agent in PLA like kenaf fibre.

The crystallinity of hybrid composites was significantly lower as compared to PLA/KF and ρPLA/KF. Similar reasoning can be offered as explained previously on the KF which act as nucleating agent for PLA. Besides, KF also crystallised at the fibre/matrix interface rather than its primary role as the reinforcing materials (Ou et al. 2011; Dong et al. 2014). This crystalline interface has improved the interfacial strength of composite which in turn leads to better transfer of applied load from matrix to fibres. The nucleating agent such as fibre may promotes the degree of orientation of polymer chains and increase crystallization rate which in the end improve the molecular entanglement between crystal grains (Jiang et al. 2007; Bikiaris 2010). Consequently, the crystal strength might increase, which indirectly increase the mechanical properties of the composites.

4 Conclusion and Future Perspective

The addition of plasticizer enhanced the fracture toughness of PLA hybrid biocomposites. The fracture toughness of plasticized PLA hybrid biocomposites is found to be highest than the other biocomposites systems. The fracture toughness recorded for plasticized PLA/KF/MMT biocomposite is 416.0 kJ/m^2. FESEM observation revealed that mechanism involved in this system includes rough surface structure, existence of fibre pull-out, fibre bridging and fibre debonding. XRD analysis shows that the crystallinity of plasticized hybrid biocomposite was slightly lowered compared to the non-plasticized hybrid biocomposite. Further improvement on the hybrid biocomposite not only derived from PLA need to be done in order to be utilized in industry.

Acknowledgments The authors wish to thank International Islamic University Malaysia for the financial support and facilities provided in making these studies a success.

References

Agubra, V.A., Owuor, P.S., Hosur, M.V.: Influence of nanoclay dispersion method on the mechanical behavior of e-glass/epoxy nanocomposites. Nanomater **3**, 550–563 (2013)

Anuar, H., Noor Azlina, H., Suzana, A.B.K., et al.: Effect of PEG on impact strength of PLA hybrid biocomposite. In: Proceeding IEEE Symposium on Business, Engineering and Industrial Applications, pp. 473–476

Ayandele, E., Sarkar, B., Alexandridis, P.: Polyhedral Oligomeric Silsequioxane (POSS)—containing polymer nanocomposites. Nanomater **2**, 445–475 (2012)

Bikiaris, D.: Microstructures and properties of propylene/carbon nanotube nanocomposites. Mater **3**, 2884–2946 (2010)

Bonnia, N.N., Ahmad, S.H., Zainol, I., et al.: Mechanical properties and environmental stress cracking resistance of rubber toughened polyester/kenaf composite. Express Polym. Lett. **4**, 55–61 (2010)

Borba, P.M., Tedesco, A., Lenz, D.M.: Effect reinforcement nanoparticles addition on mechanical properties of SBS/curaua fibre composites. Mater. Res. **17**(2), 412–419 (2014)

Chen, X., Irene, J., Beyerlein, L., et al.: Curved-fibre pull out model for nanocomposites. Part 2. Interfacial debonding and sliding. Mech. Mater. **41**(3), 293–307 (2009)

Dehbari, N., Moazeni, N., Wan Abdul Rahman, W.A.: Effects of kenaf core on properties of polylactic acid biocomposite. Polym. Compos. **35**, 1220–1227 (2014)

Domun, N., Hadavinia, H., Zhang, T., et al.: Improving the fracture toughness and the strength of epoxy using nanomaterials—a review of the current status. Nanoscale **7**, 10294–10329 (2015)

Dong, Y., Ghautara, A., Takagi, H., et al.: Polylactic acid (PLA) biocomposites reinforced with coir fibre: evaluation of mechanical performance and multifunctional properties. Compos. Part A Appl. Sci. Manuf. **63**, 76–84 (2014)

El Fattah, A.A., El Demerdash, A.G.M., Sadik, W.A.A., et al: The effect of sugarcane bagasse fibre on the properties of recycled high density polyethylene. J. Compos. Mater. 1–12

Fernandes, E.M., Correlo, V.M., Mano, J.F., et al.: Novel cork-polymer composites reinforced with short natural coconut fibre: effect of fibre loading and coupling agent addition. Compos. Sci. Technol. **78**, 56–62 (2013)

Ferrari, V.J., Arquez, A.P., De Hanai, J.B., et al.: Development of high performance fibre reinforced cement composites (HPFRCC) for application as a transition layer of reinforced beams. Ibracon Struct. Mater. J. **7**, 965–975 (2014)

Greco, F., Leonetti, L., Lonetti, P.: A two scale failure analysis of composite materials in presence of fibre/matrix crack initiation and propagation. Compos. Struct. **95**, 582–597 (2013)

Hari, J., Pukanzsky, B.: Nanocomposites: preparation, structure, properties. In: Kutz, M. (ed.) Applied PlasticsEngine Handbook: Processing Materials, pp. 109–142. Elsevier Inc., Waltham (2011)

Hsieh, T.H., Kinloch, A.J., Taylor, A.C., et al.: The Effect of Carbon Nanotubes on the fracture and fatigue performance of thermosetting epoxy polymer. J. Mater. Sci. **46**, 7525–7535 (2011)

Huang, T., Miura, M., Nobukawa, S., et al: Chain Oacking and it anomalous effect on mechanical toughness of poly(lactic acid). Biomacromoleculs 1660–1666 (2015)

Indira, K.N., Parameswaranpillai, J., Thomas, S.: Mechanical properties and failure topography of Banana Fibre PF macrocomposites fabricated by RTM and CM techniques. Polym. Sci. 1–8 (2013)

Jiang, X.L., Luo, S.J., Chen, X.D.: Effect of nucleating agents on crystallization kinetic of PET. Express Polym. Lett. **1**(4), 245–251 (2007)

Kaiser, R., Anuar, H., Samat, N., et al.: Effect of processing route on mechanical, thermal and morphological properties of PLA-based hybrid biocomposites. Iran. Polym. J. **22**, 123–131 (2013)

Kumar, M.S., Raghavendra, K., Venkataswamy, M.A., et al.: Fractographic Analysis of tensile failures of aerospace grade composites. Mater. Res. **15**(6), 990–997 (2012)

Lamon, J., R'Mili, M.: Investigation of the residual tensile behaviour of fibre bundles after static fatigue: implications for the prediction of durability of composites. Compos. Part A Appl. Sci. Manuf. **67**, 149–156 (2014)

Lazim, Y.M., Salit, S.M., Zainudin, E.S., et al.: Effect of alkali treatment on the physical, mechanical and morphological properties of waste betel nut (*Areca catechu*) husk fibre. Biores **9**(4), 7721–7736 (2014)

Li, X., Liu, W., Sun, L., Aifantis, K.A., et al.: Resin composites reinforced by nanoscaled fibres or tubes for dental regeneration. Biomed. Res. Int. 1–13 (2014)

Liao, H., Wu, Y., Wu, M., et al.: Effects of fibre surface chemistry and roughness on interfacial structures of electrospun fibre reinforced epoxy composite films. Polym. Compos. 837–845 (2011)

Liu, X., Grant, D.M., Parsons, A.J., et al.: Magnesium coated bioresorbable Phosphate Glass Fibres: Investigation of the interface between Fibre and Polyester Matrices. Biomed. Res. Int. 1–10 (2013)

Majeed, K., Jawaid, M., Hassan, A., et al.: Potential materials for food packaging from nanoclay/natural fibres filled hybrid composites. Mater. Des. **46**, 391–410 (2013)

Ou, R., Guo, C., Xie, Y., et al.: Non isothermal crystallization kinetics of kevlar fibre-reinforced wood flour/HDPE composites. Biores **6**(4), 4547–4565 (2011)

Panagiotis, I., Xidas, K., Triantafyllidis, S.: Effect of the type of alkylammonium ion clay modifier on the structure and thermal/mechanical properties of glassy and rubbery epoxy–clay nanocomposites. Eur. Polym. J. **46**(3), 404–417 (2010)

Piscitelli, F., Scamardella, A.M., Valentina, R., et al.: Epoxy composites based on amino-silylated MMT: the role of interfaces and clay morphology. J. Appl. Polym. Sci. **124**(1), 616–628 (2012)

Poletto, M., Ornaghi Junior, H.L., Zattera, A.J.: Native cellulose; structure, characterization and thermal properties. Mater **7**, 6105–6119 (2014)

Prasanth, R., Shankar, R., Anna, Dilfi., Thakur, V.K., et al.: Eco friendly fibre reinforced natural rubber green composites: a perspective on the future. In: Thakur, V.K. (ed.) Green Composites from Natural Resources, pp. 206–235. Taylor Francis (2010)

Qiu, J., Song, P., Fu, S., et al.: Compatibilization of Polypropylene/ starch plasticized with diethanol amine. Adv. Mater. Res. **610–613**, 475–479 (2013)

Ravikumar, M., Prasad, M.S.: Fracture toughness and mechanical properties of aluminum oxide filled chopped strand mat e-glass fibre reinforced-epoxy composites. Int. J. Sci. Res. Publ. **4**(7), 1–7 (2014)

Rios, C.J., Chomik, E., Balderrama, J.J., et al.: Determination of fracture toughness J on fibre-metal laminate type CARALL with sheets aluminium 6061. Proced. Mater. Sci. **9**, 530–537 (2015)

Saba, N., Md Tahir, P., Jawaid, M.: A review on potentially of nano filler/natural filled polymer hybrid composites. Polym **6**, 2247–2273 (2014)

Saba, N., Md Tahir, P., Jawaid, M.: Mechanical properties of kenaf fibre reinforced polymer composite: a review. Constr. Build. Mater. **76**, 87–96 (2015)

Sarifuddin, S., Ismail, H., Zuraida, A.: The effect of kenaf Core Fibre Loading on Properties of low cost high density polypropylene/thermoplastic sago starch/kenaf core fiber composites. J. Phys. Sci. **24**(2), 97–115 (2013)

Salleh, Z., Taib, Y.M., Koay, M.H., et al.: Fracture toughness investigation on long kenaf/woven glass hybrid due to water absorption effect. Proc. Eng. **41**, 1667–1673 (2012)

Shah, D.U., Schuber, P.J., Licence, P., et al.: Determining the minimum, critical and maximum fibre content for twisted yarn reinforced plant fibre composites. Compos. Sci. Technol. **72**(15), 1909–1917 (2012)

Shi, X., Zhang, G., Phuong, T.V., et al.: Synergistic effects of nucleating agents and plasticizers on the crystallization behavior of poly(lactic acid). Molecules **20**, 1579–1593 (2015)

Shiraz, N.Z., Enferrad, E., Monfared, A., et al: Preparation of nanocomposite based on exfoliation of montmorillonite in acrylamide thermosensitive polymer. ISRN Polym. Sci. 1–5 (2013)

Sumalla, M., Amber, L., Bawa, M.: Effect of Fibre Length on the Physical and Mechanical Properties of Random Oriented, Nonwoven Short Banana (*musa balbisiana*) Fibre/Epoxy Composite, vol. 2(1), pp. 139–149. www.Leena-luna.co.jp

Sylvester, O.N., Christopher, I.C., Celestine, O.N., et al.: Experimental investigations and statistical analysis of creep properties of hybridized epoxy-alumina-calcium silicate nanocomposite material operating at elevated temperatures. Int. J. Sci. Technol. Res. **6**, 36–45 (2012)

Tian, H.: Tagaya. H (2007) Preparation, characterization and mechanical properties of the polylactide/perlite and the polylactide/montmorillonite composites. J. Mater. Sci. **42**, 3244–3250 (2007)

Todo, M., Takayama, T.: Fracture mechanisms of biodegradable PLA and PLA/PCL blends. In: Rosario, P. (ed.) Biomaterials-Physics and Chemistry. http://www.intechopen.com/books/biomaterials-physics-and-chemistry/fracture-mechanisms-of-biodegradable-pla-and-pla-pcl-blends

Weitsman, Y.J.: Effects of fluids on mechanical properties and performance chapter 7. In: Fluid effects in Polymer and Polymeric Composites, pp. 123–147

Wong, K.J., Zahi, S., Low, K.O., et al.: Fracture characterization of short bamboo fibre reinforced polyester composites. Mater. Des. **31**, 4147–4154 (2010)

Yousif, B.F., Shalwan, C.W., Chin, K.C.M.: Flexural properties of treated and untreated kenaf/epoxy composites. Mater. Des. **40**, 378–385 (2012)

Zhao, P., Liu, W., Wu, Q., Ren, J.: Preparation, mechanical and thermal properties of biodegradable polyesters/poly(lactic acid) blends. J. Nanomater. 1–8 (2010)

Zhu, J., Zhu, H., Njuguna, J., et al.: Recent development of flax fibres and their reinforced composites based on different polymeric matrices. Mater **6**, 5171–5198 (2013)

Mechanical, Dynamic Mechanical and Vibration Behavior of Nanoclay Dispersed Natural Fiber Hybrid Intra-ply Woven Fabric Composite

M. Rajesh, P. Jeyaraj and N. Rajini

Abstract Influence of nanoclay dispersion on mechanical, dynamic mechanical and free vibration characteristics of basket type intra-ply woven banana/jute (banana yarn-weft direction, jute yarn-warp direction) hybrid polyester composite has been investigated. Results revealed that loading of nanoclay enhances the mechanical properties of the composite significantly as it improves the adhesion between fiber and matrix. Dynamic mechanical analysis result reveals that nanoclay addition significantly increases the storage modulus and glass transition temperature of the intra-ply hybrid composite. From free vibration studies, carried out on laminated beam like structures, it is found that the natural frequency of the composite laminate increases till 2 wt% of nanoclay while the modal damping increases when the wt% of nanoclay is more than 2. However, due to the agglomeration effect of nanoclay the properties are not improved for nanoclay loading beyond 2 wt%.

Keywords Nanoclay · Weaving · Dynamic mechanical analysis · Mechanical properties

M. Rajesh · P. Jeyaraj
Department of Mechanical Engineering, National Institute of Technology Karnataka, Surathkal, Mangalore 575 025, India

N. Rajini (✉)
Department of Mechanical Engineering, Kalasalingam University, Krishnankoil, Virudhunagar 626 126, Tamilnadu, India
e-mail: rajiniklu@gmail.com

N. Rajini
Laboratory of Biocomposite Technology, Institute of Tropical Forestry and Forest Products, Universiti Putra Malaysia, 43400 Serdang, Selangor, Malaysia

M. Jawaid et al. (eds.), *Nanoclay Reinforced Polymer Composites*, Engineering Materials, DOI 10.1007/978-981-10-0950-1_12

1 Introduction

Natural fibers can be used an alternative material to conventional synthetic fibers such as glass, carbon, boron and kevlar fibers. Natural fibers such as banana, jute, sisal and hemp have advantages such as low cost, less weight, easy to process and bio-degradable compared to the synthetic fibers (Joshi et al. 2004; Venkateshwaran et al. 2011). Several researchers have investigated mechanical properties of several natural fiber hybrid composites and demonstrated that hybrid composites have better mechanical strength. Jacob et al. (2004) investigated the effect of sisal/oil palm fiber in natural rubber composite on mechanical properties. They concluded that increasing the concentration of fiber in matrix reduces the tensile strength. Venkateshwaran et al. (2011) analyzed the banana/sisal/epoxy hybrid composite with different fiber length and weight percentage. They achieved better mechanical properties for 50 wt% of sisal fiber in banana/epoxy composite. Jawaid et al. (2011) hybridized the oil palm empty fruit bunch fiber mat with woven jute composite and found that hybridized composite got maximum tensile and flexural result.

Several researchers demonstrated that based nanoclay added composites increases mechanical and dynamic properties such as tensile strength, modulus and glass transition temperature of composites compared to same filler reinforced in micron scale level in petroleum based resins such as unsaturated polyester, epoxy, vinyl ester etc. and natural resins such as vegetable oil, animal facts etc. (Wang et al. 2006; Haq et al. 2009). Chan et al. (2011) found that uniform distribution of nanoclay into polymer matrix increases the mechanical property as nanoclay added in small amount increases mechanical interlocking inside the composite and reduces the crack propagation while loading. Huang and Netravali (2009) found that micro/nano-sized bamboo fibrils in soy protein resin increases the fracture stress and Young's modulus of material. Fu et al. (2008) proved that particle size, particle/matrix interface adhesion and particle loading are important parameters to define the strength of composite materials. Yasmin et al. (2006) investigated the dynamic mechanical properties of epoxy based composite with 1–10 wt% of nanoclay content. They concluded that higher concentration (10 wt%) of nanoclay composite gives better property than lower and neat resin in all types of nanoclay composite. Huang and Netravali (2007) added nano particle in yarn and woven type flax/soy protein resin biodegradable composites. Their results revealed that nanoclay added woven composite has higher mechanical property than yarn type composite. Bozkurt et al. (2007) found that the addition of nanoclay in glass/epoxy composite up to 6 % increases the flexural modulus and dynamic mechanical properties such as storage modulus and loss modulus of composite. They also found that addition of nanoclay is not showing any significant improvement in the glass transition temperature. Haq et al. (2008) analyzed reinforcement of nanoclay in polyester/soybean oil/hemp composite and found that nanoclay addition increases stiffness, ultimate tensile stress and toughness of the composite. Jo et al. (2008) analyzed addition of montmorillonite (MMT) nanoclay into polyester composite and found that the nanoclay improves mechanical properties and also enhances the internal damping of composite material (Chandradass et al. 2007; Barick and Tripathy 2011). They concluded that organically modified silicate nanoclayenhances the viscous properties of composite material. In general, the addition of nanoclay in a composite increases the storage and loss modulus in both the glassy and rubbery region as the addition of nanoclay increases the resistance of molecular movement (Liu et al. 2013; Galooyak et al. 2011; Kostopoulos et al. 1993; Haddad and Yehia 2013; Gracida et al. 2005).

From the literature survey, it is found that the addition of nanoclay improves the mechanical properties, natural frequency and dynamic behavior of the polymer composites. Different kinds of natural fibers are available abundant in nature are becoming waste material as they are not being utilized effectively. Hybrid natural composites are fabricated by orienting different fibers layer by layer or keeping them in chapped form in the polymer matrix. Some researchers demonstrated that natural fiber reinforced in woven form gives better results compared to short natural fiber in random orientation. From the previous study it is found that the basket type woven composites have better mechanical properties compared to plain, twill, stain, huckaback woven composites. So the intra-ply hybrid woven natural fiber mats are prepared in basket weaving pattern. Additionally, MMT nanoclay has been added as a secondary reinforcement in the intra-ply natural/natural fiber hybrid composite to enhance the properties without affecting mass of the composite significantly.

The present work investigates the effect of nanoclay addition in a basket type intra-ply hybrid woven banana/jute natural fiber on the mechanical properties, free vibration behavior and dynamic mechanical properties such as storage modulus, loss modulus and damping factor.

2 Experimental Details

2.1 Woven Natural Fiber Mat and Polymer Preparation

In the present work, jute/banana natural fibers intra-ply hybrid woven mat is used as primary reinforcement while montmorillonite (MMT) nanoclay is used as secondary reinforcement in a polyester resin. Initially, these natural fibers are spurned into yarn, which contains around 100–150 numbers of loose fibers. From these yarns intra-ply hybrid basket type woven natural fiber mat is prepared manually. From the previous study it is found that basket type weaving pattern has higher mechanical properties than other type of weaving pattern. So intra-ply hybrid woven mats are prepared with basket type weaving pattern in the present work.

Figure 1 shows the schematic diagram of intra-ply hybrid woven natural fiber mats used in the present work. In intra-ply hybrid woven mat, one natural fiber yarn is oriented along warp direction while the other fiber is oriented along weft direction. Initially the secondary reinforcement MMT nanoclay is added with the unsaturated polyester resin then the mixer is stirred using high speed mechanical

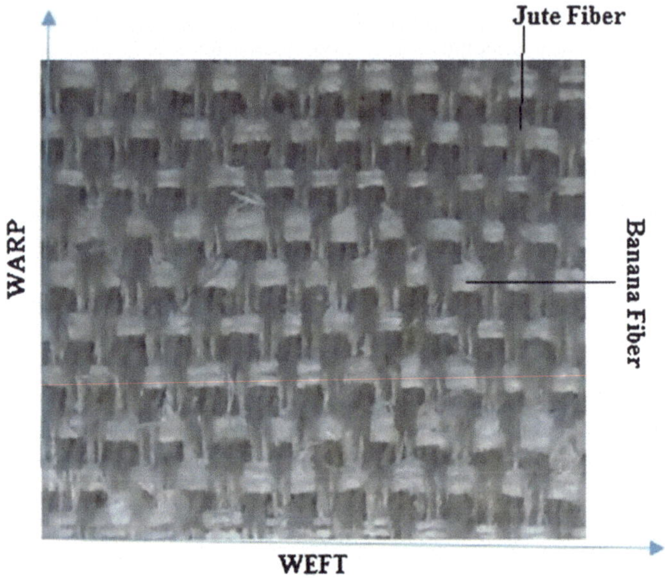

Fig. 1 Intra-ply hybrid basket type woven natural fiber mat

shear mixer at 900 rpm for 4 h for the reinforcement of nanoclay in resin with uniform dispersion. Then one wt% Methyl Ethyl Ketone Peroxide (MEKP) and cobalt naphthenate are used as catalyst and accelerator respectively and mixed with MMT nanoclay added unsaturated polyester resin in the ratio of 10:1. Weight percentage of the nanoclay is varied as 1–5 % in steps of 1 % and corresponding variations in different properties are analyzed.

2.2 *Fabrication of Composites*

Intra-ply hybrid woven composite laminates used in the present work are prepared using compression moulding technique, for that a mouldis prepared using steel of a dimension of 300 mm × 300 mm × 3 mm. The top and bottom parts are covered by stiff parallel plates made of EN steel. Firsta known amount of the prepared unsaturated, nanoclay mixture is poured then basket type intra-ply hybrid woven natural fiber mat is kept over the poured resin. In order to remove voids content from this laminate it is rolled using steel roller several times. After this mould is covered by two parallel plates and applying 20 bar pressure on the compression moulding machine. After 24 h room temperature curing, laminate is removed from the mould. Then according to ASTM standards for different test, the specimens are cut from the laminate.

2.3 Testing Standards

Tensile and flexural tests (three-point bending) are carried out using a universal testing machine while impact test is carried out using an izod impact test set-up. Tensile test is carried out as per ASTM standard D-638 with a loading speed of 2 mm/min. Three-point bending test is carried out using ASTM standard D-790. Impact test is carried out according to ASTM standard D-256 without notch. Each test is carried out on three samples and the average value is taken. For dynamic mechanical analysis, intra-ply hybrid composite specimen size of 50 mm × 50 mm × 3 mm is used and a test is carried out by the dual cantilever method using a dynamic mechanical analyzer (MODEL DMS 6100). Free vibration characteristics such as first three lowest frequencies and associated modal damping values are obtained using the impulse hammer method. The size of test specimen used for the free vibration test is 170 mm × 17 mm × 3 mm.

2.4 Material Characterization

Interfacial bonding between the fiber and matrix and morphology of fractured surface are investigated using scanning electron microscope images.

3 Results and Discussion

Several researchers investigated the natural fiber reinforced composites and reported that the mechanical properties of these composites are significantly influenced several factors such as orientation of the fiber, fiber/matrix adhesion, hybridization of natural fibers and type of the natural fiber. Some researchers found that the addition of nanoclay act as secondary reinforcement material in the polymer composite enhances the mechanical properties. The present work proposes intra-ply hybrid woven natural fiber polyester composite with different wt% of nanoclay as secondary reinforcement. Influence of basket type woven intra-ply hybrid mat and nanoclay loading on mechanical properties (tensile, flexural and impact strength), dynamic mechanical and free vibration characteristics of the intra-ply hybrid woven nano composite has been analyzed in detail.

3.1 Mechanical Properties

Influence on nanoclay loading on tensile, flexural and impact properties of the intra-ply hybrid woven mat composite is shown in Table 1. From Table 1, it is clear

Table 1 Influence of nanoclay loading on mechanical properties of the intra-ply hybrid composite (Values in bracket indicates percentage increase compared to the hybrid composite without nanoclay)

wt%	Tensile properties		Flexural properties		Impact properties
	Tensile strength (MPa)	Tensile modulus (GPa)	Flexural strength (MPa)	Flexural modulus (GPa)	Impact strength (J/m)
0	15.9	1.00	34.2	1.31	102
1	21.0 (24 %)	1.27 (26 %)	37.8 (11 %)	1.55 (18 %)	190 (86 %)
2	23.7 (33 %)	1.35 (35 %)	48.3 (41 %)	1.72 (31 %)	290 (184 %)
3	22.2 (28 %)	1.29 (29 %)	39.9 (17 %)	1.36 (4 %)	246 (114 %)
4	19.2 (21 %)	1.08 (8 %)	37.8 (10 %)	1.16 (−11 %)	134 (31 %)
5	18.5 (14 %)	1.02 (2 %)	36.7 (7 %)	1.02 (−22 %)	116 (14 %)

that the addition of nanoclay improves the mechanical properties such as tensile, flexural and impact strength of intra-ply hybrid composites. Rough surface on the fiber mat due to the addition of nanoclay increases the fiber-matrix adhesion. This enhanced fiber-matrix adhesion decreases the crack propagation rate while loading and hence increases the mechanical properties. This rough surface formed on the fiber increases the contact ratio between the fiber and matrix also. This results in increase in the mechanical interlocking capability between fiber and matrix. The nanoclay will act as small micro-pin between fiber and matrix. So it can carry high loads and transfer stress uniformly. It is evident from the SEM image. Figure 3a shows the presence of rough surface on the natural fiber after the addition of nanoclay.

Nanoclay loading in matrix always improves the mechanical properties of the hybrid composite as seen in Fig. 2 while beyond 2 wt% of nanoclay addition these properties reduce with increase in nanoclay loading. This can be attributed to the

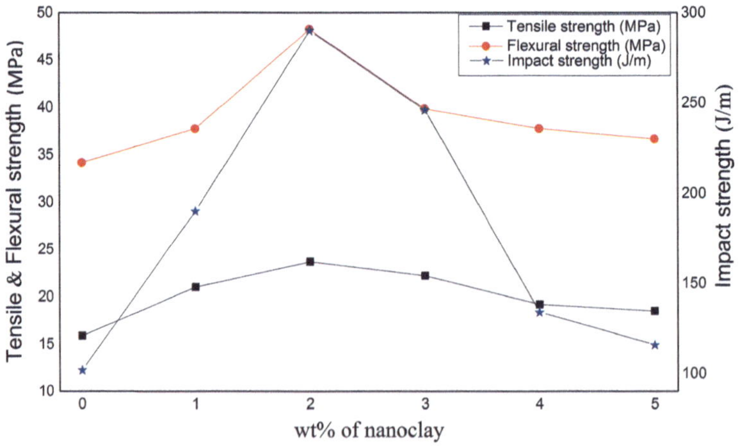

Fig. 2 Influence of nanoclay loading on mechanical properties of intra-ply hybrid composite

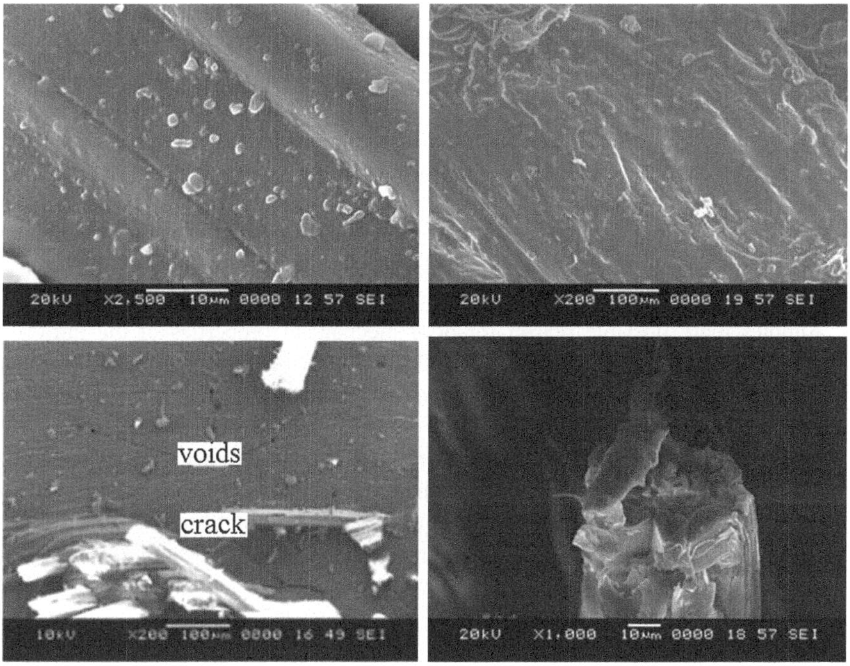

Fig. 3 SEM images to analyze influence of nanoclay loading on fractured surfaces. **a** Fiber surface after 2 wt% nanoclay addition, **b** 2 wt% nanoclay added composite under tensile load, **c** composite without nanoclay addition under tensile load, **d** fiber cross section

agglomeration of nanoclay in the matrix for the nanoclay loading beyond 2 wt%. Due to the agglomeration there will not be a uniform dispersion of nanoclay in the matrix which results in non-uniform stress distribution across the laminate and exhibits poor mechanical properties even though the amount of nanoclay loading is more than 2 wt%.

Figure 3b shows the tensile fractured surface of the intra-ply hybrid composite with 2 wt% nanoclay. Figure 3b reveals that the addition of nanoclay in the matrix increases the resistance to crack propagation while loading the specimen. But composite without nanoclay in the polymer matrix shows lowest tensile and flexural strength as they are poor in carrying high load and distribute stress uniformly between fiber and matrix while loading. This results in creation of multiple crack and propagation in all the direction under the action tensile and flexural load. It is evident from Fig. 3c shows the multiple cracks propagate, void formation and week fiber/matrix interaction while loading. Figure 3d shows the cross section of a fiber in the nanoclay reinforced matrix. It is confirmed by the appearance of the fiber shown in Fig. 3d, that the nanoclay addition in the matrix increases the resistances against the fiber pull out as the nanoclay also provides a strong interface between fiber and matrix.

Figure 2 also shows the effect of nanoclay loading on the impact strength of intra-ply hybrid composites. A trend similar to tensile and flexural strength has been observed for the impact strength also for different wt% of nanoclay addition. Toughness is an important property for a newly developed material. As it is a measure of energy stored in a material during the impact loading. Impact strength of the intra-ply hybrid woven composite also reduces beyond 2 wt% of nanoclay loading. Physical behavior of the composite with the addition of nanoclay and increasing wt% of nanoclay discussed for the tensile and flexural properties may be attributed for the variation of impact strength also. Behera et al. (2013), Kumar et al. (2015) also observed same behavior.

Another main possible reason for poor mechanical properties of the intra-ply woven hybrid composite with more than 2 wt% nanoclay is poor bonding between particle and heterogeneity property of composite material due to density variation of nanoclay and polyester resin. However, tensile strength of nanoclay composite increases by 32, 49, 39, 20, and 16 % respectively in 1–5 wt% nanoclay composites compared to composite without nanoclay. A similar trend is observed in flexural and impact strength also. The flexural strength of nanoclay added composite increases by 11, 41, 17, 11, 8 % respectively and 86, 184, 141, 31, 14 % respectively for impact strength in 1–5 wt% compared to nanoclay free composite.

3.2 *Dynamic Mechanical Analysis*

3.2.1 Storage Modulus

Effect of temperature and nanoclay loading on the storage modulus of intra-ply hybrid composite is shown in Fig. 4a. Results reveal that the addition of nanoclay improves the energy storing capacity of the intra-ply hybrid composite than the nanoclay free composite. Storage modulus of nanoclay free intra-ply hybrid composite decreases with increasing the temperature as the material losses the stiffness. So internal molecules will move freely without any restriction. Another observation found from Fig. 4a is extreme drops in the storage modulus near the glass transition temperature (T_g). This sudden dropping trend in storage modulus is distinctive of the viscoelastic actions associated with polymer based composites. In order to increase the load carrying capability of composite material, researchers proposed to incorporate the nanoclay in composites (Boumbimba et al. 2011; Abenojar et al. 2012). From Fig. 4a it is found that addition of nanoclay enriched the storage modulus substantively because of better interaction between polyester, nanoclay and fiber. So it is permissible stress between fiber and matrix is high. Nevertheless, as shown in Fig. 4a storage modulus of nanoclay composite is higher than nanoclay free composite. Higher storage modulus is observed in both glassy and rubbery regions and the nanoclay addition also increases the T_g value of composite material. This significant increment of storage modulus of the intra-ply hybrid composite

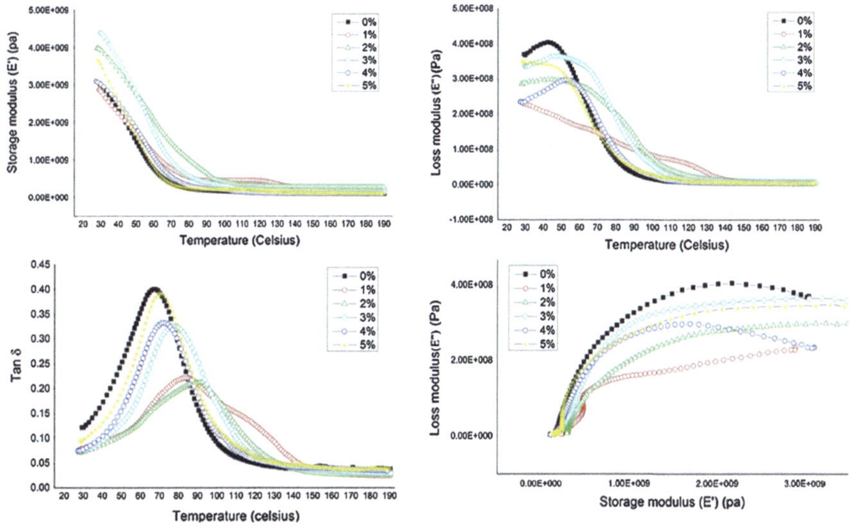

Fig. 4 Dynamic mechanical behavior of nanoclay added intra-ply hybrid composite. **a** Storage modulus, **b** loss modulus, **c** loss factor, **d** damping factor

material is due to the incorporation of nanoclay in a polymer which enhances the stress transfer by acting as a secondary reinforcement in the composite.

Another observation observed from Fig. 4a is, nanoclayupto 2 % increases the storage modulus extremely than other weight percentage but it is higher than composite without nanoclay. This is because increasing nanoclay content in the composite form cluster formation. Due to this it forms heterogeneity property and creates weak bonding between matrix and fiber, which is already discussed in Sect. 3.1. Results in Fig. 4a revealed that addition of nanoclay increases the storage modulus in both glassy and rubbery regions. This indicates that addition of nanoclay increase the thermal stability of the hybrid composite and gives better restriction to free molecular moment even at higher temperature.

3.2.2 Loss Modulus

Loss modulus is used to define the viscous nature of polymer composite. It is a measure of energy dissipation ofa material under deformation. From Fig. 4b, one can observe that variation of the loss modulus of the intra-ply hybrid woven composite with temperature and nanoclay loading is similar to the variation of storage modulus as discussed in Sect. 3.2.1. Addition of nanoclay increases the loss modulus and T_g value of intra-ply hybrid composites than composite without nanoclay. But nanoclay free composite and composite with 4 and 5 wt% has higher loss modulus in glassy region. After 45–55 °C temperature, sudden drop is observed in loss modulus peak. Lowering the loss modulus near T_g is due to the micro Brownian movement of polymer chain (Pothan et al. 2006). Normally loss modulus is dependent on frictional resistance between fiber and matrix. But due to molecule moment it reduces the frictional resistance. So it could reduce the heat loss and also it is due to poor interfacial bonding between clay/matrix/fiber. Composite with 1–2 wt% in the glassy region has less value compared to remaining wt% but composite with lower wt% shows good adhesion between fiber and matrix. Because temperatures range between 45 and 55 °C composite with lower wt% does not show sudden drop. This indicates addition of nanoclay increases the adhesion between fiber and matrix and reduces the molecular free moment and increases the frictional resistance. Due to this it increases the heat loss during deformation of materials.

3.2.3 Material Loss Factor or Tan δ

Material loss factor is a ratio of loss modulus to storage modulus and defines the material damping and it is a dimensionless property. It measures how the material dissipates energy. The variation of Tan δ peak is associated with molecule moment and amount of energy dissipation. In composite material, energy dissipation is dependent on fiber/matrix interaction, a molecule moment in the polymer chain, strength of fiber, fiber breakage and crack propagation in composite material. Peak height of the loss factor indicates the material's damping nature. From Fig. 4c it is found that composite with 1 and 2 wt% nanoclay have lower Tan δ peak, which indicates better fiber/matrix adhesion, decreases the molecular mobility and increasing load carrying capacity. This is because addition of nanoclay will act as secondary reinforcement material and also it carries more load and observe a large amount of energy. So it will increase the performance of composite material. Because the addition of nanoclay allows a small amount of energy at the interface. That means, stronger interface will dissipate less energy. So it will increase the stiffness of the material also increase the T_g value of composite material.

This indicates that addition of nanoclay provides better adhesion between the fiber and matrix due to the formation of rough surface on fiber as already discussed. Influence of nanoclay loading on the material loss factor is shown in Fig. 4c. From the results it is observed that composite with 1 and 2 wt% nanoclay has a low magnitude of loss factor compared to the nanoclay free composite. This indicates that addition of nanoclay in composite increases the restriction to molecular free moment. Because Tan δ peak height associated with molecular motion in the polymer chain and the amount of energy dissipation between fiber and matrix. But further addition of nanoclay increases the Tan δ peak high which indicates addition of nanoclay increases the interaction between fiber and matrix. Interesting result is observed from Table 2 is addition of nanoclay in intra-ply hybrid woven composite increases the glass transition temperature due to restriction of molecular free moment. From Table 2, it is found that T_g value from Tan δ is more realistic than loss modulus curve. But nanoclay free composite has the highest peak as nanoclay

Table 2 Peak height and glass transition temperature of nanoclay added composites

wt% (%)	Peak height of Tan δ	Temperature (°C)	
		T_g from Tan δ	T_g from loss modulus
0	0.4007	67.8	45.9
1	0.2219	84.3	60.5
2	0.2126	91.5	64.3
3	0.3273	77.8	58.4
4	0.3337	72.0	53.0
5	0.3916	70.9	50.0

added intra-ply woven composite has better fiber/matrix interaction. Due to this it dissipates more energy than nanoclay composites.

3.2.4 Cole-Cole Plot

Figure 4d shows the Cole-Cole plot of different weight percentage of nanoclay added intra-ply hybrid composites. Adding of nanoclay in the polymer changes the perfect semi-circle into imperfect semicircular shape. It indicates that, nanoclay changes the materiel behavior from homogeneous to heterogeneous. From the results, it can be seen that composite without nanoclay showing perfect semicircle, but composite with nanoclay shows an imperfect semi-circle. Because beginning of the curve is showing imperfect shapes. This is due to the influence of nanoclay addition in the composite. From this it is concluded that nanoclay influences the dynamic mechanical properties of material and affects the Cole-Cole plots shape and behavior of materials.

3.3 Free Vibration Analysis

Three important parameters such as storage modulus, loss modulus and the material loss factor of a material analyzed with the help of dynamic mechanical analyzer gives dynamic properties associated with a material at different temperatures. Any newly developed material will be used to develop structural members for several engineering applications. It is important to predict and analyze natural frequencies and associated modal damping of these structures for the sake of their undesirable behavior during resonance. Modal analysis performed on a structure will give these two basic and important inherent dynamic characteristics of the structure.

In the present work modal analysis is carried out on intra-ply hybrid woven nano composite beam like structure for Fixed-Free and Fixed-Fixed boundary condition. From the free vibration analysis, natural frequency and damping factor associated with the first three bending modes are obtained for different intra-ply hybrid nano composite in order to investigate the influence of nanoclay loading on the free vibration behavior. A rectangular specimen of size 170 mm × 17 mm × 3 mm is used for the free vibration analysis. The beam is excited with the help of turned impact hammer and vibration response is obtained using a less weight accelerometer. The less weight accelerometer is chosen particularly to avoid the added mass effect of the sensor on the light weight composite. The vibration response obtained for the impulse hammer excitation at five different chosen locations is given as an input to DEWETRON, modal analysis software, through DEWESoft data acquisition system. The modal analysis will give the mode shapes and its natural frequency and associated modal damping factor using circle fit method. Equation 1 used to calculate the damping ratio which is given below

$$\xi = \frac{\omega_2^2 - \omega_1^2}{2\omega_0\left[\omega_2 \tan\frac{\alpha_2}{2} + \omega_1 \tan\frac{\alpha_1}{2}\right]} \quad (1)$$

where, ω_0 refers angular frequency at resonance, ω_2, ω_1 are angular frequencies, α_1, α_2 are angle between angular frequencies.

Table 3 shows the influence of different wt% of nanoclay addition on free vibration characteristic of intra-ply hybrid woven nanocomposite for Fixed-Free and Fixed-Fixed boundary conditions. From Table 3, it is seen that the addition of nanoclay increases the natural frequencies up to 2 wt% of nanoclay for both the boundary conditions. Further addition of nanoclay in the composite decreases the frequency, but it always higher than the composite without nanoclay. Because nanoclay addition enhances the polymer composite while loading and nanoclay added polymer carry more amount of stress and allows a small amount of stress between the fiber and matrix interface. Due to this intra-ply hybrid woven composite with 2 wt% nanoclay gives lower damping value than the other combination. Because of this, energy dissipation in the nanoclay polymer composite and its strong interface is less than composite without nanoclay. Therefore, it is concluded that less energy dissipation decreases the inherent damping of the material. But higher wt% nanoclay composite increases the fiber/matrix interaction and provided the frictional resistance and mechanical interlocking due to large surface area. Because of this reason, higher wt% nanoclay composite gives better inherent damping property than composite with lower wt% (Rajini et al. 2013). In general addition of nanoclay increases the natural frequency of intra-ply hybrid composite compared to the composite without nanoclay. Because the addition of nanoclay increases the modulus of intra-ply hybrid composite than nanoclay free composite and further adding nanoclay results in decreasing natural frequency but always higher than composite without nanoclay. Because at higher wt% composite will lose their stiffness due to agglomeration and poor fiber/matrix interaction. Due to this large stiffness variance in fiber/matrix interface, higher wt% added polymer composite increases the internal damping. It is evident from Table 3.

Table 3 Influence of nanoclay loading on natural frequency and damping factor for intra-ply hybrid composite

Beam	wt% (%)	Mode 1		Mode 2		Mode 3	
		Natural frequency (Hz)	Damping factor	Natural frequency (Hz)	Damping factor	Natural frequency (Hz)	Damping factor
Fixed-free	0	34	0.173	229	0.045	600	0.025
	1	39	0.159	239	0.038	644	0.024
	2	44	0.133	253	0.043	693	0.047
	3	39	0.102	250	0.045	688	0.032
	4	35	0.162	234	0.051	644	0.053
	5	34	0.179	200	0.063	590	0.058
Fixed-fixed	0	219	0.151	565	0.093	1137	0.055
	1	234	0.094	610	0.069	1171	0.043
	2	254	0.061	678	0.063	1298	0.041
	3	252	0.096	644	0.086	1206	0.058
	4	239	0.109	639	0.089	1220	0.066
	5	215	0.197	551	0.087	1137	0.099

Natural frequencies evaluated experimentally are compared with numerical and analytical solution by assuming the composite as quasi-isotropic material. Commercial available finite element software ANSYS is used for the numerical analysis and beam is modeled using four noded structural shell element SHELL 181 in ANSYS by giving the equivalent Young's modulus (based on the flexural modulus), density (based on mass) of the laminates and assuming a poison ratio of 0.3. The expression used for analytical method given by Thomson et al. (2008) for bending mode frequency of ith mode based on Euler-Bernoulli hypothesis is

$$\omega_i = \beta_i^2 \sqrt{\frac{EI}{\rho A L^4}} \text{ rad/s}$$

where, E = Young's modulus (N/m^2), A = Cross-section area of beam (m^2), ρ = Density of composite material (kg/m^3), I = Moment of inertia (m^4), L = Length of composite specimen (m), β value for first three modes of Fixed-Free condition is 1.875, 4.694, 7.854 and Fixed-Fixed condition is 4.73, 7.8532 and 10.995 (Thomson et al. 2008). Table 4 shows comparison of experimental results with the results of analytical and finite element methods for the intra-ply woven hybrid composite with 2 wt% of the nanoclay. Experimentally obtained natural frequencies are in good agreement with the natural frequencies evaluated based on analytical and finite element methods as seen in Table 4.

Similar to the mechanical properties and dynamic mechanical properties variation with nanoclay, free vibration frequencies and modal damping factors of the intra-ply hybrid woven composite beam also reduces when wt% of the nanoclay is above 2.

Table 4 Comparison of natural frequency (Hz) of nanoclay added intra-ply hybrid composite for first three mode with analytical and finite element methods

Boundary condition	Experimental			Analytical			FEM method		
	Mode			Mode			Mode		
	1st	2nd	3rd	1st	2nd	3rd	1st	2nd	3rd
Fixed-free	44	253	693	40	247	692	40	249	695
Fixed-fixed	254	678	1298	235	647	1268	236	632	1227

4 Conclusion and Future Work

Influence of addition of the small amount of nanoclay as secondary reinforcement in intra-ply woven natural fiber hybrid composite's mechanical properties, dynamic mechanical and free vibration characteristics are analyzed. Results revealed that, inclusion of nanoclay in the intra-ply hybrid polymer composite influences the material inherent properties and hence changes mechanical behavior significantly. Results also revealed that addition of nanoclay upto 2 wt% increases the mechanical properties. Further addition of nanoclay in the composite decreases the stiffness of the composite material, but higher than composite without nanoclay. From the dynamical mechanical analysis, it is found that addition of nanoclay increases the glass transition temperature however it reduces the material loss factor. Natural frequencies associated with beam like intra-ply woven hybrid composites increases nanoclay addition till 2 wt% then it reduces however, modal damping behavior is reverse to that of natural frequency variation.

- Exploring the possibilities of using natural fiber woven fabric reinforced polymer nano composite in consumer goods and structural applications.
- Improving the properties of the composite by sandwiching the natural fiber woven fabric polymer nano composite between composite facing layers.
- Improving the mechanical behavior of the composite by replacing the weak natural fiber with glass fiber in the intra-ply woven fabric.

Acknowledgments Authors wish to thank Department of Science and Technology-India for funding through SR/FTP/ETA-64/2012 and also wish to thank Material Characterization Lab, Department of Mechanical Engineering, NITK, Surathkal for allowing us to carry out tensile and flexural tests.

Compliance with Ethical Standards

Funding: This study was funded by SERB-Department of Science and Technology, India. (SR/FTP/ETA-64/2012).

Conflict of Interest: We declare that we have no conflict of interest.

References

Abenojar, J., Martnez, M., Pantoja, M., et al.: Epoxy composite reinforced with nano and micro sic particles: curing kinetics and mechanical properties. J. Adh **88**, 418–434 (2012)

Barick, A., Tripathy, D.: Effect of organically modified layered silicate nanoclay on the dynamic viscoelastic properties of thermoplastic polyurethane nanocomposites. App. Clay. Sci. **52**, 312–321 (2011)

Behera, A.K., Avancha, S., Sen, R., et al.: Development and characterization of nanoclay-modified soy resin-based jute composite as an eco-friendly/green product. Poly. Plas. Technol. Eng. **52**, 833–840 (2013)

Boumbimba, R.M., Ahzi, S., Bahlouli, N., et al.: Dynamic mechanical properties of pmma/organoclaynanocomposite: experiments and modeling. J. Eng. Mat. Tech. **133**, 030908 (2011)

Bozkurt, E., Kaya, E., Tanoglu, M.: Mechanical and thermal behavior of non-crimp glass fiber reinforced layered clay/epoxy nanocomposites. Comput. Sci. Technol. **67**, 3394–3403 (2007)

Chan, M.I., Lau, K.T., Wong, T.T., et al.: Mechanism of reinforcement in a nanoclay/polymer composite. Compos. Part B Eng. **42**, 1708–1712 (2011)

Chandradass, J., Kumar, M.R., Velmurugan, R.: Effect of nanoclay addition on vibration properties of glass fibre reinforced vinyl ester composites. Mater. Lett. **61**, 4385–4388 (2007)

Fu, S.Y., Feng, X.Q., Lauke, B., et al.: Effects of particle size, particle/matrix interface adhesion and particle loading on mechanical properties of particulate polymer composites. Compos. Part B Eng. **39**, 933–961 (2008)

Galooyak, S.S., Dabir, B., Nazarbeygi, A., et al.: The effect of nanoclay on rheological properties and storage stability of sbs-modified bitumen. Petr. Sci. Technol. **29**, 850859 (2011)

Gracida, J., Perez-Guevara, F., Cardoso-MartNez, J.: Thermal and dynamic mechanical properties of binary blends of bacterial copolyester poly (hydroxybutyrate-co-hydroxyvalerate)(PHBHV) with poly (2-hydroxyethylmethacrylate)(PHEMA). J. Mater. Sci. **40**, 2565–2567 (2005)

Haddad, Y.M: Mechanical Behaviour of Engineering Materials: Volume 2: Dynamic Loading and Intelligent Material Systems. Springer Science & Business Media (2013)

Haq, M., Burgueno, R., Mohanty, A.K., et al.: Hybrid bio-based composites from blends of unsaturated polyester and soybean oil reinforced with nanoclay and natural fibers. Comput. Sci. Technol. **68**, 3344–3351 (2008)

Haq, M., Burgueno, R., Mohanty, A.K., et al.: Processing techniques for bio-based unsaturatedpolyester/clay nanocomposites: tensile properties, efficiency, and limits. Compos. A Appl. Sci. Manf. **40**, 394–403 (2009)

Huang, X., Netravali, A.: Characterization of flax fiber reinforced soy protein resin based green composites modified with nano-clay particles. Compos. Sci. Technol. **67**, 2005–2014 (2007)

Huang, X., Netravali, A.: Biodegradable green composites made using bamboo micro/nanofibrils and chemically modified soy protein resin. Compos. Sci. Technol. **69**, 1009–1015 (2009)

Jacob, M., Thomas, S., Varughese, K.T.: Mechanical properties of sisal/oil palm hybrid fiber reinforced natural rubber composites. Compos. Sci. Technol. **64**, 955–965 (2004)

Jawaid, M., Khalil, H.A., Bakar, A.A.: Woven hybrid composites: tensile and flexural properties of oil palm-woven jute fibres based epoxy composites. Mater. Sci. Eng. A **528**, 5190–5195 (2011)

Jo, B.W., Park, S.K., Kim, D.K.: Mechanical properties of nano-mmt reinforced polymer composite and polymer concrete. Constr. Build. Mater. **22**, 14–20 (2008)

Joshi, S., Drzal, L., Mohanty, A., Arora, S.: Are natural fiber composites environmentally superior to glass fiber reinforced composites? Compos. A Appl. Sci. Manf. **35**, 371–376 (2004)

Kostopoulos, V., Vellios, L., Philippidis, T.P., et al.: Dynamic mechanical analysis of a two-dimensional carbon-carbon composite. J. Mater. Sci. **28**, 5495–5499 (1993)

Kumar, M., Arun, S., Upadhyaya, P., et al.: Properties of pmma/clay nanocomposites prepared using various compatibilizers. Int. J. Mech. Mater. Eng. **10**, 1–9 (2015)

Liu, W., Liu, B., Wang, X.: Morphology, rheological properties, and crystallization behavior of polypropylene/clay nanocomposites. Int. J. Poly. Mater. **62**, 164–171 (2013)

Pothan, L.A., Thomas, S., Groeninckx, G.: The role of fibre/matrix interactions on the dynamic mechanical properties of chemically modified banana fibre/polyester composites. Compos. A Appl. Sci. Manf. **37**, 1260–1269 (2006)

Rajini, N., Jappes, J.W., Jeyaraj, P., et al.: Effect of montmorillonite nanoclay on temperature dependence mechanical properties of naturally woven coconut sheath/polyester composite. J. Rein. Pla. Com **32**, 811–822 (2013)

Thomson, W.T., Dahleh, M.D., Padmanabhan, C.: Theory of Vibrations with Applications, 5th edn. Pearson Education, India (2008)

Venkateshwaran, N., Elaya Perumal, A., Alavudeen, A., et al.: Mechanical and water absorption behaviour of banana/sisal reinforced hybrid composites. Mater. Des. **32**, 4017–4021 (2011)

Wang, L., Wang, K., Chen, L., et al.: Preparation, morphology and thermal/mechanical properties of epoxy/nanoclay composite. Compos. A Appl. Sci. Manf. **37**, 1890–1896 (2006)

Yasmin, A., Luo, J., Abot, J., et al.: Mechanical and thermal behavior of clay/epoxy nanocomposites. Comput. Sci. Technol. **66**, 2415–2422 (2006)

The manufacturer's authorised representative in the EU is Springer Nature Customer Service Centre GmbH, Europaplatz 3, 69115 Heidelberg, Germany. If you have any concerns regarding our products, please contact ProductSafety@springernature.com

Printed and bound by CPI Group (UK) Ltd, Croydon, CR0 4YY
29/07/2026
02180627-0005